NITROGEN FIXATION WITH NON-LEGUMES

Developments in Plant and Soil Sciences

VOLUME 79

The titles published in this series are listed at the end of this volume.

Nitrogen Fixation with Non-Legumes

Proceedings of the 7th International Symposium
on Nitrogen Fixation with Non-Legumes,
held 16–21 October 1996 in Faisalabad, Pakistan

Edited by

Kauser A. MALIK

and

M. Sajjad MIRZA
National Institute for Biotechnology and Genetic Engineering (NIBGE),
P.O. Box 577, Jhang Road, Faisalabad, Pakistan

and

J.K. LADHA
International Rice Research Institute (IRRI), P.O. Box 933,
Manila 1099, Philippines

KLUWER ACADEMIC PUBLISHERS
DORDRECHT \ BOSTON \ LONDON

A C.I.P. Catalogue record for this book is available from the Library of Congress

ISBN 0–7923–4873–7

Published by Kluwer Academic Publishers,
P.O. Box 17, 3300 AA Dordrecht, The Netherlands.

Sold and distributed in North, Central and South America
by Kluwer Academic Publishers,
101 Philip Drive, Norwell, MA 02061, U.S.A.

In all other countries, sold and distributed
by Kluwer Academic Publishers Group,
P.O. Box 322, 3300 AH Dordrecht, The Netherlands.

Printed on acid-free paper

Printed in Great Britain

Contents

Foreword — xi
J. Dobereiner

Preface — xiii

Opening lecture

1. Biological nitrogen fixation by endophytic diazotrophs in non-leguminous crops in the tropics — 3
J. Dobereiner and V.L.D. Baldani

SECTION I
Biochemistry and genetics of diazotrophs

2. Analysis of nitrogen fixation and regulatory genes in the sugar cane endophyte *Acetobacter diazotrophicus* — 11
Sunhee Lee, Myrna Sevilla, Dietmar Meletzus, Katia Teixeira, Andre De Oliveira, Lena Perlova, Roman Nawroth, Ivo Baldani and Christina Kennedy

3. Functional analysis of a cloned *nifA* homologous gene of *Azospirillum brasilense* Yu62 — 21
Gui Su, Yinsuo Zhao and Jilun Li

4. Identification of plant growth hormones produced by bacterial isolates from rice, wheat and kallar grass — 25
Ghulam Rasul, M. Sajjad Mirza, Farooq Latif and Kauser A. Malik

SECTION II
Ecophysiological studies for the improvement of plant–microbe interaction

5. The rice endophytic diazotroph and PGPR — 41
W. Song, H.L. Yang, X.L. Sun, Y.S. Wang, Y.D. Wang and Z.H. Chen

6. Effect of *Azospirillum* inoculation on growth and N_2 fixation of maize subjected to different levels of FYM using ^{15}N-dilution method 49
H.M.A. El-Komy, T.M.M. Moharram and M.S.A. Safwat

7. Contribution of non-leguminous biofertilizers to rice biomass, nitrogen fixation and fertilizer-N use efficiency under flooded soil conditions 61
Sikander Ali, Naima Hamid, Ghulam Rasul, Samina Mehnaz and Kauser A. Malik

8. Detection of inoculated plant growth-promoting rhizobacteria in the rhizosphere of rice 75
Samina Mehnaz, M. Sajjad Mirza, Uzma Hassan and Kauser A. Malik

Posters

9. Salt tolerance of diazotroph *Alaligenes faecalis* and its salt-tolerant association with host rice 87
S.Z. Ping, M. Lin, Y.N. Liu, C.B. You and Kauser A. Malik

10. Interaction of *Azorhizobium caulinodans* with different rice cultivars for increased N_2-fixation 91
Shahida N. Khokhar and Azra Qureshi

11. Response of wheat to an associative diazotroph inoculum under different rates of nitrogen fertilizer 95
A. Rashid, M.R. Sajjad, M.A. Gill, M.S. Cheema, M.S. Sindhu and M.M. Nayyar

12. Effects of high temperature on nitrogen fixation in soybean cultivars 99
M.I. Keerio and J.M. Wilson

13. Effect of plant growth regulators and *Rhizobium* inoculum on N_2-fixation and yield of chickpea 103
Zarrin Fatima, A. Bano and M. Aslam

SECTION III
Paranodulation and forced association of diazotrophs with non-legumes

14. Prospects for facilitated evolution of effective N_2-fixing associations with cereals: comparative performance of *Azospirillum brasilense* Sp7-S with various free-living diazotrophs in *para*-nodulated wheat 109
Ivan R. Kennedy, Sunietha Katupitiya, Daiguan Yu, Kate Gilchrist, Rosalind Deaker, Lily Pereg-Gerk and Craig Wood

15. Transgenic rice plants with *Parasponia* haemoglobin gene and its expression 125
Wei Guo, Shan Xueqin, Ding Qunxing, Liu Bolin and Jing Yuxiang

16. Cell structure of pseudonodules formed on the roots of barley and rice 133
Shan Xueqin and Jing Yuxiang

Posters

17. Induction of paranodulation by 2,4-D, IAA and tryptophan in wheat inoculated with various diazotrophs 145
H. Youssef, M. Monib, M. Fayez and N.A. Hegazi

18. Optimizing zinc and phosphate statuses for enhancing field colonization and contribution of the symbiotic association *Azolla–Anabaena* to rice performance 151
Y.G. Yanni

19. Enhancement of rice performance and incidence of natural fungal and insect damage with cyanobacterization, microelements, combined nitrogen and certain pesticides 159
Y.G. Yanni, M.R. Sehly and F.E. Abdallah

SECTION IV
Molecular biology of plant–diazotroph interaction

20. Colonization and *nifH* expression on rice roots by *Alcaligenes faecalis* A15 167
Hans Vermeiren, Jozef Vanderleyden and Weili Hai

21. Responses of *Azospirillum brasilense* to salinity stress 179
A.K. Tripathi and B.M. Mishra

22. Studies on the diploid indeterminate nodule-forming plant *Melilotus alba* Desr., white sweetclover 187
Ann M. Hirsch, Thomas A. LaRue, Joanna H. Norris and Norman F. Weeden

SECTION V
Molecular probes in taxonomy and ecology of diazotrophs

23. Isolation and identification of diazotrophic bacteria from rice, wheat and kallar grass 197
Uzma Hassan, M. Sajjad Mirza, Samina Mehnaz, Ghulam Rasul and Kauser A. Malik

SECTION VI
Applicability and agronomic importance of associative diazotrophs

24. Diazotrophs associated with non-legumes grown in sandy soils 209
N.A. Hegazi, M. Fayez, G. Amin, M.A. Hamza, M. Abbas, H. Youssef and M. Monib

25. Effect of inoculation with nitrogen-fixing *Klebsiella* on potato yield 223
V.N. Pishchik, I.I. Chernyaeva, A.P. Kozhemaykov, N.I. Vorobyov, A.M. Lazarev and L.P. Kozlov

26. Responses of rice plants to inoculation with *Azospirillum* sp. under field conditions 237
Ngyuen Thi Phuong Chi, Ha Hong Thanh and Nguyen Ngoc Dzung

Posters

27. BGA technology for rice cultivation in northern India 243
R.P. Gupta, N.S. Chaudhary and M.S. Pandher

28. Modified combined carbon N-deficient medium for isolation, enumeration and biomass production of diazotrophs 247
N.A. Hegazi, M.A. Hamza, A. Osman, S. Ali, M.Z. Sedik and M. Fayez

29. Contribution of integrated symbiotic and associative systems of biological nitrogen fixation to N-status of intercropped legumes and non-legumes in sandy soils of Egypt 255
M. Monib, M.E. Hassan, H.H. Abo-Taleb, A. Metwally and N.A. Hegazi

30. Some physiological studies of diazotrophic bacteria associated with maize in a temperate and a tropical soil 261
F.B. Mwaura and D. Widdowson

31. *In-vitro* propagation using phytohormones produced by associative diazotrophs 265
H.M. El-Khawas, I.A. Ibrahim, H.M. Anwar and N.A. Hegazi

32. Effect of inoculation of *Azotobacter* and PSM on fertilizer economy, plant growth and yield of winter maize 271
Puneet K. Sohal, R.P. Gupta and M.S. Pandher

33. Accumulation of radionuclides by associative bacteria and the uptake of ^{134}Cs by the inoculated barley plants 275
A.A. Belimov, A.M. Kunakova, N.D. Vasilyeva, T.S. Kovatcheva, V.F. Dritchko, S.N. Kuzovatov, I.R. Trushkina and Yu. V. Alekseyev

34. The influence of chosen fungicides on *Azospirillum* dinitrogen fixation in the soil 281
Dorota Swedrzyńska and Aleksandra Sawicka

SECTION VII
Methodologies for assessing nitrogen fixation

35. Evaluation of biological nitrogen fixation associated with non-legumes 287
Robert M. Boddey, Bruno J.R. Alves and Segundo Urquiaga

36. Use of ^{15}N isotope dilution method to quantify nitrogen fixation in legumes and its potential use for non-legumes 307
Gudni Hardarson and Christian Hera

SECTION VIII
Actinorhizal and cyanobacterial association

37. Competitivity and persistence of *Frankia* in a managed environment 315
N.R. McEwan, E.M.O. Gould, E. Lumini, J.D. McNeill and C.T. Wheeler

38. Symbiotic association between *Frankia* and actinorhizal plants 327
Nantakorn Boonkerd

39. Ecophysiological studies of cyanobacteria in paddy soils of Bangladesh 333
M.A. Hashem

40. Cyanobacterial root nodules in *Tribulus terrestris* L. (Zygophyllaceae) 345
A. Mahmood and M. Athar

Posters

41. Use of *Azolla* as biofertilizer to enhance crop yield in a rice–wheat cropping system under mild climate 353
Sikander Ali, Naima Hamid, Dilrosh Khan and Kauser A. Malik

List of papers presented at the Second Rice BNF Working Group meeting (13–15 October, 1996) organized by IRRI and NIBGE, held in conjunction with the 7th International Symposium on Nitrogen Fixation with Non-Legumes published in *Plant and Soil*, Vol. 194, Nos. 1–2, 1997 359

Foreword

Until about 25 years ago, apart from the *Rhizobium* symbioses only rhizosphere associations with diazotrophic bacteria were known. After the development of the acetylene reduction method considerable nitrogenase activity has been detected in association with roots of wetland race, maize and various tropical grasses and especially with sugar cane. This nitrogen fixation could not be explained by the presence of rhizosphere diazotrophs known until then. This nitrogenase activity was found to be maximal at low oxygen concentrations and a semi-solid medium was prepared which permitted the isolation of diazotrophic bacteria from roots and rhizosphere of these grasses and cereals. In this medium the diffusion of O_2 to the site of bacterial growth is in equilibrium with the respiration rate of the growing bacteria and the pellicles of the growing bacteria move gradually towards the surface of the medium producing a veil-like pellicle at the point where O_2 diffusion matches the O_2 uptake and consequently maintains a pO_2 enough for the nitrogenase to function. This simple but very important technique quickly led to the discovery of two new species of the genus *Azospirillum* and subsequently three additional species of this genus. In addition two new species of *Herbaspirillum* in addition to *Bacillus azotofixans* were isolated using this technique. All these diazotrophs were found in association with roots, stems and leaves of grasses and cereals. *Acetobacter diazotrophicus* in addition was found within all parts of sugar cane. All these findings on endophytic dinitrogen fixation now explain much better the high contributions non-legumes can obtain from biological N_2-fixation and their potential for sustainable agriculture.

Dr. Johanna Dobereiner
EMBRAPA, CNPAB, Brazil

Preface

The major breakthrough in agricultural production came after the advent of the Green Revolution, which was driven by technologies heavily dependent on synthetic fertilizers. Such technologies make N fertilization the major driving force in food production. Production of chemical N fertilizer consumes huge amounts of non-renewable energy sources, thus making it the most expensive input into the agricultural system. Due to the recent increases in energy costs, chemical fertilizers have become increasingly less affordable for subsistence farmers in developing countries. In addition its excessive use in the developed world has been related to environmental pollution. Biological nitrogen fixation has therefore been an attractive proposition for sustainable agriculture. Traditionally its scope has been limited to legume crops. But since the majority of crops of economic importance are non-legumes, any scientific breakthrough in this area of research will have a direct impact on agricultural production.

In the early seventies Dr. Johanna Dobereiner and her group in Brazil reported nitrogen fixation associated with the roots of grasses which initiated a number of studies by several workers around the world. This renewed interest in grasses, which also include important crops like rice, wheat and sugarcane, led to the organization of the first International Symposium on 'Nitrogen Fixation with Non-Legumes' which was held in Brazil in 1979. Ever since, the scientists working on nitrogen fixation with non-legumes have been holding such symposia every three years in different parts of the world. Up till now seven such symposia have been held, in Brazil (1979), Canada (1982), Finland (1984), Brazil (1987), Italy (1990), Egypt (1993) and Pakistan (1996). It has also been decided that the 8th International Symposium on 'Nitrogen Fixation with Non-Legumes' will be held in Australia.

Over the last few years considerable progress has been made towards understanding the mechanisms involved in plant–diazotroph interactions. Several new associations have been discovered and new lines of action have been reported. The two international institutes engaged on cereals, namely IRRI and CIMMYT, are actively evaluating these technologies. IRRI has already developed a New Frontier Project to coordinate the worldwide collaborative efforts among research centers committed to reducing dependency of rice on mineral N resources. An International Rice BNF working group was established to review and share research results. The second meeting of this working group was held at NIBGE, Faisalabad, in conjunction with the 7th International Symposium. All the invited keynote speakers of the Symposium participated in this working group meeting. Its proceedings have been published separately as a volume of *Plant and Soil* (Volume 194, Nos. 1–2, 1997: contents page reproduced at the end of this book).

During the Symposium 30 oral presentations were made and 70 posters were displayed. This information was actively discussed by 130 participants from 30 countries. The present volume comprises most of the oral presentations and extended summaries of selected poster presentations. The wealth of information presented in this symposium highlighted various developments and priorities of research in this area. The availability of various molecular biology tools has now made it possible to elucidate and manipulate the process of plant–microbe interaction. The *Azoarcus*–Kallar grass association and *Acetobacter diazotrophicus*–sugarcane system are the best examples which can further enhance our understanding of these systems.

I would like to thank all the members of the International Scientific Committee, especially Prof. Bob Burris and Dr. Claudine Elmerich whose continuous advice and support made this symposium a reality. This Symposium would not have been possible without generous financial support from UNESCO, ISESCO, TWAS, IAEA, USNSF and DSE/ZEL. In addition we acknowledge support from Pakistan Atomic Energy Commission, Pakistan Science Foundation, University Grants Commission, Engro Chemicals, Karachi and Perkin & Elmer as represented by AMS.

Dr. Kauser A. Malik
Symposium Organizer

List of Contributors

Abbas, M., Environmental Research and Studies Unit, Department of Microbiology, Faculty of Agriculture, Cairo University, Giza, Egypt

Abdallah, F.E., Sakha Agricultural Research Station, Kafr El-Sheikh 33717, Egypt

Abo-Taleb, H.H., Environmental Research and Studies Unit, Microbiology Department, Faculty of Agriculture, Cairo University and Agricultural Research Centre, Giza, Egypt

Alekseyev, Yu.V., All-Russian Institute of Chemical Melioration of Soil, Gusarskaya St.9, Pushkin, Saint-Petersburg, 189620, Russia

Ali, S., Environmental Research and Studies Unit, Microbiology Department, Faculty of Agriculture, Cairo University, Giza, Egypt

Ali, Sikander, National Institute for Biotechnology and Genetic Engineering (NIBGE), PO Box 577, Jhang Road, Faisalabad, Pakistan

Alves, Bruno J.R., National Centre for Agrobiology Research (EMBRAPA-CNPAB), Km 47, Seropédica, Itaguaí, 23851N970, Rio de Janeiro, Brazil

Amin, G., Environmental Research and Studies Unit, Department of Microbiology, Faculty of Agriculture, Cairo University, Giza, Egypt

Anwar, H.M., Tissue Culture Laboratory, Agriculture Development Systems Project, Ministry of Agriculture, Cairo, Egypt

Aslam, M., Applied Microbiology Section, National Agricultural Research Centre, Islamabad

Athar, M., P A R C, Islamabad, Pakistan

Baldani, Ivo, EMBRAPA/CNPBS, Seropedica, 23851, Rio de Janeiro, Brazil

Baldani, V.L.D., EMBRAPA:CNPAB, Seropédica 23851-970, Rio de Janeiro, Brazil

Bano, A., Department of Biological Sciences, Quaid-i-Azam University, Islamabad

Belimov, A.A., All-Russian Research Institute for Agricultural Microbiology, Podbelsky Sh.3, Pushkin-8, Saint-Petersburg, 189620, Russia

Boddey, Robert M., National Centre for Agrobiology Research (EMBRAPA-CNPAB), Km 47, Seropédica, Itaguaí, 23851N970, Rio de Janeiro, Brazil

Boonkerd, Nantakorn, School of Biotechnology, Institute of Agricultural Technology, Suranaree University of Technology, Nakhon, Ratchasima, Thailand

Chaudhary, N.S., Department of Agronomy, Punjab Agricultural University, Ludhiana, 141 004, India

Cheema, M.S., Agronomy Research Station, Bahawalpur, Pakistan

Chen, Z.H., Institute for Application of Atomic Energy, Chinese Academy of Agricultural Sciences. PO Box 5109, Beijing 100094, PR China

Chernyaeva, I.I., Laboratory of Biological Nitrogen, Research Institute for Agricultural Microbiology, Podbelsky Sh. 3, Saint-Petersburg-Pushkin 8, 189620, Russia

Christian Hera, Soil and Water Management and Crop Nutrition Section, Joint FAO/IAEA Division, Wagramerstrasse 5, A-1400 Vienna, Austria

De Oliveira, Andre, EMBRAPA/CNPBS, Seropedica, 23851, Rio de Janeiro, Brazil

Deaker, Rosalind, SUNFix Centre for Nitrogen Fixation, Department of Agricultural Chemistry and Soil Science, University of Sydney, NSW 2006, Australia

Dobereiner, J., EMBRAPA:CNPAB, Seropédica 23851-970, Rio de Janeiro, Brazil

Dritchko, V.F., Agrarian University, Petersburgskoye Sh.2, Pushkin-6, Saint-Petersburg, 189620, Russia

El-Khawas, H.M., Unit of Soil Biotechnology, Department of Microbiology, Faculty of Agriculture, Cairo University, Giza, Egypt

El-Komy, Department of Botany, Faculty of Science, Minia University, Minia 61111, Egypt

Fatima, Zarrin, Department of Biological Sciences, Quaid-i-Azam University, Islamabad

Fayez, M., Environmental Research and Studies Unit, Microbiology Department, Faculty of Agriculture, Cairo University, Giza, Egypt

Gilchrist, Kate, SUNFix Centre for Nitrogen Fixation, Department of Agricultural Chemistry and Soil Science, University of Sydney, NSW 2006, Australia

Gill, M.A., Agronomy Research Station, Karor, Pakistan

Gould, E.M.O., Bower Building, Division of Biochemistry and Molecular Biology, IBLS, Glasgow University, Glasgow G12 8QQ, UK

Gudni Hardarson, FAO/IAEA Agriculture and Biotechnology Laboratory, A-2444 Seibersdorf, Austria

Guo, Wei, Institute of Botany, Chinese Academy of Sciences, Beijing 100093, PR China

Gupta, R.P., Department of Microbiology, Punjab Agricultural University, Ludhiana, Punjab, India

Hae, Weili, Institute for Application of Atomic Energy, Chinese Academy of Agricultural Sciences, PO Box 5109, Beijing, 100094, PR China

Hamid, Naima, National Institute for Biotechnology and Genetic Engineering (NIBGE), PO Box 577, Jhang Road, Faisalabad, Pakistan

Hamza, M.A., Environmental Research and Studies Unit, Microbiology Department, Faculty of Agriculture, Cairo University, Giza, Egypt

Hashem, M.A., Department of Soil Science, Bangladesh Agricultural University, Mymensingh 2202, Bangladesh

Hassan, M.E., Environmental Research and Studies Unit, Microbiology Department, Faculty of Agriculture, Cairo University and Agricultural Research Centre, Giza, Egypt

Hassan, Uzma, National Institute for Biotechnology and Genetic Engineering (NIBGE), Faisalabad, Pakistan

Hegazi, N.A., Environmental Research and Studies Unit, Microbiology Department, Faculty of Agriculture, Cairo University, Giza, Egypt

Hirsch, Ann M., Department of Molecular, Cell and Developmental Biology and Molecular Biology Institute, 405 Hilgard Avenue, University of California, Los Angeles, CA 90095N1606, USA

Hong Thanh, Ha, Department of Soil Microbiology, Inst. of Biotechnology, NCST of Vietnam

Ibrahim, I.A., Tissue Culture Laboratory, Agriculture Development Systems Project, Ministry of Agriculture, Cairo, Egypt

Katupitiya, Sunietha, SUNFix Centre for Nitrogen Fixation, Department of Agricultural Chemistry and Soil Science, University of Sydney, NSW 2006, Australia

Keerio, M.I., Department of Physiology and Biochemistry, Faculty of Crop Production, SAU, Tandojam, Pakistan

Kennedy, Christina, Department of Plant Pathology, College of Agriculture, PO Box 210036, University of Arizona, Tucson, AZ 85721, USA

Kennedy, Ivan R., SUNFix Centre for Nitrogen Fixation, Department of Agricultural Chemistry and Soil Science, University of Sydney, NSW 2006, Australia

Khan, Dilrosh, Agriculture Research Station, Mingora, Swat, Pakistan

Khokhar, Shahida Nasreen, Agricultural Biotechnology Institute, NARC Park Rd., Islamabad, Pakistan

Kovatcheva, T.S., All-Russian Research Institute for Agricultural Microbiology, Podbelsky Sh.3, Pushkin-8, Saint-Petersburg, 189620, Russia

Kozhemaykov, A.P., Laboratory of Biological Nitrogen, Research Institute for Agricultural Microbiology, Podbelsky Sh. 3, Saint-Petersburg-Pushkin 8, 189620, Russia

Kozlov, L.P., Laboratory of Biological Plant Protection, Research Institute of Plant Protection, Podbelsky Sh. 3, Saint-Petersburg-Pushkin 8, 189620, Russia

Kunakova, A.M., All-Russian Research Institute for Agricultural Microbiology, Podbelsky Sh.3, Pushkin-8, Saint-Petersburg, 189620, Russia

Kuzovatov, S.N., Agrarian University, Petersburgskoye Sh.2, Pushkin-6, Saint-Petersburg, 189620, Russia

Larue, Thomas A., Boyce Thompson Institute, Tower Road, Ithaca, NY 14853, USA

Latif, Farooq, National Institute for Biotechnology and Genetic Engineering (NIBGE), PO Box 577, Faisalabad, Pakistan

Lazarev, A.M., Laboratory of Biological Plant Protection, Research Institute of Plant Protection, Podbelsky Sh. 3, Saint-Petersburg-Pushkin 8, 189620, Russia

Lee, Sunhee, Department of Plant Pathology, College of Agriculture, PO Box 210036, University of Arizona, Tucson, AZ 85721, USA

Li, Jilun, National Laboratories for Agrobiotechnology, Beijing Agricultural University, Beijing 100094, PR China

Ling, M., Institute for Application of Atomic Energy, CAAS, PO Box 5109, Beijing 100094, PR China

Liu, Y.N., China Agricultural University, College of Biological Sciences, Beijing 100094

Lumini, E., Bower Building, Division of Biochemistry and Molecular Biology, IBLS, Glasgow University, Glasgow G12 8QQ, UK.
Current address: Dipartimento di Scienze e Tecnologie Alimentari e Microbiologiche dell'Universita degli Studi di Firenze, Florence, Italy

Mahmood, A., Department of Botany, University of Karachi, Karachi-75270, Pakistan

Malik, Kauser A., National Institute for Biotechnology and Genetic Engineering, PO Box 577, Faisalabad, Pakistan

McEwan, N.R., Bower Building, Division of Biochemistry and Molecular Biology, IBLS, Glasgow University, Glasgow G12 8QQ, UK

McNeill, J.D., Forestry Commission, Northern Research Station, Penicuik EH25 9SY, UK

Mehnaz, Samina, National Institute for Biotechnology and Genetic Engineering (NIBGE), PO Box 577, Jhang Road, Faisalabad, Pakistan

Meletzus, Dietmar, Department of Plant Pathology, College of Agriculture, PO Box 210036, University of Arizona, Tucson, AZ 85721, USA, and Department of Gene Technology and Microbiology, Faculty of Biology, University of Bielefeld, 33501, Bielefeld, Germany

Metwally, A., Environmental Research and Studies Unit, Microbiology Department, Faculty of Agriculture, Cairo University and Agricultural Research Centre, Giza, Egypt

Mishra, B.M., School of Biotechnology, Faculty of Science, Banaras Hindu University, Varanasi-221005, India

Moharram, T.M.M., Agricultural Microbiology Department, Faculty of Agriculture, Minia University, Minia, 61111, Egypt

Monib, M., Environmental Research and Studies Unit, Microbiology Department, Faculty of Agriculture, Cairo University, Giza, Egypt

Mwaura, F.B., Botany Department, University of Nairobi, P O Box 30197, Nairobi, Kenya

Nawroth, Roman, Department of Gene Technology and Microbiology, Faculty of Biology, University of Bielefeld, 33501, Bielefeld, Germany

Nayyar, M.M., Agronomy Research Institute, Faisalabad-38000, Pakistan

Ngoc Dzung, Nguyen, Department of Soil Microbiology, Inst. of Biotechnology, NCST of Vietnam

Norris, Joanna H., Department of Biological Sciences, University of Rhode Island, Kingston, RI, 02881, USA

Osman, A., Environmental Research and Studies Unit, Microbiology Department, Faculty of Agriculture, Cairo University, Giza, Egypt

Pandher, M.S., Department of Microbiology, Punjab Agricultural University, Ludhiana, Punjab, India

Pereg-Gerk, Lily, SUNFix Centre for Nitrogen Fixation, Department of Agricultural Chemistry and Soil Science, University of Sydney, NSW 2006, Australia

Perlova, Lena, Department of Gene Technology and Microbiology, Faculty of Biology, University of Bielefeld, 33501, Bielefeld, Germany

Ping, S.Z., Institute for Application of Atomic Energy, CAAS, PO Box 5109, Beijing 100094, PR China

Pishchik, V.N., Laboratory of Biological Nitrogen, Research Institute for Agricultural Microbiology, Podbelsky Sh. 3, Saint-Petersburg-Pushkin 8, 189620, Russia

Qunxing, Ding, Institute of Botany, Chinese Academy of Sciences, Beijing 100093, PR China

Qureshi, Azra, Agricultural Biotechnology Institute, NARC Park Rd., Islamabad

Rashid, A., Soil Bacteriology Section, Ayub Agricultural Research Institute, PO Box 775, Faisalabad-38000, Pakistan

Rasul, Ghulam, National Institute for Biotechnology and Genetic Engineering (NIBGE), PO Box 577, Jhang Road, Faisalabad, Pakistan

Safwat, M.S.A., Agricultural Microbiology Department, Faculty of Agriculture, Minia University, Minia, 61111, Egypt

Sajjad Mirza, M., National Institute for Biotechnology and Genetic Engineering, PO Box 577, Jhang Road, Faisalabad, Pakistan

Sajjad, M.R., Soil Bacteriology Section, Ayub Agricultural Research Institute, PO Box 775, Faisalabad-38000, Pakistan

Sawicka, Aleksandra, Department of Soil Microbiology, Agricultural University of Poznań, ul. Wolyńska 35, 60N637 Poznań, Poland

Sedik, M.Z., Environmental Research and Studies Unit, Microbiology Department, Faculty of Agriculture, Cairo University, Giza, Egypt

Sehly, M.R., Sakha Agricultural Research Station, Kafr El-Sheikh 33717, Egypt

Sevilla, Myrna, Department of Plant Pathology, College of Agriculture, PO Box 210036, University of Arizona, Tucson, AZ 85721, USA

Sindhu, M.S. Agronomy Research Station, Khanewal, Pakistan

Sohal, Puneet K., Department of Microbiology, Punjab Agricultural University, Ludhiana, Punjab, India

Song, W., Institute for Application of Atomic Energy, Chinese Academy of Agricultural Sciences. PO Box 5109, Beijing 100094, PR China

Su, Gui, National Laboratories for Agrobiotechnology, Beijing Agricultural University, Beijing 100094, PR China

Sun, X.L., Institute for Application of Atomic Energy, Chinese Academy of Agricultural Sciences. PO Box 5109, Beijing 100094, PR China

Swedrzyńska, Dorota, Department of Soil Microbiology, Agricultural University of Poznań, ul. Wolyńska 35, 60N637 Poznań, Poland

Teixeira, Katia, Department of Plant Pathology, College of Agriculture, PO Box 210036, University of Arizona, Tucson, AZ 85721, USA, and EMBRAPA/CNPBS, Seropedica, 23851, Rio de Janeiro, Brazil

Thi Phuong Chi, Nguyen, Department of Soil Microbiology, Inst. of Biotechnology, NCST of Vietnam

Tripathi, A.K., School of Biotechnology, Faculty of Science, Banaras Hindu University, Varanasi-221005, India

Trushkina, I.R., Agrarian University, Petersburgskoye Sh.2, Pushkin-6, Saint-Petersburg, 189620, Russia

Urquiaga, Segundo, National Centre for Agrobiology Research (EMBRAPA-CNPAB), Km 47, Seropédica, Itaguaí, 23851N970, Rio de Janeiro, Brazil

Vanderleyden, Jozef, F.A. Janssens Laboratory of Genetics, Catholic University of Leuven, W. De Croylaan 42, B-3001 Heverlee, Belgium

Vasilyeva, N.D., All-Russian Research Institute for Agricultural Microbiology, Podbelsky Sh.3, Pushkin-8, Saint-Petersburg, 189620, Russia

Vermeiren, Hans, F.A. Janssens Laboratory of Genetics, Catholic University of Leuven, W. De Croylaan 42, B-3001 Heverlee, Belgium

Vorobyov, N.I., Laboratory of Biological Nitrogen, Research Institute for Agricultural Microbiology, Podbelsky Sh. 3, Saint-Petersburg-Pushkin 8, 189620, Russia

Wang, Y.D., Institute for Application of Atomic Energy, Chinese Academy of Agricultural Sciences. PO Box 5109, Beijing 100094, PR China

Wang, Y.S., Research Center of Biotechnology, Hebei University, Baoding, Hebei Province 071002, PR China

Weeden, Norman F., Department of Horticultural Sciences, Cornell University, Geneva, NY 14456, USA

Wheeler, C.T., Bower Building, Division of Biochemistry and Molecular Biology, IBLS, Glasgow University, Glasgow G12 8QQ, UK

Widdowson, D., Botany Department, University of Nairobi, P O Box 30197, Nairobi, Kenya

Wilson, J.M., School of Biological Sciences, University of Wales, Bangor, UK

Wood, Craig, SUNFix Centre for Nitrogen Fixation, Department of Agricultural Chemistry and Soil Science, University of Sydney, NSW 2006, Australia

Xueqin, Shan, Institute of Botany, Chinese Academy of Sciences, Beijing 100093, PR China

Yang, H.L., Institute for Application of Atomic Energy, Chinese Academy of Agricultural Sciences. PO Box 5109, Beijing 100094, PR China

Yanni, Y.G., Sakha Agricultural Research Station, Kafr El-Sheikh 33717, Egypt

You, C.B., Institute for Application of Atomic Energy, CAAS, PO Box 5109, Beijing 100094, PR China

Youssef, H., Environmental Research and Studies Unit, Department of Microbiology, Faculty of Agriculture, Cairo University, Giza, Egypt

Yu, Daiguan, SUNFix Centre for Nitrogen Fixation, Department of Agricultural Chemistry and Soil Science, University of Sydney, NSW 2006, Australia

Yuxiang, Jing, Institute of Botany, Chinese Academy of Sciences, Beijing 100093, PR China

Zhao, Yinsuo, National Laboratories for Agrobiotechnology, Beijing Agricultural University, Beijing 100094, PR China

Opening lecture

1
Biological nitrogen fixation by endophytic diazotrophs in non-leguminous crops in the tropics

J. DOBEREINER AND V.L.D. BALDANI

Introduction

The relatively high contributions of biological dinitrogen fixation (BNF) to non-legume crops, especially in the tropics, such as in Brazil, can now be much better explained by the discovery of the endophytic colonization of roots, stems and leaves of cereals, forage grasses, oil palms and especially sugar cane, that are colonized by various new diazotrophic bacteria described recently. Unconsciously Brazilian sugar cane varieties have been selected with N levels (60 kg N ha^{-1}) much lower than the incorporation level normally present in this plant (around 250 kg N ha^{-1} per 18 months). The same happened with most other cereals and forage grasses, even though they do not obtain sufficient N from BNF for maximal yields.

Azospirillum spp. have been found within roots and even aerial parts of various gramineous plants such as sugar cane, wheat, maize and oil palms. However, the role of this organism present within the plant has not yet been identified, except for strain Sp 245 of *A. brasilense* in wheat, where it was shown that the bacteria enhanced nitrate assimilation from soil due to their very active nitrate reductase (Baldini *et al.*, 1983). During the last 5 years, however, four new diazotrophs were described which are obligate endophytes and do not survive for longer periods in soil as *Herbaspirillum* spp. (Olivares *et al.*, 1996). Similar results were also observed for *Acetobacter diazotrophicus* (Baldani *et al.*, 1997) and for the recently proposed *Burkholderia brasilensis* (Baldani, 1996).

Identification of endophytic diazotrophs

Herbaspirillum seropedicae was the first endophytic diazotroph identified (Baldani *et al.*, 1986) even though at the time it was not recognized as an obligate endophyte. The semisolid medium used for its isolation, the JNFb medium with higher

K.A. Malik et al. (eds.), Nitrogen Fixation with Non-Legumes, 3–7.

phosphate concentrations and pH 5.5, shows pellicles identical to those of *Azospirillum* spp. in NFb medium. With malate as the best C-source, the medium turns blue due to alkalinization in the presence of the indiactor Bromothymol blue. The only way to distinguish *Herbaspirillum* spp. from *Azospirillum* spp. is under the microscope. *Herbaspirillum* cells are much smaller than *Azospirillum* cells, about 0.5 × 2–3 μm in size. In wet mounts for microscopic visualization, usually only a few cells move spiraloid, close to air bubbles, and most of the others do not move.

Herbaspirillum rubrisubalbicans, previously known as *Pseudomonas rubrisubalbicans*, the causative agent of mottled stripe disease in sugar cane, was recognized as another *Herbaspirillum* species (Gillis *et al.*, 1991; Baldani *et al.*, 1996). Experiments with sorghum and sugar cane plantlets confirmed infection of the plants by these microorganisms (Olivares *et al.*, 1996), but with the Brazilian cane genotypes no symptoms of mottled stripe disease could be seen (Pimentel *et al.*, 1991). Attempts to isolate these bacteria from field-grown sugar cane in Brazil confirmed the presence of both *Herbaspirillum* spp. colonizing roots, stems and leaves. Under the transmission electron microscope colony-like agglomerates were visualized within the xylem, but only in adult plants. These findings can much better explain the high contributions that sugar cane obtains from BNF (Table 1) (see also chapter 35 of this volume) (see Boddey *et al.* at this symposium).

Herbaspirillum seropedicae can also be isolated from rice, maize, sorghum and forage grasses (Baldani *et al.*, 1986; Olivares *et al.*, 1996). Their contribution is not as high as in sugar cane, but can reach up to 50% of the N assimilated by the plant (Table 2). The selection of plant genotype with low fertilizer levels seems to be the key to the development of endophytic plant diazotroph associations. This was unconsciously done in Brazil, turning the country today into the one which uses the lowest mean N applications in agriculture among all the countries of the world (Table 3).

Another endophytic diazotroph requiring high sugar concentrations (10%) for best growth, *Acetobacter diazotrophicus*, was found in addition to *Herbaspirillum* spp., in sugar cane and sweet potatoes (Reis *et al.*, 1994). This endophyte excretes half of the N from BNF into the plant. This was shown in mixed cultures of *Acetobacter* sp. with yeast which, right from the beginning, multiplied in numbers of cells

Table 1. Dinitrogen fixation in sugar cane varieties (BNF contribution estimated by g N/m^2 per 3 years)

Variety	Total N in plants	N balance (g N/m^2)	Percentage N balance	^{15}N dilution (g N/m^2)	Percentage
CB 45–3	84.3ab	20.9	74	17.5b	62
SP 70–1143	77.5bc	18.6	72	17.3b	67
Krakatau	102.8a	27.8	79	23.9a	70
Chunee	33.0d	3.8	34	5.6	51
Caiana	11.6	–3.4	0	2.2d	58

Means followed by the same letter are not different by *F* test ($p = 0.05$)

Table 2. Contribution of BNF to N-incorporation in rice due to inoculation with endophytic diazotrophs

Inoculant	Log no. of cells/g fresh weight			Total plant mg/plant	Percentage BNF
	Root	Stems	Leaves		
Control	1.00	1.00	1.00	0.76	0
A. brasilense SP 109	6.14	3.14	3.14	1.13	32
H. seropedicae Z 94	6.40	4.14	4.14	1.07	54
H. seropedicae Z 67	6.14	5.14	4.17	1.06	31
H. rubrisubalbicans M4	6.04	4.50	4.17	0.96	36

Table 3. Mean fertilizer application in the world (kg/ha) (Solos and Adubos, 1988)

Country	NPK	N	N/P
Europe	345	172	4.3
China	142	87	11.4
USA	105	58	5.7
Mexico	63	36	7.6
India	46	27	7.6
Brazil	51	10	1.4

similar to the bacterium and, in the total-N, twice as much N_2 was fixed as in pure cultures (Cojho *et al.*, 1993). The co-colonization of sugar cane by *Herbaspirillum* spp. and *Acetobacter diazotrophicus* within the plant, into the stems where much sucrose is available, might explain the high contribution of BNF and the maintenance of a constant pH 5.5 of the sugar cane juice. This hypothesis is based on the physiological behaviour of *Acetobacter*, which uses the sucrose and excretes organic acids. *Herbaspirillum* uses the acids, and therefore maintains the pH constant at 5.5, which is the optimal pH for growth of both diazotrophs.

Recently a fourth endophytic diazotroph was found in rice and most other cereals, for which the name *Burkholderia brasilensis* was proposed (Baldani, 1996). This diazotroph forms more diffuse pellicles in semisolid medium, grows better at pH 4.5 and uses sugars or mannitol as carbon source. Very recently it was also isolated from fruit trees such as banana and pineapple, and from oil palms, together with a probably new additional *Herbaspirillum* species.

Impacts of BNF on tropical agriculture

The best example of the impact of BNF is certainly Brazil where, among all countries in the world, the lowest N applications are used (Table 3). The reason for relatively high yields with very little N contrasts with most other developed, and also developing, countries where green revolution varieties are used. Brazil produces sufficient cereals for food, and much more cheaply, because large areas are available for agriculture, and this is not the case in many other tropical countries.

The BNF in sugar cane became the key to the viability of the Proalcool programme. Recommendations based on ^{15}N experiments include the planting of the two varieties which fix most N_2, CB 45–3 and Sp 701143, without any N fertilizer. The farmer can use the money for higher phosphate applications and spraying plant tops with Mo solutions. In addition the elimination of burning cane leaves before harvesting increases soil fertility and cane yields which, in regions which use this technology, increased yields from 60 t ha^{-1} to more than 100 t, easily paying for the more expensive cane cutting with leaves, and creating jobs on the farm. With this technology the energy balance of alcohol production in Brazil will increase from 3 times (more energy produced than used for its production) to more than 5, while other countries, where high N fertilizers are applied obtain energy balances little above 1.

Recent studies have opened new possibilities for the replacement of diesel fuel by oil from palm plants, which produce 50% more energy than sugar cane alcohol (Programa Nacional de Dendê, 1987). In roots, stems and leaves up to 10^6 diazotrophic bacteria/g have now been found, probably new ones, permitting the complete elimination of N fertilizers which, in Brazil, are rarely used on oil palms. It has been calculated that half of the already-deforested Amazon region, if planted with oil palms mixed with legume trees, could revitalize these eroded regions and produce sufficient palm oil to replace all the diesel fuel in the country. In addition, the PRONADEM programme would create thousands of jobs in these poorest areas of the country.

The now well-confirmed endophytic colonization of Gramineae and other monocotyledonous plants forming a symbiotic association can much better explain the high contributions of BNF in the tropics than rhizosphere associations. The relatively high contribution of fixed N to rice, a C3 plant, indicates a much larger extension of this symbiosis than has been thought so far.

With these new endophytic diazotrophs many new possibilities arise to reduce pollution in tropical countries, and to solve the most difficult social and environmental problems.

References

Baldani V L D 1996. Efeito da inoculaçãode *Herbaspirillum* spp. no Processo de colonização e infecção de plantas de arroz e occorrência e caracterização parcial de uma bactéria diazotrofica. PhD thesis, UFRRJ, Seropédica, RJ.

Baldani J I, Baldani V L D, Seldin L, Dobereiner J 1986. Characterization of *Herbaspirillum seropedicae* gen. nov. sp. nov. a root associated nitrogen fixing bacterium. Int. J. Syst. Bacteriol., 36, 86–93.

Baldani J I, Caruzo L, Baldani V L D, Goi S R, Döbereiner J 1996a. Recent advances in BNF with non-legumimes plants. Soil Biol. Biochem, 29(5/6), 911–922.

Baldani J I, Pot B, Kirchhof G, Falsen E, Baldani V L D, Olivares F L, Horste B, Kersters K, Hartman A, Gillis M, Dobereiner J 1996. Emended description of *Herbaspirillum*; inclusion of '*Pseudomonas*' *rubrisubalbicans*, a mild plant pathogen, as *Herbaspirillum rubrisubalbicans* comb. nov.; and classification of a group of clinical isolates (EF group) as *Herbaspirillum*. Int. J. Syst. Bacteriol., 46(3), 802–810.

Baldani V L D, Baldani J I, Dobereiner 1983. Effects of *Azospirillum* inoculation and nitrogen incorporation in wheat. Can. J. Microbiol., 29, 924–929.

Cojho E H, Reis V M, Schenberg A C G, Dobereiner J 1993. Interactions of *Acetobacter diazotrophicus* with an amylolytic yeast in nitrogen free batch culture. FEMS Microbiol. Lett., 106, 341–346.

Gillis M, Dobereiner J, Pot B, Goor M, Falsen E, Horste B, Reinhold B, Kersters K 1991. Taxonomic relationships between (*Pseudomonas*) *rubrisubalbicans*, some clinical isolates (EF Group 1), *Herbaspirillum seropedicae* and (*Aquaspirillum*) *autotrophicum*. In: Polsinelli M, Materassi R, Vincenzine M, eds. Nitrogen Fixation. Kluwer, Dordrecht, 293–364.

Olivares F L, Baldani V L D, Reis V M, Baldani J I, Dobereiner J 1996. Occurrence of endophytic diazotrophs *Herbaspirillum* spp. in roots, stems and leaves, predominantly of Gramineae. Biol. Fert. Soils, 21, 197–200.

Pimentel J P, Olivares, F L, Reis V M, Akiba F, Dobereiner J 1991. Dinitrogen fixation and infection of grass leaves by *Pseudomonas rubrisubalbicans* and *Herbaspirillum seropedicae*. Plant Soil, 137, 1–65.

PRONADEM. O óleo de dendê como recurso energético. Brasilia: EMBRAPA-DDT, 1987. 40p. (EMBRAPA:UEPAE de Belem. Documento, 1).

Reis V M, Olivares F L, Dobereiner J 1994. Improved methodology for isolation of *Acetobacter diazotrophicus* and confirmation of its endophytic habitat. World J. Appl. Microbiol. Biotechnol. 10, 401–405.

Solos, Adubos 1988. Boletin Informativo, São Paolo 21, 147p.

SECTION I

Biochemistry and genetics of diazotrophs

2
Analysis of nitrogen fixation and regulatory genes in the sugar cane endophyte *Acetobacter diazotrophicus*

SUNHEE LEE, MYRNA SEVILLA, DIETMAR MELETZUS, KATIA TEIXEIRA, ANDRE DE OLIVEIRA, LENA PERLOVA, ROMAN NAWROTH, IVO BALDANI AND CHRISTINA KENNEDY

Abstract

Acetobacter diazotrophicus, a diazotrophic endophyte of sugarcane, carries a cluster of *nif* genes similar to those characterized in other Proteobacterial diazotrophs. These include, in order, *nifA, nifB, nifHDKEN, nifUSVW* and *fixABC*. This order of *nif* and *fix* gene organization is most similar to that found in *Azospirillum brasilense*, although the deduced gene products of each gene are generally more similar to those characterized in species of *Rhizobium, Bradyrhizobium* or *Azorhizobium*. Downstream of *fixABC* is a gene encoding a protein with similarity to MCPA, a methyl-accepting chemotaxis protein which is involved in chemotatic responses to extracellular signals. *mcpA* gene may play an important role in symbiosis and plant–microbe interaction by allowing colonization of sugarcane tissues far distant from the site of *A. diazotrophicus* infection or entry.

Introduction

The application of molecular genetics to nitrogen fixation and nodulation in free-living diazotrophs and rhizobial species has led to immense understanding of both processes. The identity and genomic arrangement of approximately 20 *nif* genes encoding proteins for the conventional Mo-containing nitrogenase enzyme and its activity have been most fully established for two members of the γ group of Proteobacteria, *Klebsiella pneumoniae* and *Azotobacter vinelandii* and for the α group member *Rhodobacter capsulatus* (Dean and Jacobson, 1992; Masepohl and Klipp, 1996). Some if not all *nif* genes have also been identified in several other diazotrophs, including other Proteobacterial species in the genera *Azospirillum, Rhodospirillum, Rhizobium, Bradyrhizobium, Azorhizobium, Herbaspirillum*, as well as in some *Clostridium*, Cyanobacteria and Archaebacteria (reviewed

K.A. Malik et al. (eds.), Nitrogen Fixation with Non-Legumes, 11–19.

by Dean and Jacobson, 1992; see also Almassy *et al.*, 1986; Elmerich *et al.*, 1992; Haselkorn and Buikema, 1992; Chien and Zinder, 1996). A striking feature that emerges from the comparison of *nif* gene arrangement in these organisms is the diversity of alignment and linkage that occurs. For example, while in most organisms the *nifHDK* nitrogenase structural genes are contiguous and cotranscribed, in *Bradyrhizobium japonicum, nifH* and *nifDK* are separated by 15 kb of DNA. The differences in arrangement of *nif* genes that occur among diazotrophs are too numerous to name; the only conserved feature apparent is that the *nifDK* genes, encoding the structural subunits of the heterotetrameric MoFe protein or Component I of nitrogenase, are contiguous and cotranscribed, although in *Anabaena* the *nifD* gene is disrupted by and separated from *nifK* by an insertion element that is excised upon derepression (Haselkorn and Buikema, 1992). The diversity of arrangement is also reflected in the relative proximity of *nif* genes to each other; in some organisms, most *nif* genes are clustered in one or two segments of the bacterial chromosome, in others, there may be up to four or five genetically unlinked groups of *nif* (and/or *fix*) genes (Dean and Jacobson, 1992).

Additionally, while nitrogenase activity and *nif* gene expression are influenced in all diazotrophs by the environmental availability of fixed N supply (ammonium) and oxygen, these control mechanisms are often quite different and involve either inactivation of nitrogenase enzyme or the expression or activity of the *nifA* gene product, NifA, a transcriptional activator of expression of all other *nif* genes in Proteobacterial diazotrophs (Merrick and Edwards, 1995; Colnaghi *et al.*, 1997).

Applications of molecular tools to understanding the mechanism of rhizobial nodulation of legume plants has led to another paradigm, equally important to that describing *nif* gene identity, product function, and regulation. The general concepts and details can be found in recent reviews and represent an elegant and effective mechanism for two partners in a symbiosis to both signal specificity and trigger differentiation for plant and bacterial cells to become appropriately adapted (Denarie *et al.*, 1992; Fisher and Long, 1992; Spaink, 1994). In the rhizobium/legume case, flavonoids or related molecules released by plant roots induce expression of bacterial genes whose products direct synthesis of chito-lipo-oligosaccharides, named Nod factors, that induce root tissues to differentiate into nodule structures that house rhizobia and are the site of nitrogen fixation.

The recent identification of nitrogen-fixing bacteria with endophytic habitats raises the possibility of a new classification of symbiosis; one in which, like the rhizobium/legume symbiosis, the plant partner supplies photosynthate to support bacterial nitrogen fixation and the bacterium gives up its fixed nitrogen for plant growth. Unlike the rhizobium/legume symbiosis, there appears to be no differentiated organ analogous to a root nodule as the site of nitrogen fixation. One of the most promising, both in terms of specificity of endophytic association, and suggestive evidence for significance of bacterial nitrogen fixation for plant N nutrition is that between a Proteobacterial group member, *Acetobacter diazotrophicus*, and sugarcane (Cavalcante and Dobereiner, 1988; Gillis *et al.*,

1989; James *et al.*, 1994). Our goals in studying *A. diazotrophicus* and its endophytic association with sugarcane are: (1) to analyse the bacterial genes responsible for nitrogen fixation and its regulation, and (2) to establish that *A. diazotrophicus* colonizing sterile sugarcane plants contributes significant fixed nitrogen to the plant to enhance its growth in nitrogen-limiting growth conditions.

The first goal has contributed to the second in identifying genes to mutagenize for the construction of non-nitrogen-fixing mutants. In the longer term such understanding will lead to strategies to manipulate regulatory genes for isolation of mutants with enhanced ability to fix nitrogen. The genes we sought to isolate and characterize belong to the groups *nif*, *ntr*, and *gln* because of their established importance in nitrogen fixation and regulation in other diazotrophs (Merrick and Edwards, 1995; Colnaghi *et al.*, 1997).

This chapter reports the identification and/or genomic mapping of 14 *nif* and *fix* genes, four *ntr* genes and three *gln* genes. Another gene of possible relevance to symbiosis and lying adjacent to *nif* and *fix* regions encodes a product with similarity to a protein involved in chemotaxis. In addition, an insertion mutation of the *nifD* gene and its reintroduction into the *A. diazotrophicus* genome has resulted in the construction of a Nif mutant. Both wild type Nif^+ strain and the $nifD^-$ mutant that is unable to fix nitrogen are able to colonize the interior of sugarcane.

Materials and methods

Bacteria and growth

A. diazotrophicus strain PAL5 was chosen for this work because it was isolated from a sugarcane cultivar in Brazil that has been agriculturally productive for many years in soils not supplied with nitrogenous fertilizer (Boddey *et al.*, 1991), and also because, unlike most other isolates of *A. diazotrophicus*, this strain does not apparently contain any plasmid DNA, thus minimizing the influence of extraneous, probably irrelevant, genetic material in analysis of nitrogen fixation and/or symbiotically relevant genes. In a screening of several plasmid-containing isolates of *A. diazotrophicus*, none had *nif* genes located on identifiable plasmids (Teixeira *et al.*, 1993).

Genomic library

A. diazotrophicus was grown as described in James *et al.* (1994). Genomic libraries were prepared in: (1) a lambda cloning vector, EMBL3, to allow preparation of large amounts of DNA to be analysed by Southern blot hybridization to known *nif* and other gene probes, and (2) in pLAFR3, a wide host range cosmid that can be transferred by conjugation to other members of the Proteobacteria, for the isolation of *nif* and related genes of *A. diazotrophicus* by their ability to complement mutants of *A. vinelandii.*

Results and discussion

Identification, isolation, sequencing and mapping of nif *and* fix *genes*

The *nifHDK* genes were isolated from a lambda library carrying *A. diazotrophicus*, DNA by hybridization with the *nifHDK* region of *Azospirillum brasilense* (Sevilla *et al*., manuscript in preparation). Appropriate subclones and deletion derivatives of this region were sequenced using standard dideoxynucleotide sequencing reactions. The amino acid sequences deduced from the nucleotide sequence of the three nitrogenase structural genes are most similar to those from *R. leguminosarum* bv. *phaseoli* (*nifH*: 85% identity); *H. seropedicae* (*nifD*: 83% identity) and *B. japonicum* (*nifK:* 76% identity).

The *A. diazotrophicus nifA* gene was isolated on a cosmid from a pLAFR3 gene library able to complement to Nif$^+$ a *nifA* mutant of *A. vinelandii* (Teixeira *et al*., manuscript in preparation). Sequencing of the *nifA* region on the cosmid insert DNA, identified by the ability of subclones to complement the *A. vinelandii nifA* mutant strain and by Southern hybridization experiments using the *A. brasilense nifA* gene as probe, revealed the *nifA* ORF as well as an ORF corresponding to the N-terminal-encoding region of the *nifB* gene located just downstream of *nifA*. The *nifA* gene product of *A. diazotrophicus* is most similar to NifA of *Azorhizobium caulinodans* among the NifA's of the Proteobacterial diazotrophs (51% identity, 68% similarity). The 980 amino acids of the N-terminus of the NifB protein of *A. diazotrophicus* are most similar to those of the *R. capsulatus* gene product (76% identity, 83% similarity).

A second pLAFR3 library cosmid carrying *nif* genes was isolated by its ability to complement *nifE* and *nifV* mutants of *A. vinelandii* (the same cosmid was isolated from Nif$^+$Tet^r transconjugants of DJ35 and DJ71; Jacobson *et al*., 1989). Subcloning and sequencing from the ends of particular subcloned restriction fragments has resulted in the 'sketch' of *nif* and *fix* genes shown in Figure 1, along

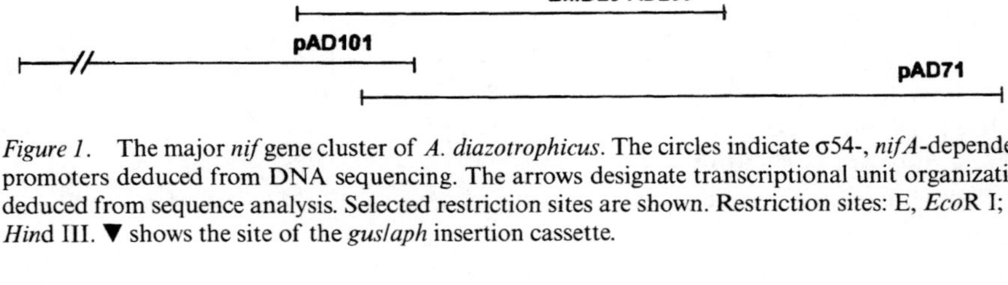

Figure 1. The major *nif* gene cluster of *A. diazotrophicus*. The circles indicate σ54-, *nifA*-dependent promoters deduced from DNA sequencing. The arrows designate transcriptional unit organization deduced from sequence analysis. Selected restriction sites are shown. Restriction sites: E, *Eco*R I; H, *Hin*d III. ▼ shows the site of the *gus/aph* insertion cassette.

with the *nifHDK* and *nifAB* genes described above. While only four of these genes and ORF's have been completely sequenced on both strands, a sufficient number of nucleotides representing a significant proportion of the amino acid sequence of deduced gene products identified from database searches have been obtained to make this map valid. The data obtained so far are summarized in Table 1. Of great interest is that these three regions isolated by different strategies are actually adjacent to each other on the PAL5 genome. This was discovered by sequencing results which indicated the overlaps between EMBL-AD200, pAD101, and pAD71 (Figure 1).

Of the five other Proteobacterial diazotrophs for which information about

Table 1. Similarity of *A. diazotrophicus* gene products to those of known *nif, fix* and other genes

Functions	Genes	Sequenced gene (%)[a]	Most similar organism[b]	Identity (%)[c]	Similarity (%)[c]
Nitrogenase structure	*nifH*	100	*R. leguminosarum*	85	91
	nifD	100	*H. seropedicae*	83	91
	nifK	100	*B. japonicum*	76	83
FeMo cofactor biosynthesis	*nifB*	71	*R. capsulatus*	76	83
	nifE	97	*B. japonicum*	63	75
	nifN	41	*R. meliloti*	61	80
	nifV	92	*R. sphaeroides*	44	55
Maturation or activation of nitrogenase	*nifS*	55	*R. sphareoides*	68	83
	nifW	78	*R. capsulatus*	44	59
Electron transfer	*fixA*	95	*R. leguminosarum*	59	72
	fixB	88	*B. japonicum*	49	79
	fixC	66	*B. japonicum*	43	64
	fixX	50	*R. leguminosarum*	55	78
Transcriptional activation	*nifA*	100	*A. caulinodans*	51	68
Other nitrogen regulation genes	*glnB*	100	*A. brasilense*	77	87
	glnZ	100	*A. brasilense*	78	89
	glnD	75	*A. vinelandii*	38	54
	ntrB	100	*A. brasilense*	47	61
	ntrC	100	*A. brasilense*	68	81
	nifR3	95	*A. brasilense*	59	71
	ntrX	100	*A. brasilense*	63	73
	ntrY	100	*A. caulinodans*	43	58
Chemotaxis response	*mcpA*	46	*C. crescentus*	59	69

[a] Percentage of gene sequence is estimated from the total base pairs obtained so far for the *A. diazotrophicus* gene compared to the known size of that gene in other organisms.
[b] The organism named is that in which the indicated gene product of *A. diazotrophicus* was found to be most similar. Organisms: *Azospirillum brasilense*; *Azorhizobium caulinodans*; *Azotobacter vinelandii*; *Bradyrhizobium japonicum*; *Caulobacter crescentus*; *Herbaspirillum seropedicae*; *Rhizobium leguminosarum*; *Rhizobium meliloti*; *Rhodobacter capsulatus*; *Rhodobacter sphaeroides*.
[c] Percentage identity and similarity are of the deduced *A. diazotrophicus* gene product sequence compared to the gene product most related, indicated in the previous column.

the clustering and arrangements of *nif* genes is known, the arrangement of *nif* genes in *A. brasilense* is most similar to that of *A. diazotrophicus* (Elmerich *et al.*, 1992). While some of the individual *A. diazotrophicus nif* gene products show more similarity to Nif proteins of other Proteobacterial species, those of rhizobial genera in particular, than to those of *A. brasilense*, the overall arrangement of *nif* and *fix* genes is more like that of the latter organism.

Two other genes were *mocA* and *mcpA*. *mcpA* was located by sequencing DNA downstream of *fixABCX* and is currently being mutagenized to determine whether it has any influence on nitrogen fixation or colonization. The *mcpA* gene product is a transmembrane protein with a perisplasmic input, or sensing domain and cytoplasmic output, or signalling domain (Parkinson, 1993). This pathway allows the transfer of signals indicating chemical gradients from the environment to the bacterial flagella motor where alterations in rotation result in movement towards attractants and away from repellents. Bacteria capable of such chemotaxic response may have a competitive advantage by detecting and migrating towards optimal conditions for growth (Zhulin and Taylor, 1995). Differences in the composition of plant root exudates may stimulate the specific chemotaxic response resulting in specificity of the bacteria/plant association (Barbara *et al.*, 1985).

Isolation and sequencing of *ntr* and *gln* genes of *A. diazotrophicus*

Genes known to be involved in regulation of *nif* gene expression and ammonium assimilation in other diazotrophs include *ntrBC, glnB*, and *glnD* (see Merrick and Edwards, 1995, for review). Homologues of these genes were isolated by the ability of a pLAFR3 cosmid, pAD101, to complement *ntrC* mutants of *A. vinelandii* for growth on nitrate (Toukdarian and Kennedy, 1986). Sequence analysis of pAD101 subclones revealed a *nifR3–ntrBC* gene cluster similar to the organization of similar genes identified in *Azospirillum brasilense, Rhodobacter capsulatus* and *Rhizobium leguminosarum* (Machado *et al.*, 1995; Fosterhartnett *et al.*, 1993; Patriarca *et al.*, 1993) (Table 1). In additon, the deduced amino acid sequences of two open reading frames downstream of *ntrC* were found to be highly similar to the *ntrYX* operon, coding for a unique two-component system of *Azorhizobium caulinodans* that regulates expression of the *nifA* gene (Pawlowski *et al.*, 1991). Mutants of *A. diazotrophicus* with lesions in these genes are currently being analysed to determine their influence on regulation of nitrogen fixation and interaction between *A. diazotrophicus* and its sugar cane host.

The PII protein encoded by *glnB* plays a major role in the ammonia sensing mechanism in enteric and other bacteria (see Colnaghi *et al.*, 1997). To isolate this gene, degenerate oligonucleotide primers encoding conserved sequences deduced from the alignment of known PII proteins were used in a PCR amplification experiment with total DNA of *A. diazotrophicus*. Amplified fragments were cloned in *E. coli* and sequenced in both orientations. The corresponding full-length clones were isolated by screening of the EMBL3-phage library. Interestingly, sequencing data revealed that *A. diazotrophicus*, like *Azospirillum brasilense* and *E. coli*, harbours two copies of *glnB*-like genes sharing 75% nucleotide base pair identity and 73% deduced amino acid sequence identity with each other, and 77% and 78% amino acid identity with the *glnB* and *glnZ* gene products of *A. brasilense* (Table 1). The *glnD* gene was isolated by complementation of the *nfrX (glnD)* mutant strain MV17 (Santero *et al.* 1988) of *A. vinelandii* for growth on N_2 with the pLAFR3 cosmid library. Introduction of deletions as well as subcloning of restriction fragments followed by DNA sequencing identified the DNA region encoding the *glnD* gene on the cosmid pAD401. Work is in progress on the molecular analysis of these genes and on the corresponding mutants with respect to their effect on regulation of *nif* gene expression.

Isolation of Nif$^-$ (*nifD*) mutants of *A. diazotrophicus*

A *nifD*: Kan *uidA* (*GUS*) insertion mutation was constructed by insertion of the Kan *uidA* cassette into the *EcoR* I site within the *nifD* gene (see Figure 1) cloned in a suicide vector, pSUP203, which is unable to replicate in *A. diazotrophicus*. Transfer of this plasmid by conjugation, mediated by RP4 *mob* genes, into *A. diazotrophicus* and selection for Kan^r transconjugants was successful. Transconjugants carrying the *nifD* mutation replacing the wild-type gene were isolated. These strains were Nif$^-$ (unable to grow on N-deficient medium and failed to reduce acetylene), but not apparently affected in other phenotypes (Sevilla, manuscript in preparation).

Colonization of sugarcane by wild-type and *nifD*$^-$ strains

Preliminary experiments in which micropropagated sugarcane plantlets were inoculated with *A. diazotrophicus* wild-type and *nifD*$^-$ mutant strains indicate that both strains are equally able to colonize the interior of the plants, whether they are grown in medium with or without mineral N. The evidence for this is from scanning electron microscopy which showed similar numbers of both wild-type and mutant strains to be present within intercellular spaces of plant roots and shoots 10 days after inoculation (data not shown). In addition, estimation of the number of bacteria by the most probable number technique showed the presence of bacteria in roots, stems and shoots of inoculated plants. Numbers were similar in plants inoculated with either the wild-type or *nifD*$^-$ strains. Experiments are in progress to determine whether sterile sugarcane plants inoculated with wild-type and mutant strains are able to grow better under conditions of ammonium excess or limitation.

Acknowledgements

Work on *Acetobacter diazotrophicus* was supported at the University of Arizona by International Collaborative Grants from NSF (INT9302788) and USDA (5319R4030), and by Hatch funds (ARZT-136701-05-114). We thank DFG (Germany) for supporting the postdoctoral work of D. Meletzus at the University of Arizona and CNpQ (Brazil) for funding collaborative visits of K. Teixeria and I. Baldani to the University of Arizona. We are also grateful to Dr Johanna Dobereiner for her seminal discovery of *Acetobacter diazotrophicus* in sugarcane and for her enthusiastic encouragement of our collaborative research programme.

References

Almassy R J, Janson C A, Hamlin R, Xuong N H, Eisenberg D 1986. Novel subunit–subunit interactions in the structure of glutamine synthetase. Nature, 323, 304–309.

Barbara R, Thomas H, Istban F 1985. Strain specific chemotaxis of *Azospirillum* spp. J Bacteriol. 162, 190–195.

Boddey R M, Urquiaga S, Reis V, Dobereiner J 1991. Biological nitrogen fixation associated with sugarcane. Plant Soil, 137, 111–117.

Cavalcante V A, Dobereiner J 1988. A new acid-tolerant nitrogen-fixing bacterium associated with sugarcane. Plant Soil, 108, 23–31.

Chien Y T, Zinder S H 1996. Cloning, functional organization, transcript studies, and phylogenetic analysis of the complete nitrogenase structural genes (nifHDK2) and associated genes in the archaeon *Methanoscarcina barkeri* 227. J. Bacteriol., 178, 143–148.

Colnaghi R, Green A, He L, Rudnik P, Kennedy C 1997. Strategies for increased ammonium production in free-living or plant associated nitrogen fixing bacteria. Plant and Soil, 194, 145–154.

Contreras C, Drummond M, Bali A, Garcia E, Bush G, Kennedy C, Merrick M 1991. The product of the nitrogen fixation regulatory gene *nfrX* of *Azotobacter vinelandii* is functionally and structurally homologous to the uridylytransferase encoded by *glnD* in enteric bacteria. J. Bacteriol., 173, 7741–7749.

Dean D, Jacobson M R 1992. Biochemical genetics of nitrogenase. In: Stacey G, Evans H J, Burris R, eds. Biological Nitrogen Fixation. Chapman & Hall, New York, 763–834.

Denarie J Debelle J, Rosenberg C 1992. Signaling and host range variation in nodulation. Annu. Rev. Microbiol., 46, 497–531.

Elmerich C, Zimmer W, Vieille C 1992. Associative nitrogen-fixing bacteria. In: Stacey G, Burris R H, Evans H J, eds. Biological Nitrogen Fixation. Chapman & Hall, New York, 212–258.

Fisher R F, Long S R 1992. *Rhizobium*–plant signal exchanges. Nature, 357, 655–659.

Fosterharnett D, Cullen P J, Gabbert K K, Kranz R G 1993. Sequence, genetic analysis and *lacZ* fusion analysis of a *nifR3–ntrB–ntrC* operon in *Rhodobacter capsulatus*. Mol. Microbiol., 8, 903–914.

Gillis M, Kersters K, Hoste B, Janssens D, Kroppenstedt R M, Stephan M P, Teixeria K R S, Dobereiner J, De Ley 1989. *Acetobacter diazotrophicus* sp. nov., a nitrogen-fixing acetic bacterium associated with sugarcane. Int. J. Syst. Bacteriol., 39, 361–364.

Haselkorn R, Buikema W J 1992. Nitrogen fixation in *Cyanobacteria*. In: Stacey G, Burris R H, Evans H J, eds. Biological Nitrogen Fixation. Chapman & Hall, New York, 166–190.

Jacobson M R, Brigle K E, Bennett L, Setterquist R A, Wilson R A, Cash V L, Beynon J, Newton W E, Dean D R 1989. Physical and genetic map of the major *nif* gene cluster from *Azotobacter vinelandii*. J. Bacteriol. 171, 1017–1027.

James E K, Reis V M, Olivares F L, Baldani J I, Dobereiner J 1994. Infection of sugar cane by nitrogen-fixing bacterium *Acetobacter diazotrophicus*. J. Exp. Bot., 45, 757–766.

Machado H B, Yates M G, Funayama S, Rigo L U, Steffens M B R, Souza E M, Pedrosa F O 1995. The *ntrBC* genes of *Azospirillum brasilense* are part of a *nifR3*-like-*ntrB–ntrC* operon are negatively regulated. Can. J. Microbiol. 41, 647–684.

Masepohl B, Klipp W 1996. Organization and regulation of genes encoding the molybdenum nitrogenase and the alternative nitrogenase in *Rhodobacter capsulatus*. Arch. Microbiol. 165, 80–90.

Merrick M J, Edwards R A 1995. Nitrogen control in bacteria. Microbiol. Rev. 59, 604–622.

Parkinson J S 1993. Signal transduction schemes of bacteria. Cell, 73, 857–871.

Patriarca E J, Riccio A, Tate R, Colonnaromano S, Iaccarino M, Defez R 1993. The *ntrBC* genes of *Rhizobium leguminosarum* are part of a complex operon subject to negative regulation. Mol. Microbiol., 9, 569–577.

Pawlowski K, Klosse U, de Bruijn F J 1991. Characterization of a novel *Azorhizobium caulinodans* ORS571 two-component regulatory system, NtrY/NtrX involved in nitrogen fixation and metabolism. Mol. Gen. Genet., 231, 124–138.

Rossbach S, Rasul G, Schneider M, Eardly B, de Bruijn F 1995. Structural and functional conservation of the rhizopine catabolism (moc) locus is limited to selected *Rhizobium meliloti* strains and unrelated to their geographical origin. Mol. Plant–Microbe Interact., 8, 549–559.

Santero E, Toukdarian A, Humphrey R, Kennedy C 1988. Identification and characterization of two nitrogen fixation regulatory regions, *nifA* and *nfrX*, in *Azotobacter vinelandii* and *Azotobacter chroococcum*. Mol. Microbiol., 2, 203–314.

Spaink H P 1994. The molecular basis of the host specificity of the *Rhizobium* bacteria. Anton Leeuwenhoek Int. J. Gen. M., 65, 31–98.

Teixeira K R S, Galler R, Kennedy C, Baldani J I 1994. Plasmid contents and *nif* gene detection in *Acetobacter diazotrophicus* strain. In: Nitrogen Fixation with Non-Legumes. Hegazi N A, Fayez M, Monib M, eds. The American University in Cairo Press, Cairo, 273–281.

Toukdarian A, Kennedy C 1986. Regulation of nitrogen metabolism in *Azotobacter vinelandii*: isolation of *ntr* and gl*nA* genes and construction of *ntr* mutants. EMBO J., 5, 399–407.

Zhulin I B, Taylor B L 1995. Chemotaxis in plant-associated bacteria: the search for the ecological niche. In: Fendrik I *et al.*, eds. *Azospirillum* VI and Related Microorganisms. Springer-Verlag, Berlin and Heidelberg, 451–459.

3
Functional analysis of a cloned *nifA* homologous gene of *Azospirillum brasilense* Yu62

GUI SU, YINSUO ZHAO AND JILUN LI

Abstract

NifA proteins are the transcriptional activators of the *nif* genes in nitrogen-fixing bacteria. A *nifA*-like gene of *Azospirillum brasilense* Sp7 had been identified and functionally analysed (Liang *et al.*, 1991, named *EnifA* below). In our laboratory a *nifA* homologous gene from *A. brasilense* Yu62 had been cloned before, but without functional analysis (Yan *et al.*, 1993, named *LnifA* below). Sequence analysis showed that these are two different genes. The *ENif*A$^-$ mutant Sp 7067 (*EnifA::Tn5*) was complemented almost 100% by *LnifA*. This suggested that *LnifA* probably participated in the activation of *nif* genes in *A. brasilense*. The *LnifA* gene was mutated by inserting a *kam*^{r} gene. Some insertion mutants lost nitrogenase activity. In one of these Nif$^-$ mutants, 78% of the nitrogenase activity was restored by the *EnifA* gene. Whether the *LnifA* acts as a *nif*A-like gene needs further study.

Introduction

nifA genes have been cloned from several nitrogen-fixing bacteria (Liang *et al.*, 1991; Yan *et al.*, 1993) which are known to play a role as transcriptional activators for *nif* or *fix* genes, though the homology between these genes was low. The expression of *nif* genes was mainly controlled by the expression and activity of NifA. *A. brasilense* is a microaerobic nitrogen-fixing bacterium associated with the roots of many important cereals and grasses. A *nifA*-like gene (*EnifA*) from *A. brasilense* Sp7 had been cloned and functionally analysed. Our laboratory had cloned a *nifA* homologous gene (*LnifA*) from *A. brasilense* Yu62. However, *LnifA* displays only 37.2% identity over 430 amino acids with *ENif*A. Southern hybridization showed that *LnifA* was located on a 18.2 kb *Eco*RI fragment in *A.*

K.A. Malik et al. (eds.), Nitrogen Fixation with Non-Legumes, 21–24.

brasilense Sp7 (data not shown), whereas *EnifA* was located on a 5.6 kb *Eco*RI fragment (Liang *et al.*, 1991). Therefore, *LnifA* and *EnifA* were different genes.

Here we show that the *LnifA* gene could complement *EnifA*$^-$ mutant. Whether the *LnifA* gene is another *nifA*-like gene remains to be further studied.

Materials and methods

A. brasilense Sp7 was grown at 30°C in LD or NfbHP medium. Triparental mating and nitrogenase activity assay were carried out as described by Zhang *et al.*, (1992). Molecular subcloning and Southern hybridization were done according to Sambrook *et al.*, (1989).

```
ORF2      mrllivgtlegyitaagkiamqrgakvshtdsiegalnalraaagadlvmidvkldiatf   60
AvNifA    mnatipqrsakqnpvelydlqlqalasiartlsreqqidelleqvlavlhndlgllhglv   60

ORF2      ieslkaeritipvvacgigtdaaaavkairagakeyiplppdaeliaavleavaees---  117
AvNifA    tisdpehgalqigaihtdseavaqacegvryrsgegvignvlkhgnsvvlgrisadprfl  120

ORF2      ------------------------------------------------------------  117
AvNifA    drlalydlempfiavpiknpegntigvlaaqpdcradehmpartrfleivanllaqtvrl  180

ORF2      ----------------------------HAIVCQDPAMLATLRLADQVAPSDASVLITGE  149
AvNifA    vvniedgreaaderdelrrevrgkygfeNMVVGHTPTMRRVFDQIRRVAKWNSTVLVLGE  240

ORF2      SGTGKELMARYIHRKSRRSDAPFVAVNCAAIPENLLESELFGHEKGAFTGAVARRLGRFE  209
AvNifA    SGTGKELIASAIHYKSPRAHRPFVRLNCAALPETLLESELFGHEKGAFTGAVKQRKGRFE  300

ORF2      EANGGTLLLDELSEMHPRLQAKLLRAIQEKEIDRIGSSQPVKVNVRLVATSNRNLENEVR  269
AvNifA    QADGGTLFLDEIGEISPMFQAKLLRVLQEGEFERVGGNQTVRVNVRIVAATNRDLESEVE  360

ORF2      AGNFREDLYFRLNVFSVAIPSLRERPADIPMIADHFLKKYAQANGLGEKrLSDDALLMLR  329
AvNifA    KGKFREDLYYRLNVMAIRIPPLRERTADIPELAEFLLGKIGRQQGRPLT-VTDSAIRLLM  419

ORF2      AHHWRGNVRELENTMHRAVLLSKSETVGPDAIMLTSqllapegsaqasiptnspvanpfa  389
AvNifA    SHRWPGNVRELENCLERSAIMSEDGTITRDVVSLTGvdnespplaaplpevnladetldd  479

ORF2      gpggtvpqgraamgsrlpatrpptrhlratprgprptatlerrpaprpsrdssgarwrrs  449
AvNifA    rerviaaleqagwvqakaarllgmtprqiayriqtlnihmrki-----------------  522

ORF2      sat                                                           452
AvNifA    ---                                                           522
```

Figure 1. Amino acid sequence alignment of *Azospirillum brasilense* NifA homologous protein and *Azotobacter vinelandii* NifA. Conserved residues are shown as white letters on a black background.

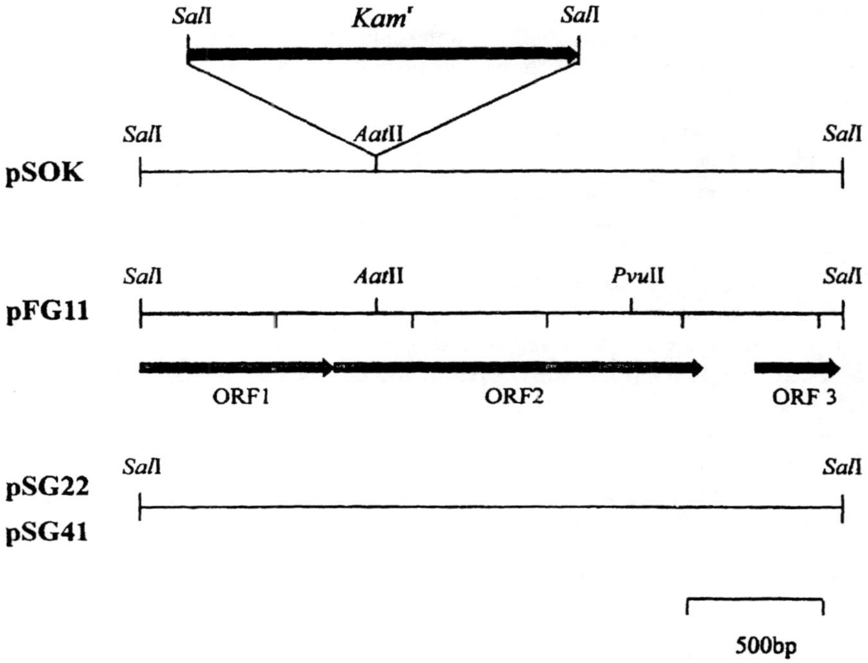

Figure 2. Construction of gene fusions pSG22, pSG41 and pSOK. pFG11 carried 2.6 kb *Sal*I fragment of ORF123 in pUC18. ORF2 is a *nifA* homologous gene. pSG22 and pSG41 are pVK100 derivatives containing the ORF123 subcloning in the *Xho*I site in the two orientations with respect to the promoter of kanamycin resistance gene. A 1.4 kb *Sal*I *kam*^r fragment from pUC4K was filled and subcloned into the filling-in *Aat*II site of ORF2 gene. The *Sal*I fragment containing the *kam*^r gene was subcloned into the *Sal*I site of pSUP202, which is named pSOK.

Results and discussion

Yan *et al.*, (1992) had screened out a *nifA* homologous gene from an *A. brasilense* Yu62 genetic library using a probe of 450 bp PCR product. The gene was searched for homologous protein using the TFAST program (Pearson and Lipman, 1988) in the NCBI GenBank. The search result showed that it is homologous to the *Caulobactor crescentus* Flbd, 64.2% identity in 371 amino acids overlap;

Table 1. The nitrogenase activity of *Azospirillum brasilense*

	Nitrogenase activity (nmol C_2H_4 h^{-1} OD^{-1} ml^{-1})			
Strains/Plasmids	No.	pSG22	pSG41	pAB57
Sp7 wild-type	89.1	51.2	48.8	50.1
Sp7067 (*nifA*-Tn5)	4.8	90.4	1.8	51.0

Azotobacter vinelandii NifA, 53.6% identity in 248 amino acids overlap; *A. vinelandii* VNifA, 52.3% identity in 237 amino acids overlap. LNifA has the conserved ATP-binding site of NifA proteins, but no conserved helix-turn-helix motif in C-terminal (Figure 1).

The *LnifA* gene was subcloned into the pVK100 vector in different orientations, designated pSG22 and pSG41 (Figure 2). They were conjugated into *A. brasilense* Sp7067 (Nif^-, *EnifA*::Tn5) and Sp7 (wild-type) respectively. The nitrogenase activity assay (Table 1) showed that pSG22 complemented Nif mutant Sp7067 almost completely, but pSG41 did not. So it could be deduced that *LnifA* did activate the expression of *nif* genes of *A. brasilense* Sp7 and it may be co-transcribed together with the upstream genes.

The Kam^r cassette was inserted into the *LnifA* gene on the suicide vector pSUP202, named pSOK (Figure 2). pSOK was conjugated into *A. brasilense* Yu62 and some Nif mutants were obtained. One of them could be complemented 78% by plasmid pAB57 containing the *EnifA* gene. However, Southern hybridization of the mutants was not successful.

Although the complementation experiment supported that *LnifA* did activate the expression of *nif* genes of *A. brasilense*, more evidence is still needed to confirm that *LnifA* is also a *nifA*-like gene. We have to discover the relationship of the two *nifA*-like genes.

References

Liang Y Y, Kaminski P A, Elmerich C 1991. Identification of a *nifA*-like regulatory gene *Azospirillum brasilense* Sp7 expressed under conditions of nitrogen fixation and in the presence of air and ammonia. Mol. Microbiol., 5, 2735–2744.

Pearson W R, Lipman D J 1988. Improved tools for biological sequence analysis. Proc. Natl. Acad. Sci. USA, 85, 2444–2448.

Sambrook J, Fritsch F, Maniatis T 1989. Molecular Cloning: A Laboratory Manual, 2nd edn. Cold Spring Harbor Laboratory Press, New York.

Yan D L, He L H, Li J L 1993. Cloning and sequencing of *ntrC* and a *nifA* homologous region from *Azospirillum brasilense*. In: New Horizons in Nitrogen Fixation. Kluwer, Dordrecht, 509.

Zhang Y P, Burris R H, Roberts G P 1992. Cloning, sequencing, mutagenesis and functional characterization of *draT* and *draG* genes from *Azospirillum brasilense*. J. Bacteriol., 174, 3364–3369.

4
Identification of plant growth hormones produced by bacterial isolates from rice, wheat and kallar grass

GHULAM RASUL, M. SAJJAD MIRZA, FAROOQ LATIF AND KAUSER A. MALIK

Abstract

Identification and quantification of the plant growth hormones indoleacetic acid and gibberellic acid, produced by plant growth-promoting rhizobacteria (PGPR), was carried out by using high-pressure liquid chromatography (HPLC). The PGPR strains were isolated from roots of rice, wheat and kallar grass and belonged to the genera *Azoarcus*, *Azospirillum*, *Enterobacter*, *Pseudomonas* and *Zoogloea*. For these studies, bacteria were grown in liquid nitrogen free malate (NFM) or combined carbon medium (CCM) containing tryptophan and combined nitrogen. Some *Azospirillum* strains produced both indoleacetic acid and gibberellic acid, while none of the *Enterobacter* spp. tested produced these growth hormones. *Azoarcus* strain K-1 produced higher amounts of gibberellic acid and *Azospirillum* strain ER-2 produced higher amounts of indoleacetic acid. Indoleacetic acid production increased with the age of bacterial cultures while a decrease in the production of gibberellic acid was noted at later growth stages. Pure indoleacetic acid and gibberellic acid in the concentration range 1–2 μg/ml increased root area and plant biomass of rice and wheat. Among PGPR strains tested, *Pseudomonas* 96–51 and its extract containing growth hormones increased root area, root length and plant biomass of rice and wheat.

Introduction

Plant growth-promoting rhizobacteria live freely in the soil or develop an association with plants, resulting in improvement of plant growth. Extensive work during the past few years has shown that inoculation of cereals and forage grasses with *Azospirillum brasilense* improved plant growth and productivity (Okon, 1985). These bacteria provide nitrogen by biological nitrogen fixation and produce

K.A. Malik et al. (eds.), Nitrogen Fixation with Non-Legumes, 25–37.

plant growth hormones, which enhance root development and improve mineral and water uptake by the system (Umali-Garcia *et al.*, 1980; Okon and Kapulnik, 1986). Different bacterial genera have been shown to possess the property of growth hormone production. *Azotobacter* species were reported to produce plant growth hormones in N-free media (Brown and Walker, 1970; Barea and Brown, 1974; Gonzalez-Lopez *et al.*, 1986). *Azospirillum* strains are also known to produce plant growth hormones such as auxins, gibberellins and cytokinins (Tien *et al.*, 1979; Hartmann *et al.*, 1983). Production of indoleacetic acid (IAA) has been reported in *Bradyrhizobium*, *Acetobacter*, *Alcaligenes*, *Enterobacter*, *Pseudomonas* and *Xanthomonas* (Fett *et al.*, 1987; Koga *et al.*, 1991; Minamisawa and Fukai, 1991; Fuentes-Ramirez *et al.*, 1993; Kobayashi *et al.*, 1993).

Colorimetric methods have long been applied for the detection of IAA produced by plants and microorganisms (Gordon and Weber, 1951; Hartmann *et al.*, 1983; Bric *et al.*, 1991; Malik *et al.*, 1994). Paper chromatography, thin-layer chromatography (TLC) and bioassays have been used for the identification of auxins, gibberellins and cytokinins (Lee *et al.*, 1970; Gonzalez-Lopez *et al.*, 1986; Harari *et al.*, 1989). The indoles, gibberellins and cytokinins were also identified by HPLC (Tien *et al.*, 1979; Crouch *et al.*, 1992; Costacurta *et al.*, 1994).

We have isolated several bacterial strains from soil and roots of kallar grass, wheat and rice. Using colorimetric methods, production of IAA by some of these strains has been reported earlier (Malik *et al.*, 1994). In the present study, production of IAA and gibberellic acid (GA) by PGPR strains, and their identification and quantification by HPLC, is reported. The effect of synthetic/pure phytohormones IAA and GA, and crude bacterial extracts containing these hormones, has been studied on rice and wheat.

Materials and methods

Identification of plant growth hormones by HPLC

Bacterial strains

A list of bacterial strains studied for growth hormone production is given in Table 1. These strains have been isolated from kallar grass, *Atriplex*, rice and wheat, and belonged to the genera *Azoarcus*, *Azospirillum*, *Enterobacter*, *Pseudomonas* and *Zoogloea* (Bilal and Malik, 1987; Bilal *et al.*, 1990; Malik *et al.*, 1991, 1994).

Growth media and growth conditions

The bacterial cultures were grown in CCM (Rennie 1981) or dye-free NFM (Okon *et al.*, 1977) liquid medium containing 1 g/L of NH_4NO_3 and 100 mg/L of tryptophan. Tryptophan was added as a precursor in IAA synthesis, while biotin and para-aminobenzoic acid (PABA) were eliminated from both the

Table 1. Production of plant growth hormones gibberellic acid and indoleacetic acid by different PGPR strains

Bacterial strain	Plant sp.	Gibberellic acid (μg/ml)		Indoleacetic acid (μg/ml)	
		10 days	22 days	10 days	22 days
Azoarcus K-1	Kallargrass	10*	–	7	8
Zoogloea Ky-1	Kallargrass	–	–	–	1
Azospirillum N-4	Rice	2	–	4	5
Azospirillum WRRN23	Rice	4	–	2	6
Flavobacterium 96–57	Rice	2	ND	ND	7
Pseudomonas 96–51	Rice	2	–	ND	15
Azospirillum ER-2	Wheat	–	–	ND	22
Azospirillum ER-20	Wheat	–	–	7	14
Azospirillum ER-201	Wheat	–	–	4	14
Azospirillum ER-24	Wheat	–	–	9	12
Enterobacter ER-23	Wheat	–	–	–	–
Enterobacter QH7	Wheat	ND	–	ND	–
E. agglomerans AX-12	*Atriplex*sp.	ND	–	ND	–

Quantities are given as μg/ml of the culture filtrate.
*= Production was also estimated at 5 days; –, not present; ND, not determined.

media. *Azospirillum* strains and *Azoarcus* K-1 were grown in NFM and *Enterobacter*, *Flavobacterium*, *Pseudomonas* and *Zoogloea* strains were grown in CCM. The 400 ml of CCM/NFM liquid medium was inoculated with 4 ml of overnight-grown culture. The cultures were harvested by centrifugation at 10,000 rpm for 10 min after 10 and 22 days.

Extraction process and identification of growth hormones by HPLC

The cell-free liquid culture medium was concentrated by freeze-drying up to 80 ml and the pH of the extract was adjusted to 2.8 with concentrated hydrochloric acid. Extraction with ethyl acetate was carried out as described by Tien *et al.* (1979). The ethyl acetate extract was evaporated to dryness and the residue was dissolved in one millilitre ethanol.

The samples were analysed on HPLC using Turbochom software (Perkin-Elmer, USA). A Licosorb-C18 column was used for elution of growth hormones. Methanol : acetic acid : water (30 : 1 : 70) was used as mobile phase at the rate of 1.5 ml/min.

For identification, 20 μl samples, filtered through a 0.45 μm filter, were injected into the column. The growth hormones were identified on the basis of retention time of the standard IAA and GA by using a refractive index detector (RI). The concentration of each acid was calculated on the basis of peak height and peak area.

Effect of pure IAA and GA, PGPR and extracts of PGPR on rice and wheat growth

Seeds of wheat variety PAK-81 and rice variety NIAB-6 were germinated on water agar plates and 2–5-day-old seedlings were transferred to glass tubes

(20 × 2 cm, length × diameter) containing vermiculite and nitrogen containing Hoagland solution (¼ strength). One seedling was grown in each tube. Different concentrations of pure IAA and GA (0.25–4.0 μg/ml) were applied both on rice and wheat. Four seedlings were used for each treatment. The tubes were kept in a controlled-temperature room at 28 ± 2 °C with 16 h light and 8 h dark. Light intensity was 400 $\mu E/m^2$. The plants without inoculation/hormone treatment were used as control. The plants were harvested after 3 weeks and compared for their root area, root length and plant dry weight.

Bacterial strains, namely *Azoarcus* K-1, *Azospirillum* N-4 and *Pseudomonas* 96–51, and their extracts, were applied on both rice and wheat. For rice 1 ml of cell suspension (10^8–10^9 cells/ml) was used to inoculate each plant. For growth hormones the concentration of the extract in ethanol was calculated on the basis of IAA contents in the extract and this extract was applied to make a final concentration of 1, 2 and 4 μg/ml in each tube.

To study the effect of inoculation of PGPR and application of 2 μg/ml ethyl acetate extract of PGPR strains on wheat, seeds of variety Inqlab were grown in plastic buckets (1.5 kg capacity), half-filled with vermiculite, Hoagland liquid medium (750 ml) was provided in every bucket. Five millilitres of respective cell suspension (10^8–10^9 cells/ml) were inoculated into each bucket. The extract concentration 2 μg/ml was adjusted in the buckets, in the same way as for rice. Two buckets were used for every treatment and five plants were grown in each bucket. The plants without inoculation and without extract application were used as control.

Estimation of root area and plant dry weight

The root area and root length were measured with the help of a computer and a scanner by using the Root Image Analysis program; this program has been developed by Washington State University Research Foundation, Washington State University, USA. For plant dry weight, plants were kept at 70°C until no change in their weight was noted.

Results and discussion

Identification of growth hormones by HPLC

Gibberellic acid and indoleacetic acid were identified by HPLC on the basis of retention time of pure GA and IAA, which was 8.0 and 10.8 min, respectively, under the assay conditions (Figure 1). Out of 13 PGPR strains which were studied for growth hormone production, 10 produced both IAA and GA (Table 1). *Azoarcus* K-1 produced higher amounts of GA (10 μg/ml) as well as IAA (8 μg/ml). *Azospirillum* strain ER-2 also produced higher amounts of IAA, followed by *Pseudomonas* strain 96–51 and *Azospirillum* ER-20 and ER-201. *Enterobacter*

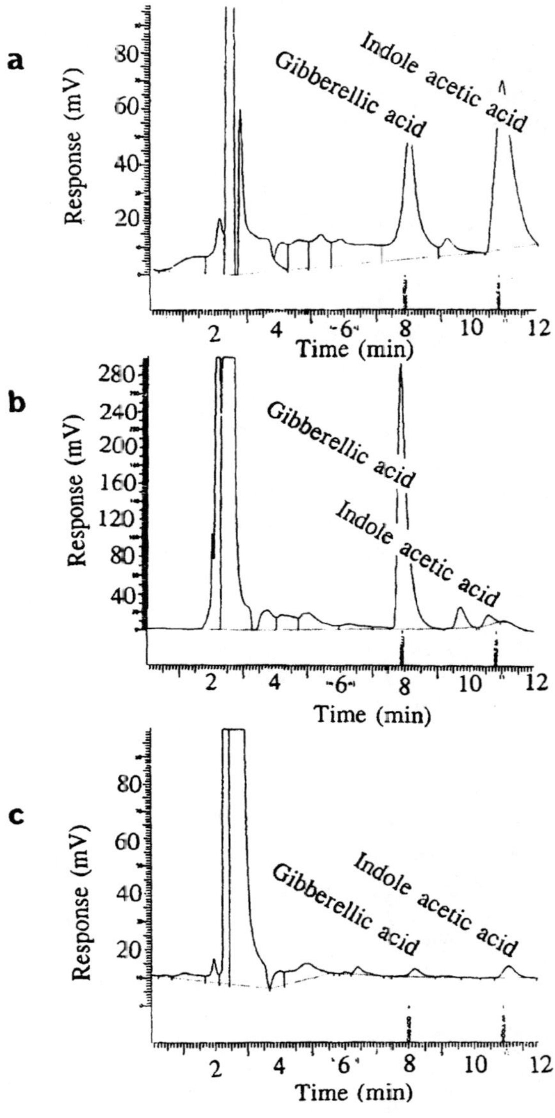

Figure 1. Identification of indoleacetic acid and gibberellic acid production by different PGPR strains by HPLC. Indoleacetic acid and gibberellic acid standards (**a**), *Azoarcus* K-1 (**b**) and *Flavobacterium* 96–57 (**c**).

strains QH7, ER-23 and Ax-12 did not produce IAA or GA (Table 1). GA production was higher at 10 days and IAA was maximum after 22 days of cultures growth (Table 1).

Azospirillum, Azotobacter, *Pseudomonas* and many other genera have already been reported for production of phytohormomes (Reynders and Vlassak, 1979; Tien *et al.*, 1979). Production of IAA has been reported with and without tryptophan, under nitrogen fixing and non-fixing conditions and with and without combined nitrogen. Hartmann *et al.* (1983) described that there was no difference in IAA production by *Azospirillum lipoferum* under nitrogen-fixing and non-fixing conditions. It was also observed that IAA production increased in the presence of 0.1 mg/ml of tryptophan in the medium. In the present studies IAA and GA production was studied in the presence of combined nitrogen source because we have already reported that IAA production is increased in the presence of combined nitrogen (Malik *et al.*, 1994). However, Lee *et al.* (1970) reported that IAA production by *Azotobacter vinelandii* was reduced in the presence of combined nitrogen. Tien *et al.* (1979) reported that the combined nitrogen had little effect on the production of IAA. Gonzalez-Lopez *et al.* (1986) reported reduction in auxin production and increase in GA production (from 4 to 6 μg/ml) in the presence of combined nitrogen. Hartmann *et al.* (1983) described a 10-fold increase in excretion of IAA by *A. brasilense* in the presence of 10 mM ammonia.

In the present study in which HPLC was used, relatively lower levels of IAA production by PGPR strains were detected as compared to those reported previously (Malik *et al.*, 1994) in which the colorimetric method was employed. This indicates an overestimate of IAA production by the colorimetric method in which IAA conjugates like IAA alanine, IAA aspartate, IAA glycine and other compounds like indolebutyric acid, indolepyruvic acid and indolepropionic acid give a colour reaction with Salkowski reagent (Gordon and Weber, 1951).

The production of IAA was affected by the age of bacterial culture. GA production decreased with increase in incubation time (Table 1). Gonzalez-Lopez *et al.* (1986) also observed an increase in auxin production by bacterial culture for 3 days which became constant for up to 15 days. GA production was maximum after 7 days of growth and then became constant up to the 15th day.

It was observed that *Azospirillum* spp. produced up to 22 μg/ml of IAA in culture medium. Hartmann *et al.* (1983) observed up to 16 μg/ml of IAA in the culture medium. In the present study, considerable variation was observed in the production of GA by PGPR strains (Table 1). Vancura (1961) reported 20 μg/ml of GA3, and Brown and Burlingham (1968) described 0.03 μg GA3 equivalent/ml. *Azoarcus* strain K-1 produced up to 10 μg/ml of GA in culture medium. The other strains produced relatively lower amounts of GA. IAA and GA production was not detected in *Enterobacter* strains. Hennequin *et al.* (1966) also reported negative results for growth hormone production with some strains of *Azotobacter vinelandii* and *Azotobacter chroococcum*.

Effect of pure IAA and GA on rice and wheat growth

Different concentrations of GA and IAA were tested for their effect on rice and wheat growth. These hormones stimulated rice and wheat growth at their lower concentrations and inhibited growth at higher concentrations (Tables 2–5).

In rice 2 μg/ml of IAA and 1 μg/ml of GA produced a higher plant biomass

Table 2. Effect of indoleacetic acid on the growth of rice plant

Concentration of IAA (μg/ml)	Shoot dry weight (mg/plant)	Root dry weight (mg/plant)	Total plant dry weight (mg/plant)
0.0	23.1±2.7	5.2±1.2	28.5±2.5
0.5	18.6±3.3	6.0±1.0	24.6±1.6
1.0	21.4±3.7	7.1±1.1	28.8±2.2
2.0	25.1±1.5	8.7±0.9	34.0±3.5
4.0	21.5±2.1	7.3±0.9	29.0±2.9
8.0	21.1±2.3	6.9±1.0	27.8±3.6

The values given in the table are averages of six replicates; ± values indicate the standard deviation from means.

Table 3. Effect of different concentrations of gibberellic acid on rice growth

Concentration of GA (μg/ml)	Root area (cm^2)	Root length (cm)	Plant dry weight (mg)	Shoot length (cm)
Control	32^{bc}	264^{b}	270^{abc}	36^{d}
0.25	42^{a}	326^{a}	290^{ab}	43^{c}
0.50	39^{a}	331^{a}	280^{ab}	44^{c}
1.00	38^{ab}	330^{a}	330^{a}	48^{b}
1.50	28^{cd}	270^{b}	213^{c}	51^{b}
2.00	24^{d}	261^{b}	265^{bc}	52^{ab}
3.00	27^{cd}	288^{ab}	275^{abc}	48^{b}
4.00	25^{cd}	271^{b}	250^{bc}	56^{a}

Figures followed by the same letters are not significantly different at the 5% level as determined by the DMR test.

Table 4. Effect of different concentrations of indoleacetic acid on wheat growth.

Concentration of IAA (ppm)	Root area (cm^2)	Root length (cm)	Plant dry weight (mg)
Control	13.5^{a}	146^{b}	99^{ab}
0.25	12.4^{a}	112^{b}	79^{b}
0.50	ND	ND	92*
1.00	15.0^{a}	160^{ab}	91^{ab}
2.00	18.8^{a}	216^{a}	98^{ab}
3.00	19.8^{a}	217^{a}	107^{a}
4.00	13.9^{a}	175^{ab}	90^{a}

* Not included in statistical analysis. Figures followed by the same letters are not significantly different at the 5% level as determined by the DMR test

(Tables 2 and 3). Root area and root length were higher at 0.25–1.00 μg/ml of GA and the value for shoot height was higher at 4 μg/ml of GA (Table 3). In wheat a higher value for root area, root length and total plant biomass was recorded at 2–3 μg/ml of IAA. For GA, root area, root length and plant dry weight were higher at 0.5–1.5 μg/ml (Tables 4 and 5).

Similar effects of growth hormone application on pearl millet and *Panicum*

Table 5. Effect of different concentrations of gibberellic acid on wheat growth

Concentration of GA (µg/ml)	Root area (cm^2)	Root length (cm)	Plant dry weight (mg)	Shoot length (cm)
Control	11.7[ab]	117[b]	72[b]	34[a]
0.25	9.5[ab]	113[b]	65[b]	34[a]
0.50	14.5[a]	167[a]	90[a]	31[a]
1.00	11.5[ab]	136[ab]	72[b]	33[a]
1.50	14.0[ab]	146[ab]	73[b]	31[a]
2.00	11.3[ab]	125[ab]	59[b]	33[a]
3.00	9.0[b]	116[b]	59[b]	33[a]
4.00	8.8[b]	100[b]	40[c]	31[a]

Figures followed by the same letters are not significantly different at the 5% level as determined by the DMR test

miliaceum have been reported (Tien *et al.*, 1979; Harari *et al.*, 1989). In pearl millet total fresh plant biomass was higher at 0.01 µg/ml IAA in the medium as compared to when a 0.05 µg/ml concentration of IAA was used. For different concentrations of GA tested, the higher fresh biomass value was observed at 0.05 µg/ml (Tien *et al.*, 1979). It has been reported that root proliferation was maximum at 10^{-9} M IAA (Harari *et al.*, 1989).

GA affected shoot formation as well as root area and root length of both rice and wheat (Tables 3 and 5). In rice the root formation was improved with a lower concentration of GA (1 µg/ml) but shoot length increased up to 4 µg/ml. There was a rapid increase in shoot length after 24 h of application of 1–4 µg/ml of GA which continued for 4–5 days (Table 3). In wheat the rate of increase in shoot length was slower as compared to that of rice plants, but wheat plants inoculated with GA were taller than the uninoculated wheat plants. The difference in shoot length of inoculated and uninoculated plants decreased with increase in the age of wheat plants (Table 5).

Effect of PGPR and extracts from PGPR strains on rice and wheat

In rice, inoculation with *Azoarcus* K-1 increased the root area, root length and dry biomass as compared to uninoculated control (Figure 2). All three concentrations of extract of the *Azoarcus* K-1, equivalent to 1, 2 and 4 µg/ml of IAA, inhibited root area, root length and dry plant biomass. Inoculation with *Azospirillum* N-4 and its extracts, except 2 µg/ml, gave lower values for root area and root length as compared to control. The application of 2 µg/ml extract of *Azospirillum* N-4 increased the root area and root length. Plant dry weight values for N-4 inoculation and its extracts (1, 2 and 4 µg/ml) applications, were also lower than control. *Pseudomonas* strain 96–51 and its 1 µg/ml extract application increased root area, root length and plant dry weight while other concentrations (2 and 4 µg/ml) of the extract inhibited the growth (Figure 2).

For wheat, bacterial inoculation and 2 µg/ml extracts from different PGPR

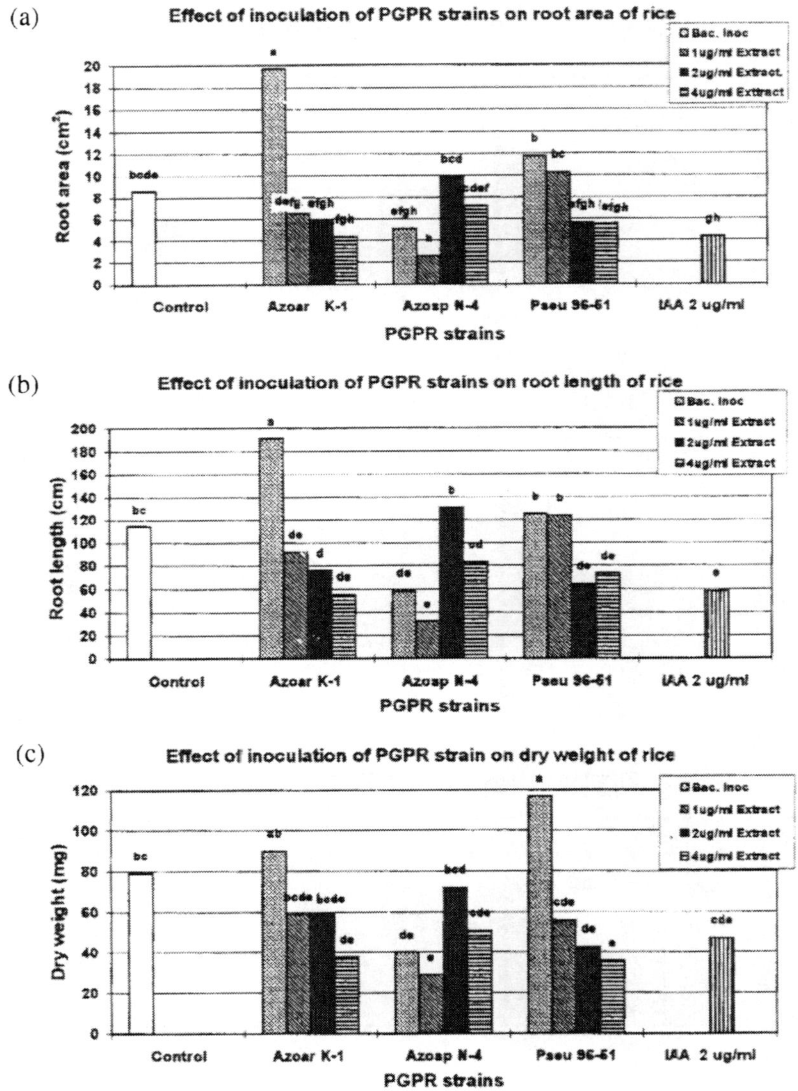

Figure 2. Effect of inoculation of PGPR strains and application of extracts from PGPR strains on root area (**a**), root length (**b**) and dryweight (**c**) of rice. Azoar = *Azoarcus* K-1, Azosp = *Azospirillum* N-4, Pseu = *Pseudomonas* 96–51. Bars followed by the same letters are not significantly different at the 5% level as determined by the DMR test.

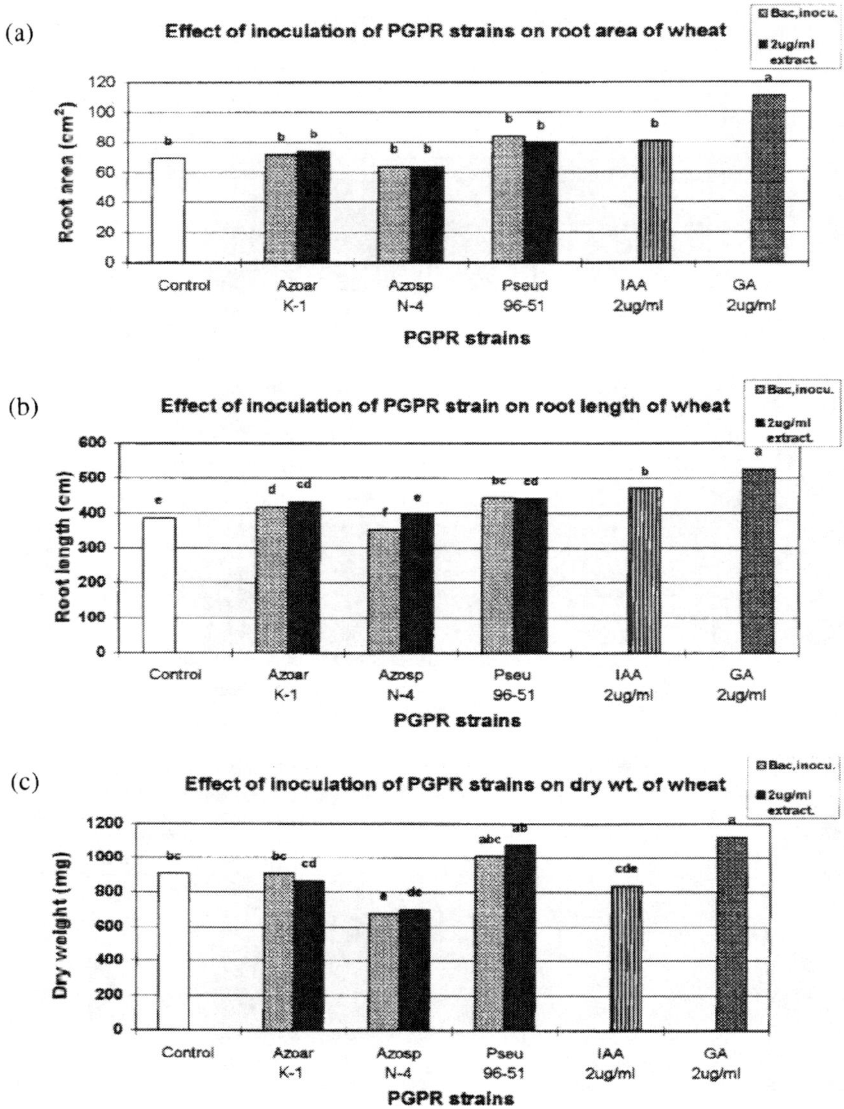

Figure 3 Effect of inoculation of PGPR strains and application of extracts from PGPR strains on root area (**a**), root length (**b**) and dry weight (**c**) of wheat. Azoar = *Azoarcus* K-1, Azosp = *Azospirillum* N-4. Pseu = *Pseudomonas* 96–51. Bars followed by the same letters are not significantly different at the 5% level as determined by the DMR test.

strains were applied (Figure 3). *Azoarcus* K-1 and its extract caused a minor increase in root area and root length as compared to control. The plant biomass values for K-1 inoculation and its extract were the same as for uninoculated control. For the treatment in which there was inoculation with *Azospirillum* N-4, the values for root area, root length and plant biomass were lower as compared to control. Application of the extract on wheat increased root length and inhibited root area and plant dry weight. Inoculation with *Pseudomonas* 96–51 and application of its extract increased root area, root length and plant biomass as compared to control (Figure 3).

The results of these studies in rice and wheat indicated that the bacterial inoculation or application of extract improved or inhibited plant growth. Similar effects have been observed by Tien *et al.* (1979) in pearl millet, and Harari *et al.* (1989) in *Panicum miliaceum*. The effects of plant growth may be due to single or combined action of growth hormones present in the extracts, and also depend on the relative concentrations of these hormones. Therefore, to understand the growth stimulation mechanism, in addition to correct identification and quantification of growth hormones, separation of growth hormones present in the extract, and study of different concentrations of each component is required. Work on these lines is in progress.

Conclusions

Azoarcus K-1, *Azospirillum* strains, *Zoogloea* Ky-1, *Flavobacterium* 96–57 and *Pseudomonas* 96–51 produced both indoleacetic acid and gibberellic acid in the liquid media. *Enterobacter* strains did not produce these hormones. A significant increase in root area as well as root and shoot length was observed in rice and wheat by application of lower concentrations ($\leq$ 2 µg/ml) of indoleacetic acid and gibberellic acid. Inoculation with *Azoarcus* K-1 and *Pseudomonas* 96–51 increased root area, root length and total plant dry weight. The lower concentration of extracts from these PGPR strains which contained growth hormones (2 µg/ml from *Azospirillum* N-4 and 1 µg/ml from *Pseudomonas* 96–51) improved, and the rest of the higher concentrations inhibited plant growth.

Acknowledgements

Financial support for this work was partially provided by PSF by means of PSF project P-NIBGE/Agr (153).

References

Barea J M, Brown M E 1974. Effects on plant growth produced by *Azotobacter paspali* related to synthesis of plant growth regulating substances. J. Appl. Bacteriol., 37, 583–593.

Bilal R, Malik K A 1987. Isolation and identification of a N_2 -fixing zoogloea-forming bacterium from kallar grass histoplane. J. Appl. Bacteriol., 62, 289–294.

Bilal R, Rasul G, Qureshi J A, Malik K A 1990. Characterization of *Azospirillum* and related diazotrophs associated with roots of plants growing in saline soils. World J Microbiol. Biotechnol., 6, 46–52.

Bric J M, Bostock R M, Silverstone S E 1991. Rapid *in situ* assay for indoleacetic acid production by bacteria immobilized on a nitrocellulose membrane. Appl. Environ. Microbiol., 57, 535–553.

Brown M E, Burlingham S K 1968. Production of plant growth substances by *Azotobacter chroococcum*. J. Gen. Microbiol., 53, 135–144.

Brown, M E, Walker N 1970. Indolyl-3-acetic acid formation by *Azotobacter chroococcum*. Plant Soil, 32 250–253.

Costacurta A, Keijers V, Vanderleyden J 1994. Molecular cloning and sequence analysis of an *Azospirillum brasilense* indole-3-pyruvate decarboxylase gene. Mol. Gen. Genet., 243, 463–472.

Crouch I J, Smith M T, Van Staden J, Lewis M J, Hoad G V 1992. Identification of auxins in a commercial seaweed concentrate. J Plant Physiol., 139, 590–594.

Fett W F, Osman S F, Dunn M F 1987. Auxin production by plant pathogenic *Pseudomonas* and *Xanthomonas*. Appl. Environ. Microbiol., 53, 1839–1845.

Fuentes-Ramirez L E, Jimenez-Salgado T, Abaca-Ocampo I R, Caballero-Mellado J 1993. *Azotobacter diazotrophicus*, an indoleacetic acid producing bacterium isolated from sugar cane cultivars of Mexico. Plant Soil, 154, 145–150.

Gonzalez-Lopez J, Salmeron V, Martinez-Toledo M V, Ballesteros F, Ramos-Cormenzana A 1986. Production of auxin, gibberellins and cytokinins by *Azotobacter vinelandii* ATCC 12837 in chemically defined media and dialysed soil media. Soil Biol. Biochem., 18, 119–120.

Gordon S A, Weber R P 1951. Colorimetric estimation of indoleacetic acid. Plant Physiol., 26, 192–195.

Harari A, Kigel J, Okon Y 1989. Involvement of IAA in the interaction between *Azospirillum brasilense* and *Panicum miliaceum* root. In: Skinner F A, Body R M, Fendrik J, eds. Nitrogen Fixation with Non-Legumes. Kluwer, Dordrecht, Academic Publishers. 227–234.

Hartmann A, Singh M, Klingmuller W 1983. Isolation and characterization of *Azospirillum* mutants excreting high amounts of indoleacetic acid. Can. J. Microbiol., 29, 916–923.

Hennequin J R, Blachere H, Tabareau A M 1966. Recherches sur la synthese de phytohormones et de composes phenoliques par *Azotobacter* et des bacteries de la rhizosphere. Ann. Inst. Pasteur Suppl. 3, 111, 89–101.

Kobayashi M, Izui H, Nagasawa T, Yamada H 1993. Nitrilase in biosynthesis of the plant hormones indole-3-acetic acid from indole-3-acetonitrile: cloning of *Alcaligenes* gene and site directed mutagenesis of cysteine residues, Proc. Natl. Acad. Sci. USA, 90, 247–251.

Koga J, Adachi T, Hidaka H 1991. IAA biosynthetic pathway from tryptophan via indole-3-pyruvic acid in *Enterobacter cloacae*. Agric. Biol. Chem., 55: 701–706.

Lee M, Breekenridge C, Knowles R 1970. Effect of some culture conditions on the production of indole-3-acetic acid and a gibberellin-like substance by *Azotobacter vinelandii*. Can. J. Microbiol., 16, 1325–1330.

Malik K A, Bial R, Rasul G, Mahmood K, Sajjad M I 1991. Associative N2-fixation in plants growing in saline sodic soils and its quantification based on ^{15}N natural abundance. Plant Soil, 137, 67–74.

Malik K A, Rasul G, Hassan U, Mehnaz S, Ashraf M 1994. Role of N2- fixing and growth hormones producing bacteria in improving growth of wheat and rice. In: Hegazi N A, Fayez M, Monib M, eds. Nitrogen Fixation with Non-Legumes. Proceedings of The Sixth International Symposium with Non-Legumes, Ismailia, Egypt, 6–10 September 1993. Cairo University, Giza, Egypt, 409–422.

Minamisawa K, Fukai K 1991. Production of indole-3-acetic acid by *Bradyrhizobium japonicum*: a correlation with genotype grouping and rhizobitoxine production. Plant Cell Physiol., 32, 1–9.

Okon Y 1985. *Azospirillum* as a potential inoculant for agriculture. Trends Biotechnol., 3, 223–228.

Okon Y, Kapulnik Y 1986. Development and function of *Azospirillum* inoculated roots. Plant Soil, 90, 3–16.

Okon Y, Albercht S L, Burris R H 1977. Methods for growing *Spirillum lipoferum* and for counting it in pure culture and in association with plants. Appl. Environ. Microbiol., 33, 85–88.

Rennie R J 1981. A single medium for the isolation of acetylene reducing (dinitrogen fixing) bacteria from soil. Plant Soil, 27, 8–14.

Reynders L, Vlassak K 1979. Conversion of tryptophan to indoleacetic acid by *Azospirillum brasilense*. Soil Biol. Biochem., 11, 557–558.

Tien T M, Gaskins M H, Hubbell D H 1979. Plant growth substances produced by *Azospirillum brasilense* and their effect on the growth of pearl millet (*Pennisetum americanum* L.) Appl. Environ. Microbiol., 37, 1016–1024.

Umali-Garcia M, Hubbell D H, Gaskin M H, Dazzo F B 1980. Association of *Azospirillum* with grass roots. Appl. Environ. Microbiol., 39, 219–226.

Vancura V 1961. Detection of gibberellic acid in *Azotobacter* culture. Nature, 192, 88–89.

SECTION II

Ecophysiological studies for the improvement of plant–microbe interaction

5
The rice endophytic diazotroph and PGPR

W. SONG, H.L. YANG, X.L. SUN, Y.S. WANG, Y.D. WANG AND
Z.H. CHEN

Abstract

The selective pressure of the rhizoplane of rice plants for colonized beneficial rhizobacteria has been investigated. The rhizobacteria were isolated from the rhizoplane of rice variety Yue Guang (with a middling disease-resistant trait) and Yue Fu (with a high yield trait). The *in-vitro* antagonistic assay to 14 pathogens, including three major rice plant pathogens, *Xanthomonas oryzae*, *Corticium sasakii* and *Piricularia oryzae* Cavara, showed that the ratio of positive reactions for the strains isolated from rice variety Yue Guang were much greater than that of Yue Fu. The antagonistic spectra of those rhizoplane bacterial strains from Yue Guang were much wider than those from Yue Fu. The endophytes of rice Yue Fu plants were isolated during the whole rice-growing season in 1996. There were 10^3–10^6 cfu/g fw (plant) of endophytes present in rice plant roots, stems and leaves. A considerable part of the endophytic population showed acetylene reduction activity and $^{15}N_2$-fixing activity in pure culture.

Introduction

In the past decade non-legume nitrogen fixation has attracted much attention. The inoculation of associative diazotrophs such as PGPR increased the yields of cereal crops extensively (Döbereiner, 1992). However, the main problem in the application of PGPR as bio-fertilizer or biopesticide inoculant was inconsistent performance in the field. Many factors could contribute to the inconsistent performance of PGPR, including complex interactions among host, pathogen, rhizobacteria and the soil environment. One of the most important factors is variability in root colonization by PGPR strains. Therefore studies on the adaptation of PGPR strains to their host plant would be very important.

K.A. Malik et al. (eds.), Nitrogen Fixation with Non-Legumes, 41–48.

Materials and methods

Rice varieties *Oryza sativa L. japonica* Yue Fu and Yue Guang were used in pot and field experiments.

Variety Yue Fu is with higher yield trait while Yue Guang shows middling disease resistance trait.

Acetylene reduction activity (ARA) assay

Pure bacterial colonies were selected from N-free solid medium and five colonies were transferred to ARA assay bottles containing semi-solid N-free medium. The bottles were kept at 30°C for 24 h. C_2H_4 formation was tested on a Varian Vista-6000 gas chromatograph.

$^{15}N_2$-fixing activity ($^{15}N_2FA$) assay

Bacterial strains with ARA values more than 1 nmol C_2H_4 formation/ml h were assayed for their $^{15}N_2$-fixing activities. Assay bottles containing 40 ml of semi-solid N-free medium were used for the $^{15}N_2FA$ assay. $^{15}N_2$ with approximately 20% of ^{15}N abundance was used for the gas phase. The assay bottles were flushed with argon three times and kept under a vacuum line; 92% of the gas volume of $^{15}N_2$ and 8% of the gas volume of sterilized O_2 were injected into the assay bottles, respectively. The assay bottles were kept at 30°C for 3–4 days. The culture media with bacterial colonies were taken out and baked at 80°C until thoroughly dried. The total N values of dried ^{15}N samples were determined with the Kjeldahl method and their ^{15}N abundance was determined on MU-1305 mass spectrograph.

Isolation of the PGPR from rice rhizoplane

Rice roots were washed thoroughly with sterile water and then shaken at 200 rpm for 30 min in sterile water containing glass beads. The serial dilutions of the water suspension were made for inoculating the N-free and N-containing solid medium in the plates, respectively. The bacterial colonies were observed and counted after 4 days.

Isolation of endophytes of rice plant

1. Rice seeds were sterilized with 95% alcohol for 1 min and 8–13% of NaOCl for 45 min, followed by rinsing with sterile distilled water four times. The final rinsed solutions were plated to serve as checks. Samples were discarded if bacterial colonies occurred from the sterile checks within 48 h. The sterile seeds were crushed in a sterile mini-milling bowl with a proper amount of sterile water. The seed homogenates were diluted with phosphate buffer solution in a gradient series and plated, then incubated at 30°C for 4 days. The bacterial colonies were observed and counted.

2. Rice plants with roots were carefully taken out from paddy soil in the field. Aliquots of 1 g of plant material (roots, stems and leaves respectively) were washed with tap water, and dried with filter papers. The plant materials were immersed in 70% alcohol for 3–5 min, transfered to 2–3% of NaOCl for 3–10 min, and rinsed with sterile phosphate buffer solution more than four times. The final rinsed solutions were plated to serve as checks. Samples were discarded if bacterial colonies occurred from the sterile checks within 48 h. The sterile plant materials were crushed in a sterile mini-milling bowl with 9 ml of sterile distilled water. The homogenates were diluted with phosphate buffer solution in a gradient series and plated, then incubated at 30°C for 4 days. The bacterial colonies were observed and counted.

Pathogenic strains used:

21: *Corticium sasakii* (shirai) Massumoto	ACCC 30021
42: *Piricularia oryzae* Cavara	ACCC 30042
02: *Alternaria longipes* (Ellis et Everhart) Tisdale et Wadkins	ACCC 30002
10: *Cochliobolus sativus*	ACCC 30010
11: *Colletotrichum agaves* Cavara	ACCC 30011
16: *Colletotrichum lagenarium* (Passerini) Ellis et Halsted	ACCC 30016
17: *Colletotrichum panacicola* Uyeda et Takimoto	ACCC 30017
24: *Fusarium oxysporum* f. *niveum*	ACCC 30024
33: *Helminthosporium turcicum* Passerini	ACCC 30033
41: *Phytophthora parasitica* Destur	ACCC 30041
53: *Verticillium albo-atrum* Reinke Berthold	ACCC 30053
254: *Phytophthora melonis* Kasura	ACCC 30254

PS: *Pseudomonas solanacearum* (Smith) (Institute of Plant Protection, CAAS)
XO: *Xanthomonas oryzae pv. oryzae* ZHE 173 (Institute of Crop Breeding and Cultivation, CAAS)

The in-vitro antagonistic activity assay for rhizobacteria to phytopathogens

A suspension of pathogenic strains was prepared in 5 ml sterile water and added to 150 ml of PDA medium kept at 40°C. After shaking thoroughly, the medium was plated for testing. Four pieces of discoid sterile filter papers (dI: 5 mm) were carefuly set up on the surface of the medium in those plates in a diagonal arrangement. The plates were incubated at 28°C for 48 h; 5 μl of rhizobacterial strain suspention (10^8 cfu/ml) were added to the four discoid filter papers on plates, and incubated at 30°C. The *in vitro* antagonistic activities were checked daily.

Results and discussion

The rhizobacteria with ARA or in-vitro *antagonistic activity isolated from rice rhizoplane of different varieties at different growth stages*

Several reports had appeared on the selective pressure of plant roots on rhizobacteria colonizing the roots (Neal *et al.*, 1970; Mavingui *et al.*, 1992). The population

Table 1. The population (cfu/g f.w. root) of rhizobacteria isolated from rhizoplane at different rice growth stages

Growth stages	N-containing medium		N-free medium	
	Yue Guang	Yue Fu	Yue Guang	Tue Fu
Seedling	1.50×10^6	2.16×10^6	2.30×10^6	2.40×10^6
Tillering	1.38×10^7	1.09×10^7	2.23×10^7	7.40×10^6
Booting	1.31×10^7	2.16×10^7	5.00×10^6	1.20×10^7
Heading	1.33×10^7	7.80×10^6	1.24×10^7	9.30×10^6

of rhizobacteria colonizing rhizoplane of rice plants varied with rice varieties and growth stages (Table 1). During the seedling stage populations of colonized rhizobacteria were approximately 10^6 cfu/g f.w. root. The population increased with the age of rice plant. The maximum population for variety Yue Guang was up to 1.38–2.22 × 10^7 (cfu/g f.w. root) during tillering stage. For variety Yue Fu the maximun population was up to 1.20–2.16 × 10^7 (cfu/g f.w. root) during booting stages.

Table 2. The rhizobacteria* with ARA and *in-vitro* antagonistic activity isolated from rice rhizoplane at different growth stages.

Rice variety	Seedling		Booting		Heading	
	I	II	I	II	I	II
Yue Guang	1AG2		2BG1	2AG1	3AG2	3AG2
	1BG2		2BG2	2AG4	3AG3	3AG3
	1BG4		2BG4	2AG5	3AG5	3AG7
	1BG5		2BG5	2AG7	3AG10	3AG8
			2BG6	2AG8	3AG11	3AG9
			2BG7	2AG9	3AG12	3AG10
			2BG8	2AG10	3AG13	3AG11
			2BG9	2AG11	3BG7	3AG12
			2BG10		3BG12	3AG13
			2BG11			3AG14
						3AB15
						3BG7
						3BG12
Yue Fu	1AF1	1AF3	2AF8	2AF5	3AF5	3AF1
	1AF2	1AF6	2BF1	2AF7	3AF7	3AF4
	1AF4	1AF7	2BF2	2AF8	3AF8	3AF5
	1AF7		2BF3	2AF10	3AF10	3AF6
	1BF1		2BF4	2AF11		3AF7
	1BF4		2BF5			3AF8
			2BF6			3AF10
			2BF7			3BF5
			2BF8			3BF6
			2BF9			3BF8
			2BF10			3BF10
			2BF11			

I: with ARA more than 1 nmol C_2H_4 formation/vial.h. II: with *in-vitro* antagonistic activity. *The populations listed were more than 1 x 10^5 cfu/g f.w root

Table 3. The *in-vitro* antagonistic activity of some rhizobacteria isolated from rice rhizoplane during rice growth stages

Rice variety	Growth stage	Strains	Pathogen strains*														Inhibition zones
			XO	21	42	02	10	11	16	17	24	33	41	53	254	PS	
Yue Guang	Booting	2AG1	+					+			+		+	+			5
		2AG4	+	+	+	+	+	+	+	+	+	+	+	+	+	+	14
		2AG5			+	+			+			+					4
		2AG7						+			+		+				3
		2AG8		+				+									2
		2AG9			+			+				+					3
		2AG10						+					+				2
		2AG11					+										1
	Heading	3AG2														+	1
		3AG3														+	1
		3AG7					+										1
		3AG8														+	1
		3AG9														+	1
		3AG10														+	1
		3AG11	+	+	+		+								+	+	6
		3AG12														+	1
		3AG13					+										1
		3AG14		+			+								+		3
		3AG15					+										1
		3BG3														+	1
		3BG7	+				+							+	+		4
		3BG12	+												+	+	3
	Total	22	5	4	4	2	8	6	2	1	3	3	4	3	5	10	60
Yue Fu	Seedling	1AF3					+										1
		1AF6					+										1
		1AF7					+										1
	Booting	2AF5			+			+									2
		2AF7			+			+									2
		2AF8						+									1
		2AF10						+									1
		2AF11			+			+									2
	Heading	3AF1														+	1
		3AF4			+												1
		3AF5													+		1
		3AF6													+		1
		3AF7		+	+			+								+	4
		3AF8														+	1
		3AF10			+										+		2
		3BF5					+										1
		3BF6														+	1
		3BF8														+	1
		3BF10					+										1
	Total	19	0	1	6	0	5	6	0	0	0	0	0	0	3	5	26

A: Isolated from N-containing medium. B: Isolated from N-free medium. +, The diameters of the inhibition zone are more than 20 mm. * The names of pathogen strains are described in the text.

The results of *in-vitro* antagonistic activity shown in Table 2 and Table 3 indicated that 60 inhibition zones were present against 14 pathogen strains by 22 tested strains isolated from Yue Guang. The positive reaction rate was 272%. For 19 tested strains isolated from Yue Fu, only 26 inhibition zones were observed. The positive reaction rate was 136%. The ratio of positive reactions for the strains isolated from Yue Guang was almost twice that from Yue Fu.

The *in-vitro* antagonistic spectra of rhizobacteria strains isolated from rice variety Yue Guang were much wider than that from Yue Fu. There were nine strains from Yue Guang to which more than three positive reactions were observed, while there was only one strain isolated from Yue Fu to which three positive reactions were recorded. The strains 2AG4 and 3AG11 were with even wider *in-vitro* antagonistic spectra than others, to which 14 and 6 pathogen strains were inhibited, respectively.

The *Xanthomonas oryzae*, *Corticium sasakii* and *Piricularia oryzae* Cavara are the main pathogens for rice plant diseases. There were 13 and 7 positive reactions observed for the strains isolated from rice varieties Yue Guang and Yue Fu, respectively. The positive reaction rate for the strains from Yue Guang was almost twice that from Yue Fu.

In conclusion, there may be some interrelations between rice varieties with different plant disease resistance and the species of rhizobacteria colonizing the rice rhizoplanes.

Isolation and population dynamics of endophytic diazotrophs associated with rice plants at different growth stages

A number of endophytic diazotroph strains have been isolated from strictly surface-sterilized rice Yue Fu plants in the rice field during the whole growing season. As the results show in Table 4, there were 10^3–10^6 cfu/g f.w. root of endophytes present in roots, stems and leaves. The population of endophytes with *in-vitro* ARA and ^{15}NFA increased until the tillering stage. The population of root endophytes with ARA and ^{15}NFA was more stable during the whole growing season. Within stems the population of endophytes with ARA and ^{15}NFA increased during the tillering to booting stages, followed by a decrease at the heading stage. Within leaves the population of endophytes with ARA and ^{15}NFA

Table 4. The endophytic bacteria population (cfu/g f.w. plant) isolated from rice Yue Fu plants

Rice growing season	Seeds	Roots	Stems	Leaves
Seed	7.23×10^5			
Seedling		2.04×10^4	1.07×10^5	4.32×10^4
Tillering		6.42×10^5	9.31×10^4	4.87×10^4
Booting		6.52×10^3	1.14×10^5	3.45×10^4
Heading		7.23×10^4	3.51×10^6	1.59×10^6

Table 5. The endophytic bacteria with ARA and $^{15}N_2FA$ isolated from rice Yue Fu plants during rice growing stages

Rice Growing					Roots				Stems				Leaves			
Season	Str	Pop	ARA	$^{15}N_2FA$	Str	Pop	ARA	$^{15}N_2FA$	Str	Pop	ARA	$^{15}N_2FA$	Str	Pop	ARA	$^{15}N_2FA$
Seed	SD13	2.82×10^5	+	3.087												
Seedling					SR3	1.52×10^3	+	2.291	SSml	1.56×10^4	+	4.804				
					SR10	7.58×10^2	+	0.542	SSm13	1.56×10^3	+	2.981				
					SR11	1.52×10^2	+	2.291								
Tillering					TR7	2.00×10^3	+	0.233	TSm5	1.00×10^3	+	0.490	TL5	1.00×10^2	+	4.094
					TR9	2.00×10^4	+	3.237	TSm13	4.00×10^3	+	0.493				
					TR10	2.00×10^4	+	0.442	TSm16	2.00×10^3	+	1.097				
									TSm18	1.00×10^3	+	0.853				
Booting					BR2	1.18×10^3	+	7.089	BSm6	2.34×10^3	+	5.065	BL1	3.68×10^3	+	5.853
					BR13	4.70×10^2	+	0.167	BSm7	3.51×10^3	+	3.810	BL2	6.13×10^3	+	0.328
									BSm8	8.18×10^3	+	5.517	BL7	1.23×10^3	+	1.739
									BSm17	9.36×10^2	+	0.161	BL11	2.45×10^2	+	4.933
									BSm18	4.68×10^2	+	0.034	BL12	6.14×10^2	+	0.048
									BSm19	4.01×10^2	+	0.470				
Heading					HR2	3.85×10^3	+	4.154	HSm6	1.04×10^5	+	1.889	HL5	2.78×10^4	+	0.113
					HR10	3.85×10^2	+	0.200	HSm18	4.11×10^5	+	2.049	HL7	2.78×10^4	+	4.136
													HL8	8.33×10^4	+	3.092
													HL11	1.67×10^5	+	3.924
													HL13	1.39×10^5	+	1.785

Str: strain; Pop: population; ARA +: > 1 nmol C_2H_4 formation/ml of culture.h; $^{15}N_2FA$: $^{15}N\%$ a.e.; SD: seed, S: Seedling, T: Tillering, B: Booting, H: Heading, R: Roots, Sm: Stems (2 cm above basal part), L: Leaves.

increased during the booting stage until heading stage (Table 5). It appears that with aging of rice plants, endophytic diazotrophs may move to the plant organs with vigorous metabolisms. The identification of these strains and *in-situ* assays for inoculating the strains on rice plants to check their biocontrol effects to rice plant diseases are in progress.

Acknowledgements

This work was supported by the National Natural Sciences Fundation of China (Project No. 39570025), Natural Science Foundation of Hebei Province (Project No. 396073) and the China Agriculture Ministry (Project No. 85 Agri-08–04–12). We thank Professor Su Baolin for providing rice varieties and the experimental field. We also thank Professor Cai Miaoyin for guidance on classification and identification of rhizobacteria.

References

Dobereiner J 1992. History and new prospectives of diazotrophs in association with non-leguminous plants. Symbiosis, 13, 1–13.

Mavingui P, Laguerre G, Berge O, Heulin T 1992. Genetic and phenotypic diversity of *Bacillus polymyxa* in soil and in the wheat rhizosphere. Appl. Environ. Microbiol., 58, 1894–1903.

Neal J L, Atkinson Jr T G, Larson R I 1970. Changes in the rhizosphere microflora of spring wheat induced by disomic substitution of a chromosome. Can. J. Microbiol., 16, 153–158.

6

Effect of *Azospirillum* inoculation on growth and N_2 fixation of maize subjected to different levels of FYM using ^{15}N-dilution method

H.M.A. EL-KOMY, T.M.M. MOHARRAM AND M.S.A. SAFWAT

Abstract

Using the ^{15}N dilution method, a pot experiment was conducted to study the effect of *Azospirillum* inoculation under different levels of farmyard manure (0, 5, and 10 t/ha) on N_2 fixation, growth and N yield of two maize cultivars. The percentage of nitrogen fixation (%Ndfa) by *Azospirillum* inoculation varied from 6% to 31% according to *Azospirillum* species and plant cultivars. Application of FYM at the rate of 5 t/ha decreased the nitrogen fixation in both cultivars and it was completely inhibited at a higher level (10 t/ha). Significant increases in the N yield were recorded in both cultivars irrespective of the FYM level (up to 10 t/ha). Such increases were mainly attributed to increases in shoot dry weights, as well as increases in the percentage of N derived mainly from the soil and fertilizer, not from N_2 fixation. *Azospirillum* spp. showed a higher degree of host specificity, and the degree of specificity occurred at plant cultivar level.

Introduction

One of the major problems of Egyptian soils is a deficiency in organic matter content, not exceeding 1–2% in all cultivated soils and a small fraction of 1% in sandy and newly reclaimed soils (El-Hadidy, 1960; Abd-El-Malek *et al.*, 1961). Therefore, the application of organic fertilizers seems to be of a great value for improving soil biological, chemical and physical properties, thus improving its productivity (Hegazi *et al.*, 1983; Peoples *et al.*, 1995).

In Egypt, farmyard manure (FYM) represents the basic organic fertilizer, as it has been known as long as agriculture itself; other organic fertilizers are used on a small scale. The interaction between FYM amendment and soil microflora, especially with nitrogen-fixing bacteria, was studied a long time ago (Bright and

K.A. Malik et al. (eds.), Nitrogen Fixation with Non-Legumes, 49–59.

Conn, 1919; Winogradsky, 1924). The addition of manure caused an increase in the total number of microflora in general within few days (Hegazi *et al.*, 1979, 1983), and nitrogen fixers in particular (Roper and Halsall, 1986; Roper and Smith, 1991; Roper *et al.*, 1994; Roper and Ladha, 1995).

Azospirillum spp. are frequent inhabitants of the soil and rhizosphere of a wide variety of plants. The nitrogen fixation and plant growth-promoting abilities of *Azospirillium* spp. have aroused interest in its use as a bacterial fertilizer (Dart, 1986; Okon and Kapulnik, 1986; Dobereiner and Pedrosa, 1987).

The objective of the present study was to determine the response of two different maize genotypic cultivars (H.II-Giza 215 and H.III-Giza 310) to inoculation with *Azospirillum* spp. and/or FYM amendment at different levels (0, 5 and 10 t/ha) and its effect on biological fixation of nitrogen by maize cultivars as well as the degree of host-strain specificity, using the ^{15}N-dilution method.

Materials and methods

Organisms and growth conditions

The bacterial strains used are listed in Table 1. *Azospirillum* strains were grown in liquid DN-medium (Day and Dobereiner, 1976) for 3 days at 30°C on a shaker (110 rpm). Bacterial cells were harvested by centrifugation (7 000 rpm, 9–10 min) and then washed twice with phosphate buffer pH 7.0 and used as inocula (1 ml = 10^7 cfu).

Pot experiment (under greenhouse conditions)

Homogenized mixtures of pure quartz, washed sandy soil and FYM at the rate of 0, 5 and 10 ton FYM/ha (the chemical analysis of FYM is listed in Table 2) were prepared, to fill pots of 3 kg capacity. Pots were arranged in a complete randomized design with six replicates.

Seeds of *Zea mays* (H.II Giza 215 and H. III Giza 310) cultivars were surface

Table 1. Bacterial strains used

Strains	Origin or reference
Azospirillum lipoferum st. 137 (No.27)	Kindly supplied by Prof. Dr L.F. Vassuyk, Academy of Science of Russia, Sanktbeterburg
Azospirillum lipoferum st. $R_{3/3}$	Isolated from rice-rizosphere by El-Komy (1992).
Azospirillum brasilense	Obtained from Ministry of Agriculture, Agricultural Research Center, ARC Giza, Egypt

Table 2. Chemical analysis of the used FYM

Organic C%	Total N%	C/N ratio	Organic matter	pH
15.57	1.30	11.97	26.84	6.5

sterilized according to El-Komy (1992), by immersion in a mixture solution of ethanol 96% and H_2O_2 (1 : 1, v/v) for 3 min, followed by several washings with sterile distilled water, and germinated on wet sterile filter papers in Petri dishes for 2 days. Five germinated seeds were transplanted into each pot. Each germinated seed was inoculated with 1 ml of inoculum containing 10^7 cfu of *Azospirillum* strain. After reaching 12 cm in length the plants were thinned down to two per pot. Pots were equally fertilized by $(^{15}NH_4)_2SO_4$ (1% atom excess) at the rate of 100 kg N/ha dissolved in water. Densities of *Azospirillum* were determined in the histosphere. Roots were washed and shaken for 2 min in 96% ethanol. Bacterial population was determined by the MPN method, using DN-semisolid medium (Day and Dobereiner, 1976) and the statistical tables of Cohran (1950). At 60 days after sowing, shoots and roots were harvested. They were dried overnight at 70°C and then weighed, ground and total nitrogen was determined. The atom% ^{15}N excess (N^{15}/N^{14} ratio) was determined using a mass spectrometer (Fiedler and Proksch, 1975) at the IAEA Seibersdorf Laboratory, Vienna, Austria. The percentage of N derived from atmosphere (% Ndfa) was calculated by the equation of Fried and Middleboe, 1977.

$$\%\text{Ndfa} = 1 - \frac{\%^{15}\text{N atom excess (Fs)}}{\%^{15}\text{N atom excess (nFs)}}$$

Where (Fs) and (nFs) indicate inoculated and uninoculated system respectively. N_2 fixed was calculated by the equation:

$$N_2 \text{ fixed (mg/pot)} = \frac{\%\text{Ndfa} \times \text{N yield (mg/pot)}}{100}$$

Results

Isotope determinations without FYM amendment

Considering the initial enrichment of 1% atom excess, the atomic percentage ^{15}N excess (Table 3) in all treatments was relatively low. These low values could be attributed to the residual ^{15}N associated with soil and plant roots. *Azospirillum* inoculation resulted in different ^{15}N atom excess; the highest values were recorded with the uninoculated (control) treatments. Less significant values were obtained with *Azospirillum* inoculation in both maize cultivars. This decrease in the ^{15}N enrichment of inoculated compared to uninoculated treatments indicates that the plants were receiving N_2 from the atmosphere. Inoculation with different *Azospirillum* spp. resulted in different amounts of plant N derived from air in both

Table 3. Effect of *Azospirillum* inoculation and/or FYM amendment on %^{15}N atom excess of maize cultivars

	Maize hybrid II (Giza 215)			Maize hybrid III (Giza 310)		
FYM (t/ha)	0	5	10	0	5	10
Control	0.182^{a}*	0.135^{a}	0.074^{a}	0.146^{a}	0.092^{a}	0.084^{a}
A. lipoferum (st. 137)	0.125^{b}	0.094^{b}	0.085^{b}	0.116^{b}	0.082^{b}	0.094^{b}
A. lipoferum (st. R3/3)	0.133^{c}	0.101^{c}	0.075^{c}	0.120^{c}	0.082^{c}	0.095^{bc}
A. brasilense	0.150^{d}	0.115^{d}	0.091^{d}	0.137^{d}	0.090^{d}	0.091^{d}

*Means within a column followed by the same letter do not differ significantly ($p < 0.05$) by Duncan.

maize cultivars. Inoculation with *A. lipoferum* st. 137 and st. R3/3 resulted in the accumulation of fixed N_2 and the %Ndfa, varying from 31% and 26% respectively (equal to 198 and 70 mg fixed N_2) with maize hybrid II (Tables 4 and 6) to 20% and 17% (equal to 131 and 65 mg fixed N_2) with hybrid III respectively (Tables 5 and 7). However, inoculation with *A. brasilense* strain resulted in the accumulation of fixed N_2 and %Ndfa varying from 17% to 6% (which is equal to 36 and 18 mg fixed N_2) with maize hybrid II and III respectively. Tables 4 and 5 also show that when less N_2 derived from fixation was available to the plant, more labelled fertilizer N (%Ndff) or soil N (%Ndfs) or both were assimilated by the plant. The percentage of plant N derived from ^{15}N fertilizer (%Ndff) of the uninoculated treatments varied between 18 for maize hybrid II and 15 for hybrid III, indicating uniform isotopic labelling of the soil within different plant varieties. The quantity of N obtained from the soil and fertilizer in plants inoculated with *A. brasilense* strain was higher than that obtained from the same sources with *A. lipoferum* strains.

Isotope determinations at different FYM levels

Higher ^{15}N abundance in control treatments as compared to the inoculated ones was recorded up to the level of 5 t/ha in both maize cultivars (Table 3). This is indicative of isotopic dilution as the result of the uptake of fixed nitrogen. However, the opposite was obtained at higher FYM level (10 t/ha), indicating that N_2 fixation was completely inhibited and higher N uptake from soil and fertilizer (up to 90.5%) was recorded under such condition (Tables 4 and 5).

Effect of Azospirillum *inoculation on growth and N_2 fixation of maize subjected to different levels of FYM*

Data in Tables 6 and 7 show that the growth of the two experimental maize hybrids was positively affected by *Azospirillum* inoculation irrespective of the

Table 4. Effect of *Azospirillum* inoculation and/or FYM amendment on %Ndfa, %Ndff and %Ndfs of maize hybrid II (Giza 215)

FYM (t/ha)	0			5			10		
	Ndfa	Ndff	Ndfs	Ndfa	Ndff	Ndfs	Ndfa	Ndff	Ndfs
Control	0.0[a]*	18.2[a]	81.8[a]	0.0[a]	18.5[a]	81.5[a]	0.0[a]	7.4[d]	92.6[d]
A. lipoferum (st. 137)	31.4[b]	12.5[c]	56.1[d]	30.3[b]	9.4[b]	60.3[b]	–14.6[a]	8.5[b]	106.1[b]
A. lipoferum (st. R3/3)	26.5[c]	13.5[d]	60.0[cd]	25.1[c]	10.3[c]	64.6[c]	–1.7[a]	7.5[cd]	94.2[cd]
A. brasilense	17.5[d]	15.2[bc]	67.3[b]	14.8[d]	11.8[d]	73.4[d]	–25.9[a]	9.1[a]	116.8[a]

*Means within a column followed by the same letter do not differ significantly ($p < 0.05$) by Duncan.

Table 5. Effect of *Azospirillum* inoculation and/or FYM amendment on %Ndfa, %Ndff and %Ndfs of maize hybrid III (Giza 310)

FYM (t/ha)	0			5			10		
	Ndfa	Ndff	Ndfs	Ndfa	Ndff	Ndfs	Ndfa	Ndff	Ndfs
Control	0.0[a]*	14.6[a]	85.4[a]	0.0[a]	9.5[a]	90.5[a]	0.0[a]	8.4[b]	91.6[b]
A. lipoferum (st. 137)	20.6[b]	11.6[b]	67.8[ab]	14.0[b]	8.2[d]	77.8[d]	–12.9[a]	9.5[a]	103.4[a]
A. lipoferum (st. R3/3)	17.4[c]	12.0[c]	70.6[b]	13.0[c]	8.3[cd]	78.7[cd]	–13.6[a]	9.5[a]	104.1[a]
A. brasilense	6.1[d]	13.7[d]	80.2[ab]	5.2[d]	9.0[b]	85.8[b]	–8.5[a]	9.1[a]	99.4[a]

*Means within a column followed by the same letter do not differ significantly ($p < 0.05$) by Duncan.

Table 6. Effect of *Azospirillum* inoculation and/or FYM amendment on dry weight, %N, N yield and fixed nitrogen of maize hybrid II (Giza 215).

	Dry shoot-mass (g/pot) × 10			%N			N yield (mg N/pot) × 100			Fixed nitrogen (mg N/pot) × 100		
FYM (t/ha)	0	5	10	0	5	10	0	5	10	0	5	10
Control	1.6^{d}	2.7^{d}	3.2^{c}	1.3^{a}	1.1^{d}	1.3^{a}	2.0^{d}	2.9^{bc}	4.0^{cd}	0.0^{a}	0.0^{d}	0.0
A. lipoferum (st. 137)	3.3^{a}	4.7^{a}	4.7^{a}	1.9^{bc}	1.5^{a}	1.3^{a}	6.3^{a}	6.9^{c}	6.2^{a}	1.9^{b}	2.6^{a}	0.0
A. lipoferum (st. R3/3)	2.6^{b}	3.5^{bc}	4.4^{ab}	1.0^{c}	1.2^{cd}	1.1^{bc}	2.6^{c}	4.2^{d}	4.7^{bc}	0.7^{c}	0.6^{c}	0.0
A. brasilense	2.1^{c}	3.2^{cd}	3.6^{b}	1.3^{a}	1.2^{bc}	1.2^{c}	2.6^{bc}	3.9^{a}	4.7^{d}	0.4^{d}	0.5^{bc}	0.0

*Means within a column followed by the same letter do not differ significantly ($p < 0.05$) by Duncan.

Table 7. Effect of *Azospirillum* inoculation and/or FYM amendment on dry weight, %N, N yield and fixed nitrogen of maize hybrid III (Giza 310).

	Dry shoot-mass (g/pot) × 10			%N			N yield (mg N/pot) × 100			Fixed nitrogen (mg N/pot) × 100		
FYM (t/ha)	0	5	10	0	5	10	0	5	10	0	5	10
Control	1.5^{d}	2.2^{d}	2.8^{d}	1.3^{d}	1.1^{c}	1.1^{a}	2.0^{d}	2.4^{d}	3.2^{d}	0.0^{a}	0.0^{a}	0.0
A. lipoferum (st. 137)	3.8^{a}	4.6^{a}	4.9^{a}	1.6^{c}	1.1^{d}	1.1^{a}	6.3^{a}	5.2^{a}	5.4^{a}	1.3^{b}	0.7^{b}	0.0
A. lipoferum (st. R3/3)	2.5^{b}	3.5^{b}	3.8^{bc}	1.5^{a}	1.3^{b}	1.0^{a}	3.7^{bc}	4.5^{bc}	3.7^{bcd}	0.6^{c}	0.6^{c}	0.0
A. brasilense	1.8^{cd}	2.8^{c}	3.3^{cd}	1.7^{b}	1.5^{a}	1.1^{a}	3.1^{c}	4.2^{c}	3.7^{cd}	0.2^{d}	0.2^{d}	0.0

*Means within a column followed by the same letter do not differ significantly ($p < 0.05$) by Duncan.

FYM level. This was evident from the significant to highly significant increases in dry shoot mass and total plant N yield over the control treatments, even at higher FYM level (10 t/ha).

The survival of Azospirillum *spp. in the histosphere of maize plants*

To ensure the establishment of *Azospirillum* spp. inocula within the plant histosphere (crushed root which had been washed and surface sterilization; Plazinsiki and Rolfe, 1985), and the effect of FYM amendment, the total number of *Azospirillum* spp. within the root tissue was estimated by the MPN method. As indicated in Figures 1 and 2, *Azospirillum* successfully colonized the histosphere and survived to the end of the experiment (2 months). *Az. lipoferum* strains (st. 137 and st. R3/3) were recorded in higher numbers (up to 10^7 and 10^5/g dry root respectively), while *A. brasilense* strain survived in lower numbers (up to 10^3/g dry root) and disappeared more rapidly. The amendment of FYM at both levels (5 and 10 t/ha) clearly enhanced the *Azospirillum* histosphere colonization.

Discussion

It has been established that the addition of organic matter to the soil has a beneficial effect in improving its productivity by several mechanisms; one of these

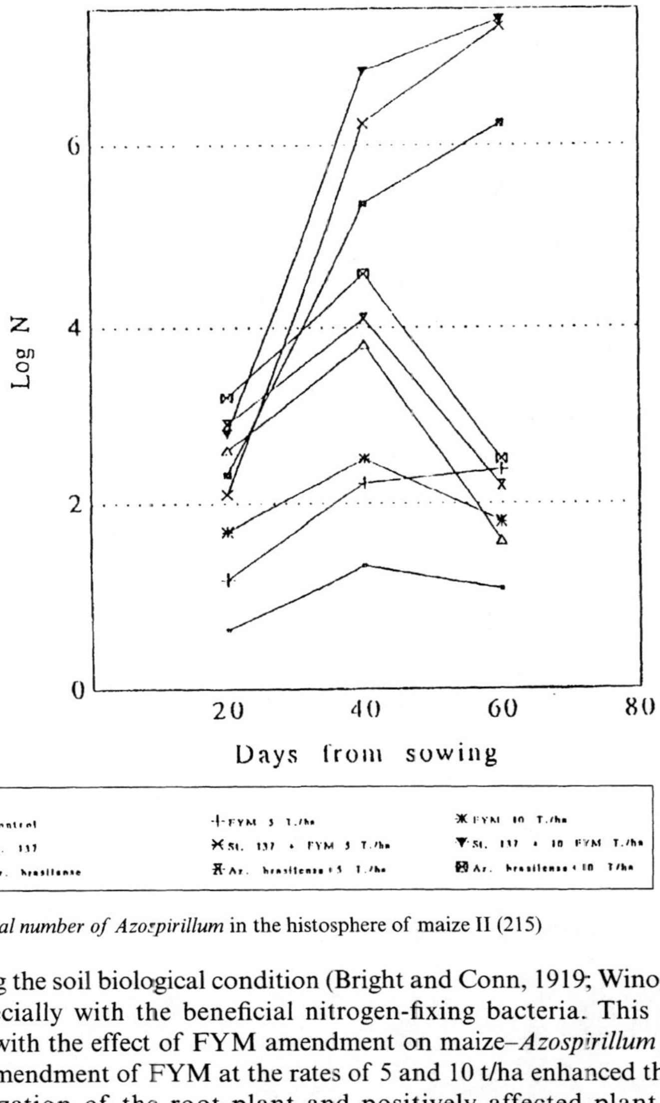

Figure 1. Total number of Azospirillum in the histosphere of maize II (215)

is improving the soil biological condition (Bright and Conn, 1919; Winogradsky, 1924), especially with the beneficial nitrogen-fixing bacteria. This study is concerned with the effect of FYM amendment on maize–*Azospirillum* association. The amendment of FYM at the rates of 5 and 10 t/ha enhanced the bacterial colonization of the root-plant and positively affected plant growth parameters, but not the nitrogen-fixing capacity and %Ndfa (measured by ^{15}N-isotope). This can be explained on the basis that, at higher levels of FYM, organic-matter mineralization resulted in higher inorganic nitrogen accumulation that inhibited the N_2-ase activity. The inhibitory effect of higher N levels on N_2-ase activity was previously established by many authors (Mertens and Hess, 1984; Millet and Feldman, 1986; Da Silva *et al.*, 1993).

Significant increases in total N yield over the control treatments by *Azospirillum* inoculation, even at high FYM level (10 t/ha), might be attributed to

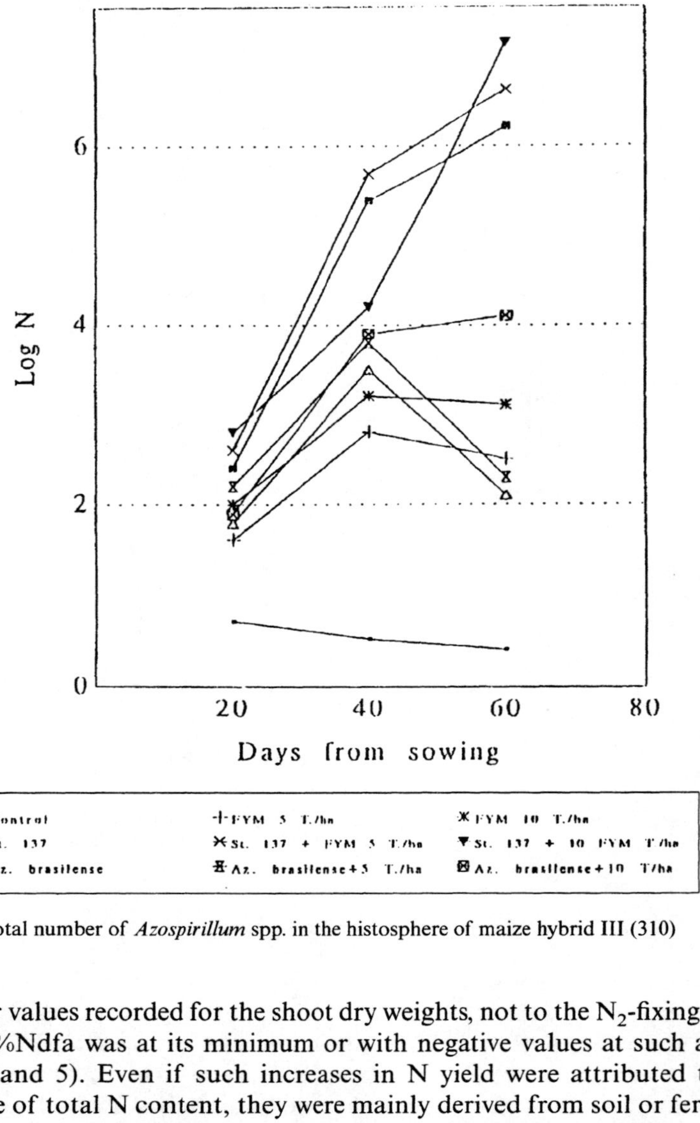

Figure 2. Total number of *Azospirillum* spp. in the histosphere of maize hybrid III (310)

the greater values recorded for the shoot dry weights, not to the N_2-fixing activity, since %Ndfa was at its minimum or with negative values at such a level (Tables 4 and 5). Even if such increases in N yield were attributed to the percentage of total N content, they were mainly derived from soil or fertilizer N, or from both. These results are in agreement with many investigators, e.g. Caballero-Mellado *et al.*, 1992; Fages, 1994. These authors concluded that significant responses of *Azospirillum* inoculated cereals were not due to N_2 fixing activities, but were mainly attributed to increased uptake of soil and fertilizer N. In other words, *Azospirillum*-inoculated plants assimilate mineral N from soil and fertilizer more than uninoculated plants (Caballero-Mellado *et al.*, 1992).

One of the most outstanding questions concerning the *Azospirillum* and host plant association is the degree of specificity required for effective association; specific differences between responses of C-3 and C-4 photosynthetic type plants was suggested; *A. lipoferum* was the predominant species colonizing C-4 plants and *A. brasilense* was the predominant species associated with C-3 plants (Baldani and Dobereiner, 1980; Baldani *et al.*, 1986). Thus when the bacterial species was inoculated onto the respective plant species, success was more frequent when the proper plant–bacterial species combination was used (Reynders and Vlassak, 1982; Garcia de Salamone, 1993). Specificity can occur at plant cultivar level (Jagnov, 1990). On the other hand, studies with *A. brasilense* strain cd/sp7 (the type strain for *A. brasilense* which is commonly used as a reference), indicated no plant specificity, and this strain positively affected many winter and summer cereals (Bashan *et al.*, 1989). Moreover, studies with *Azospirillum* inoculation of wheat and maize respectively indicated no differences between different cultivars in growth yield under field conditions (Zambre *et al.*, 1984). Our results recommended the first suggestion, that inoculation of maize plant (C-4 type plant) with two different *Azospirillum* spp., *A. lipoferum* strains (137 and R3/3) resulted in higher plant response (dry weight, %Ndfa, N-yield and higher host-colonization capacity) than *A. brasilense* inoculation. Moreover, maize hybrid II (var. 215) was more responsive for inoculation than hybrid III (var. 310), indicating the role of plant cultivar in such association.

In conclusion, like *Rhizobium* strain selection (Rolfe *et al.*, 1980) the importance of *Azospirillum* strain selection is a crucial consideration in determining specificity. Numerous strains must be isolated from different origins, characterized, and tested under various environmental conditions for each plant cultivar.

Acknowledgements

This contribution was supported by the FAO/IAEA, Vienna, Austria, to the regional African project on BNF (RAF 151010).

References

Abd-El-Malek Y, Abd-El-Salam M A, Monib M, El-Hadidy T T 1961. Bacteriological and chemical changes resulting from addition of certain organic materials to calcareous soils of the Mediterranean coastal region. Bull. Inst. Desert, Egypt. 11, 31–40.

Baldani V L D, Dobereiner J 1980. Host-plant specificity in the infection of cereals with *Azospirillum* spp. Soil Biol. Biochem., 12, 433–439.

Baldani V L D, Alvarez M A De B, Baldani J I, Dobereiner J 1986. Establishment of inoculated *Azospirillum* spp. in the rhizosphere and in roots of field grown wheat and sorghum. Plant Soil, 90, 35–46.

Bashan Y, Ream Y, Levanony H, Sade A 1989. Nonspecific responses in plant growth, yield and root colonization of noncereals crop plants to inoculation with *Azospirillum brasilense* Cd. Can. J. Bot., 67, 1317–1324.

Bright J W, Conn H J 1919. Influence of stable manure upon soil population. N.Y. Agric. Exp. Stat. Bull., 67.

Caballero-Mellado J, Carcano Mantiel M G, Mascarua-Esparza M A 1992. Field inoculation of wheat (*Triticum aestivum*) with *Azospirillum brasilense* under temperate climate. Symbiosis, 13, 243–253.

Cohran W G 1950. Estimation of bacterial densities by means of most probable number. Biometric., 6, 105–116.

Dart P J 1986. Nitrogen fixation associated with non-legumes in agriculture. Plant Soil, 90, 303–334.

Day J M, Dobereiner J 1976. Physiological aspects of N_2 fixation by *Spirillum* from *Digitaria* roots. Soil. Biol. Biochem., 8, 45–50.

Da Silva P M, Tsai S M, Bonetti R 1993. Response to inoculation and N fertilization for increased yield and biological nitrogen fixation of common bean. Plant Soil, 152, 123–130.

Dobereiner J and Pedrosa F O 1987 In: Nitrogen Fixing Bacteria in Non-leguminous Crop Plants. Springer, Berlin, 1–155.

El-Hadidy T M T 1960. Bacteriological and chemical changes due to addition of certain materials to Mediterranean calcareous soils. MSc thesis, Faculty of Agriculture, Cairo University, Egypt.

El-Komy H M A 1992. Studies on the genus *Azospirillum* from the rhizosphere of maize and rice plants. PhD thesis, Institute of Agricultural Microbiology, Academic Science of Russia, Sanktbeterburg

Fages J 1994. *Azospirillum* inoculants and field experiments. In: Okon Y, ed. *Azospirillum* Plant Association, 87–110.

Fiedler R Proksch G 1975. The determination of N-15 by emission and mass spectrometry in biological analysis. A review. Anal. Chem. Acta, 78, 1–62.

Fried M, Middelboe V 1977. Measurement of amount of nitrogen fixed by a legume crop. Plant Soil, 47, 713–715.

Garacia de Salamone Y 1993. MSc thesis, University of Buenos Aires.

Hegazi N A, Vlassak K, Monib M 1979. Effect of amendments, moisture and temperature on acetylene reduction in Nile Delta soils. Plant Soil, 51, 27–37.

Hegazi N A, Monib M, Amer H A, Shokr E 1983. Response of maize plants to inoculation with *Azospirilla* and/or straw amendment in Egypt. Can. J. Microbiol., 29, 888–894.

Jagnow G 1990. Differences between cereal crop cultivars in root-associated nitrogen fixation, possible causes of variable yield response to seed inoculation. Plant Soil, 123, 255–259.

Mertens T, Hess D 1984. Yield increases in spring wheat inoculated with *Azospirillium lipoferum* under greenhouse and field conditions of a temperate region. Plant Soil, 82, 87–99.

Millet E, Feldman M 1986. Yield response of a common spring wheat cultivar to inoculation with *Azospirillum brasilense* at various levels of nitrogen fertilization. Plant Soil, 80, 255–259.

Okon Y, Kapulnik Y 1986. Development and function of *Azospirillum*-inoculated roots. Plant Soil, 90, 3–15.

Peoples M B, Herridge D F, Ladha J K 1995. An efficient source of nitrogen for sustainable agricultural production. In: Ladha J K, Peoples M B, eds. Management of Biological Nitrogen Fixation for the Development of More Productive and Sustainable Agricultural Systems. Kluwer, Dordrecht, 3–28.

Plazinski J, Rolfe G 1985. Analysis of the pectolytic activity of *Rhizobium* and *Azospirillum* strains isolated from *Trifolium repens*. J. Plant Physoil., 120, 181–187.

Reynders L, Vlassak K 1982. Use of *Azospirillum brasilense* as a biofertilizer in intensive wheat cropping. Plant Soil, 66, 217–223.

Rolfe B G, Gresshoff P M, Shine J 1980. Rapid screening for symbiotic mutants of *Rhizobium* and white clover. Plant Sci. Lett., 19, 277–284.

Roper M M, Halsall M 1986. Use of products of straw by N_2-fixing populations of bacteria in three soils from wheat-growing areas. Aust. J. Agric. Res., 37, 1–9.

Roper M M, Ladha J K 1995. Biological N_2 fixation by heterotrophic and phototrophic bacteria in association with straw. Plant Soil, 174, 221–224.

Roper M M, Smith N A 1991. Straw decomposition and nitrogenase activity by free-living microorganisms from soil: effect of pH and clay content. Soil Biol. Biochem., 23, 275–283.

Roper M M, Turpin J A, Tompson J P 1994. Nitrogenase activity by free-living bacteria in soil in a long-term tillage and stubble management experiment on a vertisol. Soil Biol. Biochem., 26, 1087–1091.

Winogradsky S 1924. Influence of organic matter upon soil population. Compt. Rend. Acad. Sci., 178, 1236 (cited in Waksman, S A 1931).

Zambre M A, Konde B K, Sonar K R 1984. Effect of *Azotobacter chrococcum* and *Azospirillum brasilense* inoculation under graded levels of nitrogen on growth and yield of wheat. Plant Soil, 79, 61–67.

7

Contribution of non-leguminous biofertilizers to rice biomass, nitrogen fixation and fertilizer-N use efficiency under flooded soil conditions

SIKANDER ALI, NAIMA HAMID, GHULAM RASUL,
SAMINA MEHNAZ AND KAUSER A. MALIK

Abstract

A field experiment was conducted to study the effect of root-associated diazotrophic rhizobacteria and *Azolla*, alone and in combinations, on rice grown in a flooded saline soil. ^{15}N labelled ammonium sulphate at the rate of 30 kg N/ha was used to monitor its recovery and computation of biological nitrogen fixation (BNF) using an isotopic dilution method. The bacterial population of nitrogen-fixing bacteria generally increased more for inoculated treatments compared with uninoculated ones. The rice biomass (straw + grain) was maximum for *Azolla* cover followed by the bacterial treatment. The fertilizer-N recovery in rice was maximum for *Azolla* cover + bacteria followed by *Azolla* cover treatment, which may be due to lowering of floodwater pH by *Azolla*. The maximum amount of BNF (46 kg N/ha) was obtained for bacterial treatment, followed by bacteria + *Azolla* incorporated treatment. The study indicated that use of biofertilizers, along with a low input of chemical-N fertilizer, is useful for increasing rice yield, fertilizer-N use efficiency and BNF in rice grown in flooded soil conditions.

Introduction

Among fertilizer inputs, nitrogen is the major limiting nutrient for crop production. It can be supplied through chemical or biological means, but the production of chemical nitrogen fertilizer is energy intensive as 2 tonnes of fuel oil are consumed for production of 1 tonne of fertilizer nitrogen (Regan, 1988). Both the legume and non-legume nitrogen biofertilizers can convert atmospheric nitrogen to plant-usable form through biological nitrogen fixation. Despite the high N_2-fixing potential and positive effect on soil properties, the use of green

K.A. Malik et al. (eds.), Nitrogen Fixation with Non-Legumes, 61–73.

manure legumes for lowland rice has declined dramatically world-wide over the past 30 years, and at present they cover less than 5 million ha in tropical Asia (Becker *et al.*, 1995). Most of the rice in the world (128 million ha rainfed lowland and irrigated vs. 19 million ha upland) is grown in flooded conditions (IRRI, 1994). The legume green manures have certain weaknesses for the rice ecosystem, as most of them do not grow or fix atmospheric nitrogen in flooded conditions. Unlike legumes, non-leguminous biofertilizers such as *Azolla* and rhizobacteria are adapted to flooded conditions of lowland rice and do not need separate time, water and land for their multiplication, as *Azolla* can be grown on the water surface as an intercrop with rice, while rhizobacteria can rapidly multiply in the rhizosphere.

Based on data from various workers, Peoples *et al.* (1995) estimated that *Azolla* can fix 22–40 kg N/ha per month, while driving 52–99% of its nitrogen from the atmosphere. Its growth rate is very high (3–5 days under optimal conditions), and its long-term use not only increased rice yield but also improved soil fertility (Ventura and Watanabe, 1993). *Azolla* has been used traditionally as green manure for rice production in Southeast Asia and is still considered an important biofertilizer for rice crops (Lumpkin and Plucknett, 1982; Roger *et al.*, 1993; Peoples *et al.*, 1995). The application of *Azolla* has been reported to increase rice yield by 0.4–1.5 t/ha over the control, in most of the experimental sites in China, Vietnam, India, Thailand, Philippines and USA (Kikuchi *et al.*, 1984).

According to Frankenberger and Arshad (1995), many of the root-associated bacteria not only fix nitrogen but also produce some metabolites which influence the growth of the plant. Okon (1981) inoculated corn with nitrogen-fixing bacteria and found 77 kg more N/ha than in uninoculated plants. Considering all the data of inoculation with nitrogen-fixing and plant growth-promoting rhizobacteria, an average increase in rice grain yield was estimated to be 28% for pot and 14% for field conditions (Roger *et al.*, 1993).

About 14 million acres of land are salt-affected in Pakistan. Since rice can grow under varying degree of flooding, and has shown some salt tolerance, most of the salt-affected soils, whether reclaimed through chemical or biological means, are invariably sown to rice as the first crop. These soils are usually saline sodic, and due to a high floodwater pH there is a significant loss of applied fertilizer-N through ammonia volatilization (Hussain and Malik, 1983). The loss of applied mineral N in lowland rice, calculated on ^{15}N balance by various researchers, was up to 62% (Becker *et al.*, 1995) and a recovery of only 30–40% of applied N by rice has been reported in paddy soils (Cooke, 1982; Crawswell and Vlek, 1979).

Although sufficient work is reported on the nitrogen fixation by *Azolla* and root-associated nitrogen fixing bacteria only meagre information is available on the conjoint use of these biofertilizers and particularly in saline soils. We therefore tried to use these bacteria and *Azolla* alone and in different combinations to study the effect on rice yield, nitrogen fixation and fertilizer-N use efficiency in a flooded saline soil using ^{15}N labelled fertilizer.

Materials and methods

In this field study a marginally saline soil was used in the cemented microplots, and the chemical analysis of this soil is presented in Table 1. The total N was estimated with the regular Kjeldahl method, and KCl extractable NH_4 and $NO_{3-}N$ by using MgO-Devarda alloy steam distillation methods (Keeney and Nelson, 1982).

The soil was put into 1.5 × 1.5 × 0.6 m (LxWxD) cemented plots and was flooded for a few days before rice transplanting. The field experiment was carried out in a randomized complete block design, with four replications of 11 treatments (Table 2), having ^{15}N-labelled ammonium sulphate in T6–T11.

For bacterial inoculation a mixture of previously isolated root-associated bacteria (Malik *et al.*, 1994); comprising *Azoarcus* (K-1), *Flavobacterium* (96–57), *Pseudomonas* (96–51) and *Azospirillum* (N-4) was used. These bacteria were grown in semisolid nitrogen-free culture medium, and suspension of each strain was prepared at 10^8 cells/ml. Then an equal volume of these suspensions was mixed for inoculation of rice. The inoculum was applied to rice by dipping the roots of the seedling for ½ h before transplanting.

One-month-old seedlings of rice (NIAB-6) were transplanted at two seedlings per hill with 20 × 20 cm spacing. Seven days after rice transplanting (DAT), mixed culture of *Azolla* spp.; comprising *A. pinnata var. pinnata, Rong Ping* (hybrid *Azolla*), *A. microphylla, A. caroliniana, A. filiculoides,* and *A. pinnata* (local); was inoculated at 177 g (f.w.)/m^2.

To enhance *Azolla* growth superphosphate was applied at 5 kg P_2O_5/ha in *Azolla*-inoculated plots and also in all the other plots. ^{15}N-labelled ammonium sulphate (5% abundance) was applied in a single dose, in solution form at 3–4 cm below the soil surface 7 DAT at 30 kg N/ha for treatments 6 to 10, and at 60 kg n/ha for treatment 11, which was kept as the non-fixing reference treatment.

To study the colonization of inoculated diazotrophic bacteria on rice roots the most probable number (MPN) and direct plate count methods, using combined carbon medium (CCM), in vials and plates were used (Rennie, 1981). For enumeration of these bacteria in soil and associated with rice roots, serial dilutions in saline solution (0.9%) were prepared and 100 µl from every dilution was inoculated into vials and plates (Bilal *et al.*, 1990).

Table 1. Chemical properties of the experimental soil.

EC (saturation extract)	4.87 ds/m
pH (soil paste)	7.8
K (saturation extract)	0.15 mEq/L
Na (saturation extract)	66.0 mEq/L
Ca (saturation extract)	4.1 mEq/L
Total N	0.04%
Available $NH_{4-}N$	11.5 mg/kg
Available $NO_{3-}N$	13.4 mg/kg

Table 2. Effect of biofertilizers on dry matter and nitrogen yield of rice

Treatment	Dry matter yield (kg/ha)			Nitrogen yield (kg/ha)		
	Straw	Grain	Straw+grain	Straw	Grain	Straw+grain
T1. Control	7556^{b}	7583	15139^{ab}	40^{ab}	112	152^{a}
T2. *Azolla* cover	7265^{b}	7691	14956^{ab}	37^{bc}	108	145^{ab}
T3. *Azolla* incorporated	7786^{ab}	8181	15967^{ab}	41^{ab}	109	150^{a}
T4. Bacteria	7392^{b}	7656	15048^{ab}	37^{b}	107	144^{ab}
T5. *Azolla* cover + bacteria	7470^{b}	7752	15223^{ab}	40^{ab}	108	148^{ab}
T6. 30 kg N/ha	7657^{ab}	7884	15541^{ab}	41^{ab}	110	151^{a}
T7. 30 kg N + *Azolla* cover	8433^{a}	7781	16214^{a}	43^{a}	110	152^{a}
T8. 30 kg N + bac. + *Azolla* incorporated	7814^{ab}	7878	15691^{ab}	43^{a}	110	153^{a}
T9. 30 kg N + bac. + *Azolla* cover	7619^{ab}	7710	15329^{ab}	42^{a}	111	153^{a}
T10 30 kg N + bacteria	8449^{a}	7753	16202^{a}	44^{a}	113	157^{a}
T11. 60 kg N/ha	7044^{b}	7732	14776^{b}	33^{c}	103	136^{b}
		N.S.			N.C.	

^{15}N-labelled ammonium sulphate was applied to soil for treatments T6–T11.
Means followed by the same letter are not statistically different a 5% *P*.

For identification of inoculated bacteria, morphological methods (shape, size, colour of colony) and biochemical tests (using QTS strips), were used (Bilal *et al.*, 1990). In addition to these tests, strain-specific fluorescent antibody (FA) stains for the four bacterial strains were prepared in rabbits according to Somasegaran and Hoben (1985). The quality and specifity of FA was tested according to Schmidt (1974), and used for identification of inoculated strains. To minimize non-specific binding, the root was covered with gelatine/rhodamine/isothiocynate conjugate as described by Bohlool and Schmidt (1968).

In case of *Azolla* cover treatments *Azolla* was allowed to grow without incorporation, while in *Azolla*-incorporated treatments approximately half of the *Azolla* was incorporated into soil when it grew to full cover, and in total two incorporations were made, being at 55 and 91 DAT. When plots were completely covered with *Azolla*, the fresh weight of *Azolla* was recorded by using the quadrate method. At 1 week after first incorporation, pest attack was observed on *Azolla* and Furadan (1 kg carbofuran/ha) as granules was broadcast onto all *Azolla* as well as other plots to kill the insects. After 4 days of Furadan application, superphosphate (5 kg P_2O_5/ha) was again applied to all the plots, to enhance *Azolla* growth.

At maturity, 36 central rice hills of each plot were harvested, and one outermost line on each side was left as border. Total fresh weight of straw and grain was recorded and then dried at 70°C for dry weight. The ^{15}N analysis was performed by converting ammonium to dinitrogen with alkaline sodium hypobromite (Hauck, 1982) using a mass spectrometer with an inlet system (Mat GD150).

Results and discussion

Bacterial population of rhizosphere and rice roots

The population of nitrogen-fixing bacteria in rhizosphere-soil of rice, estimated by the ARA-based MPN method, increased more in inoculated treatments (T8 and T10) than in uninoculated (30 kg N/ha) treatment (T6) (Figure 1). The bacterial number increased with rice growth period for nitrogen (T6), and nitrogen+*Azolla* incorporated+bacteria (T8), nitrogen+bacteria (T10) treatments; but decreased for nitrogen+*Azolla* cover+bacteria (T9) treatment (Figure 1), which may be due to differences in bacterial ecology such as lesser diffusion of air/nitrogen into soil, through the *Azolla* mat.

The bacterial population on rice roots, also enumerated by the ARA-based MPN method, showed higher numbers for inoculated treatments than only nitrogen, at the initial stage, but it decreased with time during the rice growth period (Figure 1). The decline in population after inoculation has also been reported by other workers (Albrecht *et al.*, 1983; Chan *et al.*, 1963).

The population of inoculated bacteria, as shown by FA staining, showed that in rhizosphere soil it was generally higher for inoculated treatments than uninoculated control at panicle as well as at maturity stage of rice (Table 3).

The FA staining of roots of rice showed higher numbers for inoculated than uninoculated treatments for K-1, N-4 and 95–57 strains, whereas the population of 96–51 was not found at panicle and maturity stage of rice (Table 3). The absence of *Pseudomonas* strain 96–51 on rice roots may be due to its poor survival, or competition, or the antagonistic effect of other microorganisms in the soil (Chan *et al.*, 1963).

The above-mentioned bacterial population and ARA studies indicated that inoculation of nitrogen-fixing bacteria into rice and rice–*Azolla* culture helped in increasing the desirable bacterial population in rice ecosystem.

Azolla *biomass and N content*

Periodic visual observations indicated that the growth of inoculated *Azolla* increased after application of phosphorus. The fresh *Azolla* biomass, recorded after 49 days of its inoculation, was 0.4–0.8 kg/m^2 (4–8 t/ha), whereas the average *Azolla* biomass was reported to be 1.1 kg/m^2 in INSFER trials of various countries (Watanabe, 1987). In this study nitrogen content in the *Azolla* mat was estimated to be 6–15 kg/ha, and slightly higher N contents were observed for treatments having *Azolla* alone, or for *Azolla*+ nitrogen than for *Azolla*+bacteria. Kikuchi *et al.* (1984) have reported that in open fields one layer of *Azolla* containing 10 tonnes of green matter may have about 25–30 kg N/ha. The lower values for fresh biomass and nitrogen recorded in the present study may be due to the adverse effect of high summer temperature and high sodium or salts in this soil, as high temperature (Cary and Weerts, 1992) and salinity reduces the growth and nitrogen fixation in *Azolla* (Ali *et al.*, 1990).

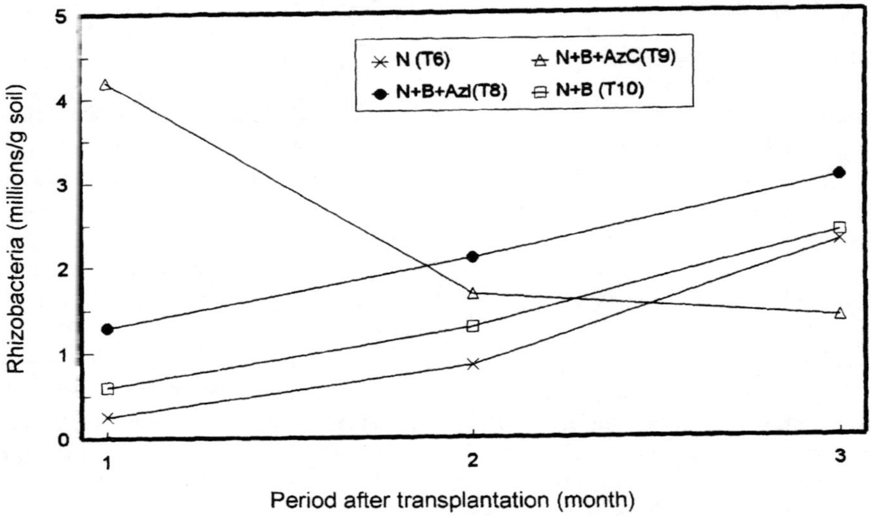

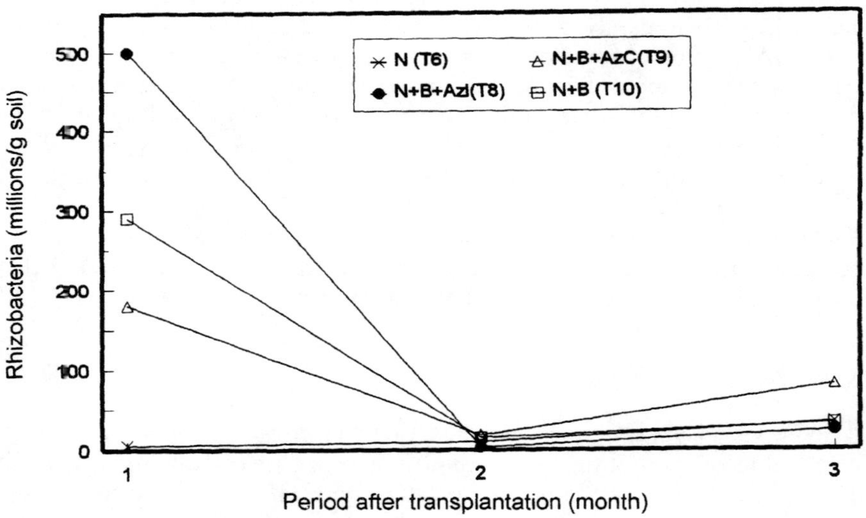

Figure 1. Diazotrophic bacterial population in the rhizosphere soil (top) and on the rice roots (bottom). (N=30 kg N/ha, B=rhizobacteria, Azl=*Azolla* incorporated, AzC=*Azolla* cover)

Rice yield

The rice yield (straw + grain) was maximum for nitrogen+*Azolla* cover (T7) and nitrogen+bacteria (T10) and minimum for 60 kg N/ha (T11), while a higher grain yield was observed for T3 (Table 2). The higher rice yield for the above-mentioned biofertilizer + 30 kg N/ha treatments indicates the usefulness of *Azolla*

Table 3. FA staining of rice roots from control and PGPR-inoculated plots at different stages of rice growth

Rice growth stage	Bacterial FA stain used			
	K-1	N-4	96–51	96–57
Rhizosphere soil				
Tillering				
Control	ND	ND	ND	ND
Inoculated	5–10	5–10	5–10	5–10
Panicle				
Control	ND	ND	5–10	5–10
Inoculated	5–10	11–50	5–10	11–50
Maturity				
Control	ND	ND	5–10	ND
Inoculated	11–50	11–50	11–50	11–50
Rice roots				
Tillering				
Control	ND	ND	ND	ND
Inoculated	5–10	5–10	5–10	5–10
Panicle				
Control	5–10	5–10	ND	5–10
PGPR	11–50	11–50	ND	5–10
Maturity				
Control	5–10	ND	ND	ND
PGPR	100–200	50–100	ND	5–10

Number of bacteria/microscopic field is given; at least 10–12 fields were observed (ND = not detected)

and bacteria for enhancing rice biomass and grain yield. The positive effect of *Azolla* on rice yield has been observed in various countries (Kikuchi *et al.*, 1984, Kumarasinghe and Eskew, 1993) and its use with chemical-N fertilizer is also considered feasible (FAO, 1988). The total nitrogen accumulated in rice plants was slightly higher for nitrogen+bacteria (T10), indicating a useful effect of bacteria on nitrogen uptake by rice.

The benefit of *Azolla* for increasing rice yield was reported to be positively correlated with the number of its incorporations and the amount of incorporation, and a greater increase in rice yield was observed for *Azolla* incorporated before rice transplanting than after transplanting (Kikuchi *et al.*, 1984; Watanabe, 1987). The very low positive effect of *Azolla* on rice yield in our experiment may be due to suboptimum *Azolla* growth, and only two incorporations made during the later stage of rice growth.

There was not much increase in rice yield even for fertilizer-N application at both 30 and 60 kg N/ha as normally observed for fertilizer application. The overall low response to biofertilizers and chemical-N fertilizer may be due to the high levels of available NH_4 and NO_{3}-N in this soil (Table 1). Recently, Hussain *et al.* (1994) have estimated the N availability index of 50 different local soils, and they have concluded that 18 and 21 mg N/kg soil, of KC1-extractable NO_{3}-N and (NH_4+NO_3)-N respectively; were the critical levels. In our soil NO_{3}-N was 13 and (NH_4+NO_3)-N was 25 mg/kg, thus it is very likely that this soil was not much deficient in nitrogen and therefore a very significant fertilizer response was not observed. The high total N content of rice (152 kg N/ha) and a very high rice grain yield (7.6 t/ha), even in control (Table 2), also indicates that sufficient nitrogen was available in this soil; as removal of about 123 kg N/ha has been reported by a similar crop (7.9 t/ha) in a tropical country (De Datta, 1981). The high level of available-N in this soil may be due to the fact that soil used in this experiment was left uncultivated for some period.

Fertilizer-N recovery and losses

As traced with ^{15}N-labelled fertilizer applied in T6–T11, the fertilizer-N recovery in rice straw, grain and straw + grain was maximum for T9 followed by T7 (Table 4). The higher fertilizer-N recovery for these two treatments indicates the usefulness of *Azolla* for increasing fertilizer-N use efficiency. Similarly, a 10–92% increase in ^{15}N recovery in rice from urea has been reported from various countries due to *Azolla* inoculation (Kumarasinghe and Eskew, 1993). The presence of *Azolla* also helped in increasing fertilizer-N retention in soil, as it was up to 14% higher for T9 as compared to T6. Thus, due to *Azolla*, the total fertilizer-N recovery in rice plant and soil was maximum (51%) for T9, and this recovery was significantly lower for non-*Azolla* treatments.

The fertilizer-N losses were 49% for T9, and 66% and 72% for 30 and 60 kg N/ha treatments, respectively (Table 4). Thus the use of biofertilizers reduced N losses by 17% and 23% as compared to low and high amounts of fertilizer application, respectively, Reduction in urea-N losses (11–19%) due to *Azolla* were also noted in a pot study by Ali and Malik (1993).

Table 4. Fertilizer N recovery of straw, grain and soil in rice field

Treatment	Fertilizer N recovery				
	Straw (%)	Grain (%)	Soil (%)	Total (%)	N lost (%)
T6. 30 kg N/ha	3.35^{b}	10.32^{bc}	20^{c}	34^{c}	66^{b}
T7. 30 kg N/ha + *Azolla* cover	4.17^{a}	11.53^{ab}	25^{b}	41^{b}	59^{c}
T8. 30 kg N/ha + bacteria. + *Azolla* incorporated	3.36^{b}	9.33^{c}	23^{b}	36^{c}	64^{b}
T9. 30 kg N/ha + bacteria + *Azolla* cover	4.33^{a}	12.43^{a}	34^{a}	51^{a}	49^{d}
T10 30 kg N/ha + bacteria	2.79^{c}	7.56^{d}	18^{cd}	28^{d}	72^{a}
T11. 60 kg N/ha	3.18^{bc}	9.30^{c}	16^{d}	28^{d}	72^{a}

^{15}N-labelled ammonium sulphate was applied to soil for treatments T6–T11.
Means followed by the same letter are not statistically different at 5% *p*.
Total = (straw + grain + soil).

Floodwater pH

The floodwater pH showed a large variation during daytime. It was low during the morning (7 am) reached maximum value in the afternoon (2 pm) and again decreased in the evening (6 pm) in all treatments (Figure 2). The increase in pH at noon was attributed to algal photosynthesis, during which the concentration of carbon dioxide and bicarbonates decreased in flood water (Mikkelsen *et al.*, 1978); Tuan and Thuyet, 1979). In general the pH remained lower in *Azolla* cover treatments than in the fertilizer treatments. The lowering of midday floodwater pH (Tuan and Thuyet, 1979; Mabbayad, 1987) as well as of floodwater temperature (Ali and Malik, 1988) has also been reported due to *Azolla*.

In this study the improvement in fertilizer-N use efficiency in rice plant and retention of higher fertilizer-N for *Azolla* treatments may be due to lowering of floodwater pH by *Azolla* cover (Figure 2), and reduction in N losses due to *Azolla* cover has also been observed in some *Azolla*–rice studies (Kumarasinghe and Eskew, 1993). The ammonia volatilization losses have been widely recognized as an important mechanism for loss of mineral N, due to high temperature and elevated floodwater pH (Mikkelsen and De Datta, 1979; Becker *et al.*, 1995), and maximum ammonia concentration and losses of applied fertilizers were reported during the daytime (10–16 hs) in a flooded soil (Ferdrazzini and Tarsitano, 1987).

Nitrogen fixation

The amount of fixed nitrogen in rice plants, computed using the ^{15}N isotope dilution technique and the high N treatment (T11) as non-fixing control, indicated

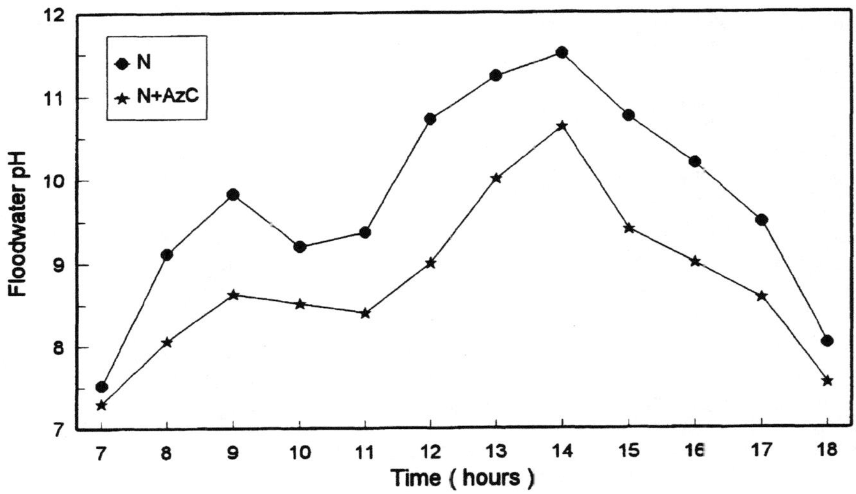

Figure 2. Kinetics of floodwater pH during daytime in rice–*Azolla* culture (N=Nitrogen treatments, average of T6 and T11; N+AzC=nitrogen+*Azolla* cover treatments, average of T7 and T9).

significantly higher values for bacterial treatments than control. Maximum BNF in rice (46 kg N/ha) was obtained for T10 where bacteria were inoculated along with 30 kg N/ha of ammonium sulphate, followed by T8 having 30 kg N/ha+bacteria+*Azolla* incorporation (Figure 3). A stimulation in the associative nitrogen fixation due to starter nitrogen fertilizer has also been reported in maize (Vlassak and Reynders, 1979). The figure shows that a greater amount of nitrogen fixed by bacteria was assimilated by rice than from *Azolla*. The higher availability of fixed N from bacteria may be due to active nitrogen fixation and an early release of N from microbial biomass than *Azolla*. Since *Azolla* has some woody tissue, and may have 18–30% lignin (Watanabe *et al.*, 1991), therefore after incorporation into soil its decomposition and mineralization may take 5–8 weeks to release 70% of its total N (Watanabe *et al.*, 1977). As *Azolla* was incorporated at later stages of rice growth (19 and 54 days before rice harvest) and it also needs a longer time for decomposition and release of its fixed N than microbes, almost no fixed N by *Azolla* was available to rice. Variable estimates of BNF have been given for nitrogen-fixing bacteria by various workers. In India seed inoculation, with *Azospirillum*, of various grasses and oats, gave a response equivalent to 20 and 40 kg N/ha respectively, and in Nepal inoculation of rice

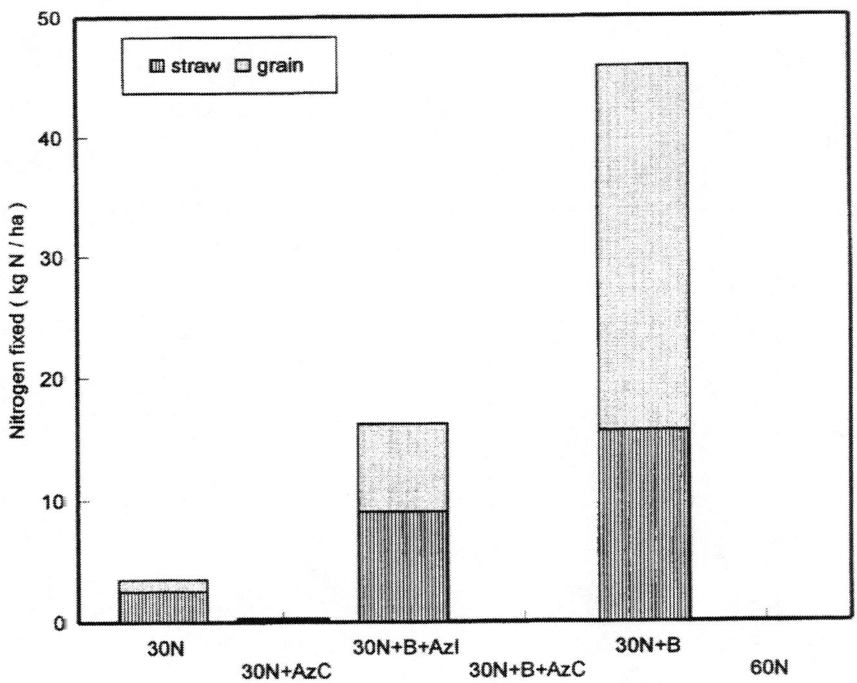

Figure 3. Amount of nitrogen fixed during rice crop, estimated by ^{15}N dilution technique. (N=nitrogen, Az=*Azolla*, B=bacteria, C=cover, I=incorporated)

seedlings increased rice yield comparable to 20 kg N/ha (FAO, 1988). Okon (1981) inoculated corn with nitrogen-fixing bacteria and found 77 kg more N/ha than in uninoculated plants. The 46 kg nitrogen fixation/ha for T10 (Figure 3) indicated that the inoculated bacteria were able to fix sufficient amounts of nitrogen even in the presence of 30 kg N/ha of fertilizer-N. Similar findings for lower sensitivity of associative nitrogen fixation to N-fertilizer application have been reported by Roger and Watanabe (1986).

Practical significance

This study indicated that the use of biofertilizers such as *Azolla* and nitrogen-fixing rhizobacteria can not only help in increasing rice yield and nitrogen fixation, but also enhance fertilizer-N use efficiency and reduce N losses of applied fertilizer, under flooded soil conditions. As they continued fixing nitrogen even in the presence of low amounts of fertilizer-N, as well as a high level of available-N in the soil, these biofertilizers can be used along with low doses of chemical-N fertilizer, to minimize the depletion of soil-N reserves.

References

Albrecht S L, Gaskins M H, Milan J R, Schank S C, Smith R L 1983. Ecological factors affecting the survival and activity of *Azospirillum* in the rhizosphere. Experientia, Suppl., 48, 136–148.

Ali S, Malik K A 1988. Effect of *Azolla* on the rice yield, fertilizer N-15 recovery, soil N budget and floodwater properties in an *Azolla*–rice intercropping system. In: Proceedings, First International Symposium on Paddy Soil Fertility, Chiangmai, Thailand, 301–314.

Ali S, Malik K A 1993. Utilization of *Azolla-Anabaena* symbiosis as nitrogen biofertilizer in rice-wheat cropping system. In: Malik K A, Nasim A, Khalid A M, eds. Proceedings, International Symposium on Biotechnology for Sustainable Development. NIBGE, PO Box 577, Faisalabad, Pakistan, 249–253.

Ali S, Malik K A, Ahmad I, Rasul E 1990. Effect of salinity on the morphology, nitrogen fixation and biomass production in *Azolla pinnata*. In: Proceedings, Indo-Pakistan Workshop on Soil Salinity and Water Management, vol. II, Pakistan Agricultural Research Council PO Box 1031, Islamabad, Pakistan, 509–520.

Becker M, Ladha J K, Ali M 1995. Green manure technology: potential, usage, and limitations. A case study for lowland rice. Plant Soil, 174, 181–194.

Bilal R, Rasul G, Mahmood K, Malik K A 1990. Nitrogenase activity and nitrogen fixing bacteria associated with the roots of *Atriplex* spp. growing in saline sodic soils of Pakistan. Biol. Fertil. Soils, 9, 315–320.

Bohlool B B, Schmidt E L 1968. Nonspecific staining: its control in immunofluorescence examination of soil. Science, 162, 1012–1014.

Cary P R, Weerts P G J 1992. Growth and nutrient composition of *Azolla pinnata* R. Brown and *Azolla filiculoides* Lamarck as affected by water temperature, nitrogen and phosphorus supply, light intensity and pH. Aquat. Bot., 43, 163–180.

Chan E C S, Katznelson H, Rouatt J W 1963. The influence of soil and root extracts on the associative growth of selected soil bacteria. Can. J. Microbiol., 9, 187–197.

Cooke G W 1982. Fertilizing for Maximum Yield. ELBS, London, 465 p.

Craswell E T, Vlek P L G 1979. Fate of fertilizer nitrogen applied to wetland rice. In: IRRI, eds. Nitrogen and Rice, International Rice Research Institute (IRRI), PO Box 933, Manila, Philippines, 175–192.

De Datta S R 1981. Principles and Practices of Rice Production. John Wiley & Sons, New York, 618 p.

FAO 1988. Bio and Organic Fertilizers: Prospects and Progress in Asia. RAPA Publication 1988/10. FAO Regional Office, Bangkok, Thailand.

Ferdrazzini F R, Tarsitano R 1987. Ammonia volatilization from flooded soils following urea application. Plant Soil, 91, 101–107.

Frankenberger Jr W T, Arshad M 1995. Phytohormones in Soils: Microbial production and function. Marcel Dekker, New York.

Hauck R D 1982. Nitrogen-isotope-ratio analysis. In: Page A L, ed. Methods of Soil Analysis. Part 2. Amer. Soc. Agron. Inc., Madison, Wis. USA, 735–779.

Hussain F, Malik K A 1983. Ammonia volatilization from a flooded rice soil system. Pak. J. Agric. Res., 4, 126–130.

Hussain F, Malik K A, Naqvi M H 1994. A comparative study of different methods for obtaining an index of nitrogen availability in upland soils. Pak. J. Sci. Ind. Res., 37, 258–264.

IRRI 1994. Rice Facts. International Rice Research Institute (IRRI), PO Box 933 Manila, Philippines (pamphlet).

Keeney D R, Nelson D W 1982. Nitrogen-inorganic forms. In: Page A L, ed. Methods of Soil Analysis. Part 2. American Society of Agronomics, Inc., Madison, WI, 643–698.

Kikuchi M, Watanabe I, Haws L D 1984. Economic evaluation of *Azolla* use in rice production. In: IRRI, eds. Organic Matter and Rice. International Rice Research Institute (IRRI), PO Box 933, Manila, Philippines, 569–592.

Kumarasinghe K S, Eskew D L 1993. Isotopic Studies of *Azolla* and Nitrogen Fertilizer on Rice. Kluwer, Dordrecht.

Lumpkin T A, Plucknett D L 1982. *Azolla* as a Green Manure: Use and Management in Crop Production. Westview Press, Boulder, CO.

Mabbayad B B 1987. The *Azolla* program of the Philippines. In: IRRI, eds. *Azolla* Utilization. Proceedings, Workshop on *Azolla* use; March 31–April 5 1985, Fuzhou, China, pp 101–108. International Rice Research Institute (IRRI), PO Box 933, Manila, Philippines.

Malik K A, Rasul G, Hassan U, Mehnaz S, Ashraf M 1994. Role of N_2-fixing and growth hormones producing bacteria in improving growth of wheat and rice. In: Hegazi N A, Fayaz M, Monib M, eds. Nitrogen Fixation with Non-Legumes. American University in Cairo Press, Cairo, Egypt, 409–422.

Mikkelsen D S, De Datta S K 1979. Ammonia volatilization from wetland rice soils. In: IRRI, eds. Nitrogen and Rice. International Rice Research Institute (IRRI), PO Box 933, Manila, Philippines, 135–156.

Mikkelsen D S, De Datta S K, Obcemea W N 1978. Ammonia volatilization losses from flooded rice soils. Soil Sci. Soc. Am. J., 42, 725–730.

Okon Y 1981. Field inoculation of grasses with *Azospirillum*. In: Graham P H, Harris S C, eds. BNF Technology for Tropical Agriculture. CIAAT, Colombia, 459–468.

Peoples M B, Herridge D F, Ladha J K 1995. Biological nitrogen fixation: an efficient source of nitrogen for sustainable agricultural production? Plant Soil, 174, 3–28.

Regan D L 1988. Other micro-algae. In: Borowitzka M A, Borowitzka L J, eds. Microalgal Biotechnology. Cambridge University Press, Cambridge, 135–150.

Rennie R J 1981, A single medium for the isolation of acetylene reducing (dinitrogen fixing) bacteria from soil. Can. J. Microbiol., 27, 8–14.

Roger P A, Watanabe I 1986. Techniques for utilizing biological nitrogen fixation in wetland rice: potentialities, current usage, and limiting factors. Fert. Res., 9, 37–77.

Roger P A, Zimmerman W J, Lumpkin T A 1993. Microbial management of wetland rice fields. In: Meeting F B, ed. Soil Microbial Ecology. Marcel Dekker, New York, 417–455.

Schmidt E L 1974. Quantitative and ecological study of microorganisms in soil by immunofluorescence. Soil Sci., 118, 141–149.

Somasegaran P, Hoben H J 1985. Methods in Legume–*Rhizobium* Technology. NifTAL and MIRCEN, University of Hawaii, USA.

Tuan D T, Thuyet T Q 1979. Use of *Azolla* in rice production in Vietnam. In: IRRI, eds. Nitrogen and Rice. International Rice Research Institute (IRRI), PO Box 933, Manila, Philippines, 385–405.

Ventura W, Watanabe I 1993. Green manure production of *Azolla microphylla* and *Sesbania rostrata* and their long-term effects on the rice yields and soil fertility. Biol. Fert. Soils, 15, 241–248.
Vlassak K, Reynders L 1979. Factors affecting biological nitrogen fixation by associative symbiosis in temperate region. In: Isotopes and Radiation in Research on Soil Plant Relationships. IAEA Vienna, 137–147.
Watanabe I 1987. Summary report of the *Azolla* programme of the international network on soil fertility and fertilizer evaluation for rice. In: IRRI, eds. *Azolla* Utilization. International Rice Research Institute (IRRI), PO Box 933, Manila, Philippines, 197–205.
Watanabe I, Espinas C R, Berja N S, Alimagno B V 1977. Utilization of the *Azolla–Anabaena* complex as a nitrogen fertilizer for rice. IRRI, Research Paper Series no. 11. International Rice Research Institute (IRRI), PO Box 933, Manila, Philippines.
Watanabe I, Padre B Jr, Ramirez C 1991. Mineralization of *Azolla* N and its availability to wetland rice. Part 1. Nitrogen mineralization of different *Azolla* species as affected by their chemical composition. Soil Sci. Plant Nutr., 35, 575–584.

8
Detection of inoculated plant growth-promoting rhizobacteria in the rhizosphere of rice

SAMINA MEHNAZ, M. SAJJAD MIRZA, UZMA HASSAN AND KAUSER A. MALIK

Abstract

A mixture of five plant growth-promoting rhizobacterial strains belonging to the genera *Azospirillum, Azoarcus, Pseudomonas* and *Zoogloea* was used to inoculate rice seedlings growing in microplots. The population of these bacteria, as well as the indigenous bacterial population in the rice rhizosphere, was determined at 4-week intervals throughout the growth season. In rhizosphere soil a continuous increase in total bacterial population (CCM-plate counts) in both inoculated and non-inoculated plots was noted until 3 months after transplantation, while the maximum population of diazotrophic bacteria (ARA-based MPN counts) was observed 1 month after transplantation. A larger number of bacteria was found associated with plant roots (10^7/g) as compared to those present in rhizosphere soil (10^6/g). The total bacterial population (plate counts) in plant roots showed a continuous increase until harvest, while the population of diazotrophs (MPN counts) started declining 1 month after transplantation. Considerably higher levels of nitrogenase activity (ARA) were noted in the roots of inoculated plants as compared to those of non-inoculated plants, where only a small number of diazotrophic bacteria were detected. The population of one *Azospirillum* strain (N-4) used as inoculum was determined by using selective media containing antibiotics, morphological characteristics, reaction with fluorescent antibodies and ARA of re-isolates. The maximum number of cells (1.4×10^5/g) of this strain were detected in the roots 1-month after seedling transplantation from the nursery, constituting about 0.4% of the total bacterial population and 2% of the diazotrophic population.

Introduction

Nitrogen fixation carried out by associative and free-living microorganisms in the rhizosphere of plants has been recognized to play an important role in

K.A. Malik et al. (eds.), Nitrogen Fixation with Non-Legumes, 75–83.

nitrogen nutrition of plants (Boddey *et al.*, 1996). Diazotrophs belonging to diverse bacterial genera such as *Azospirillum, Azotobacter, Acetobacter, Alicaligens, Bacillus, Enterobacter, Herbaspirillum, Klebsiella* and *Pseudomonas* appear to be frequent colonizers of important cereal crops including rice, wheat, sugar cane and maize (Baldani *et al.*, 1986: Roger and Watanabe, 1986; Cavalcante and Dobereiner, 1988; Berge *et al.*, 1991; You *et al.*, 1991; Malik *et al.*, 1993). Beneficial effects of these plant growth-promoting rhizobacteria (PGPR) have been attributed to biological nitrogen fixation and production of phytohormones that promote root development and proliferation, resulting in efficient uptake of water and nutrients (Tien *et al.*, 1979; Hartmann *et al.*, 1983; Haahtela *et al.*, 1990). Interest in the beneficial rhizobacteria associated with cereals has increased recently due to their potential use as biofertilizers (Okon and Labandera-Gonzalez, 1994; Bashan and Levanony, 1990). Application of bacterial inoculants as biofertilizers has resulted in improved growth and increased yield of cereal crops (Kapulnik *et al.*, 1981; Boddey *et al.*, 1986; Pereira *et al.*, 1988; Kennedy and Tchan, 1992). A number of immunological and DNA-based techniques are being developed and used to study root colonization by the inoculated bacteria and their survival in the rhizosphere (Levanony and Bashan, 1989; Hartmann *et al.*, 1993; Schloter *et al.*, 1993; Vermeiren *et al.*, 1996).

Here we report on the detection of inoculated PGPR and indigenous bacteria present in the rice rhizosphere during different stages of growth. In addition, beneficial effects of PGPR inoculation on growth and yield of rice are given.

Materials and methods

Inoculum preparation

In the present study, a mixture of five PGPR strains (Malik *et al.*, 1994) was used as inoculum for rice. *Azospirillum lipoferum* (strain N-4), *Azospirillum brasilense* (strain WB-3), *Azoarcus* (strain K1), *Pseudomonas* (96–51) and *Zoogloea* strain (Ky-1) were grown in Luria Bertani (LB) medium with shaking at 30°C. After 24 h growth, bacterial cells were harvested by centrifugation at 10 000 rpm for 5 min. Pellets were washed and resuspended in 0.85% saline to get approximately 1×10^9 cells/ml.

Inoculation of rice seedlings

Inoculum was prepared by mixing equal volumes of the cultures of five PGPR strains (10^9 cells/ml). Roots of the rice seedlings (variety NIAB-6) grown in the nursery were kept for ½ h in the inoculum. The seedlings were transplanted in cemented plots (1.5×1.5×0.6m) filled with saline soil (Ece=4.87 mS/cm, pH=7.8, K=0.15 mEq/L, Na=66 mEq/L, Ca=4.1 mEq/L; Ali *et al.*, 1995). In each plot, seedlings were transplanted in eight rows, with eight hills in each row separated by 20 cm. Two seedlings were planted on each hill. Ammonium sulphate was applied in solution form (3–5 cm below the soil surface) 7 days after transplantation in all plots. For both treatments (un-inoculated and inoculated) five replicates were used.

ARA of roots

For estimation of ARA, root samples were taken at 4-week intervals throughout the growth season. Roots were thoroughly washed with sterile water and excess water was removed with blotting paper. The root samples were incubated at 30°C with 10% C_2H_2 in vacutainers. The ARA was carried out on a gas chromatograph (Gasukuro kogyo Model 370) using Porapak N column.

Enumeration of bacteria from roots and soil

From each plot a 1 g soil sample (obtained by mixing 10 random samples) was shaken with 9 ml of saline for 30 min at 200 rpm. Serial dilutions were prepared from this suspension and spread on combined carbon medium (CCM plates; Rennie, 1981). These plates were incubated at 30°C for 48 h. The number of bacterial colonies was recorded with the help of a colony counter. For MPN counts (Alexander, 1965), dilutions were added in vials containing semi-solid CCM. These vials were incubated for 72–96 h at 30°C and then subjected to ARA. Vials showing positive results were used to calculate MPN counts.

Root samples were washed thoroughly with sterile water. One gram roots (fresh weight) were homogenized with a pestle and mortar in 9 ml saline and serial dilutions were prepared. These dilutions were porcessed as described above for plate counts and MPN counts.

For selective retrieval of *Azospirillum* strain N-4, antibiotic gentamycin (100 µl/ml) was added in plates containing LB. Serial dilutions of the crushed root suspension were prepared and two dilutions (10^{-2} and 10^{-3}) were spread on plates in triplicate. Ten colonies, resembling N-4 in their colony morphology, from each of these plates were randomly selected for detailed studies to confirm as being *Azospirillum* N-4. The colonies were used to inoculate vials containing semi-solid CCM, to observe formation of characteristic sub-surface pellicle, typical *Azospirillum*-like cell morphology and motility, ARA and reaction with specific fluorescent antibodies.

Plant biomass and grain yield

Plants were harvested after 3 months at the completion of the rice growth season. For estimation of plant biomass and grain yield, whole plants, along with the roots, were harvested from an inner 1 m^2 area, leaving one row of plants near the borders of plots. Dry weight of the roots, aerial parts and grains was measured by keeping plant material at 70°C until no change in weight was noted. For statistical analysis the *t*-test was applied.

Results and discussion

In the microplots the bacterial population was determined at 4-week intervals for 3 months after transplantation of the seedlings. In the rhizosphere of both inoculated and non-inoculated plants considerably lower numbers of bacteria were recorded as compared to plant roots (Figures 1–4). At the nursery stage the total number of bacteria (plate counts; indigenous population) colonizing roots was approximately 1×10^8 cells/g of root (Figure 1). The presence of a relatively high number of bacteria in roots is most probably the result of accumulation of root exudates from densely populated rice seedlings at the nursery stage. The PGPR are known to be attracted to root exudates of a number of plants (De Troch and Vanderleyden, 1996). From this observation it can be concluded that application of inoculum at the nursery stage is not useful as the bacterial population is already high. Application of inoculum at the time of transplantation in the field, where the population of indigenous rhizobacteria is low, seems more adequate.

After seedling transplantation the bacterial population started building up again and reached its maximum after 12 weeks. However, more diazotrophic bacteria (ARA-based MPN counts) were detected in roots 1 month after transplantation, and this number started decreasing until crop harvest (Figure 2). In the roots, 1 month after transplantation, the nitrogen-fixing bacteria comprised about 30% of the total bacterial population. At the time of harvesting (3 months after transplantation), the proportion of nitrogen fixers declined to only less than 1% of the total bacterial population in inoculated plots. This decline in MPN counts at later stages of plant growth indicates that nitrogen fixers are out-competed by non-fixers. The total bacterial population, as well as

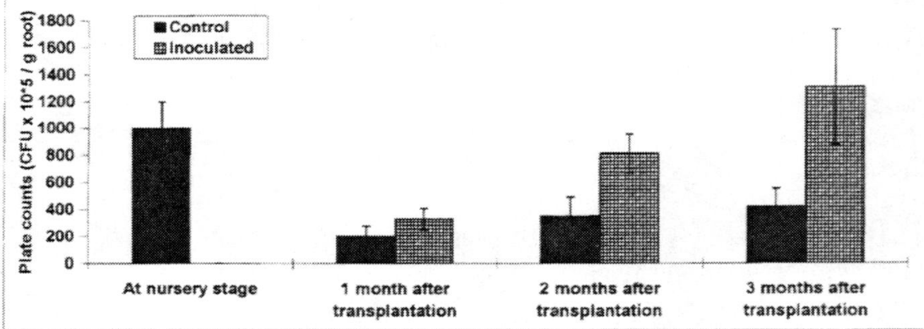

Figure 1. Detection of bacterial population associated with the roots of rice. Plate counts on combined carbon medium (CCM) were used to determine the bacterial population.

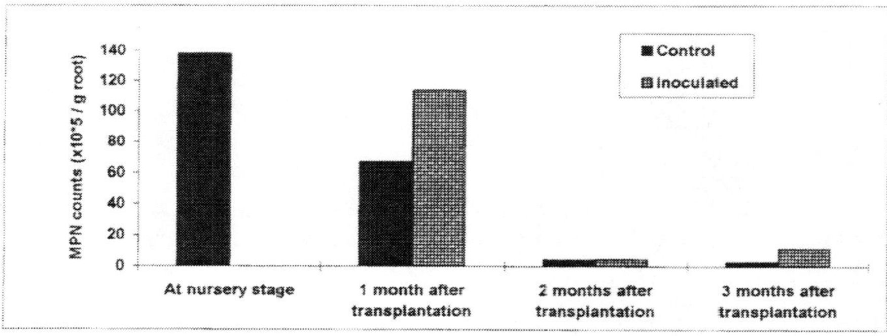

Figure 2. Detection of nitrogen-fixing bacteria in roots of rice. ARA-based MPN counts were used to determine the bacterial population.

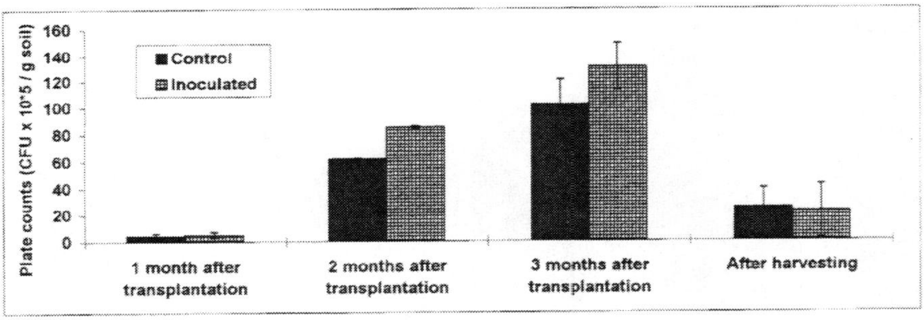

Figure 3. Detection of bacterial population in the rhizosphere soil of rice. Plate counts on combined carbon medium (CCM) were used to determine the bacterial population.

diazotrophs, were found in much greater numbers in inoculated plots as compared to non-inoculated plots (Figures 1 and 2). In the rhizosphere soil the same trend was observed (Figures 3 and 4). Total bacterial population (CCM plate counts) gradually increased 1 month after transplantation, but the diazotrophic population (ARA-based MPN counts) started decreasing with the age of the crop. A similar decrease in bacterial population at harvest in rice rhizosphere has been reported by Van Holm *et al.* (1993) where *Azorhizobium caulinodans* was used as inoculum.

Nitrogenase activity was detected in roots of both inoculated and non-inoculated plots. This ARA activity of non-inoculated roots reflects the presence of an indigenous diazotrophic population. Maximum nitrogenase activity in association with plant roots was detected 1 month after transplantation of the seedlings from the nursery (Figure 5). This high activity level coincides with the maximum number of diazotrophic bacteria found associated with the plant roots at this stage of plant growth.

The population of one inoculated strain (*Azospirillum* N-4) was determined by identifying these bacteria initially on the basis of their colony morphology,

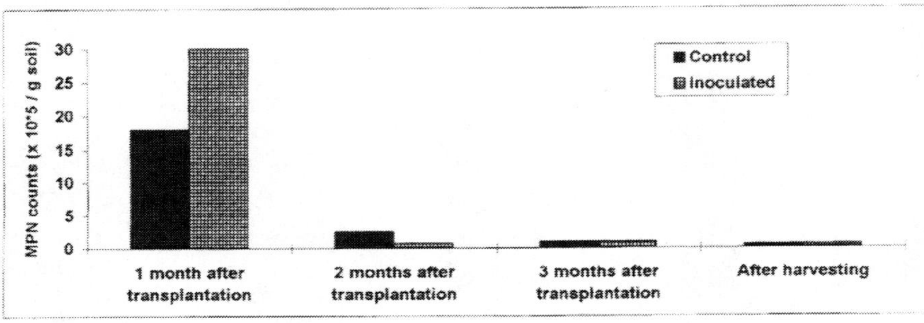

Figure 4. Detection of nitrogen-fixing bacteria in the rhizosphere soil of rice. ARA-based MPN counts were used to determine the bacterial population.

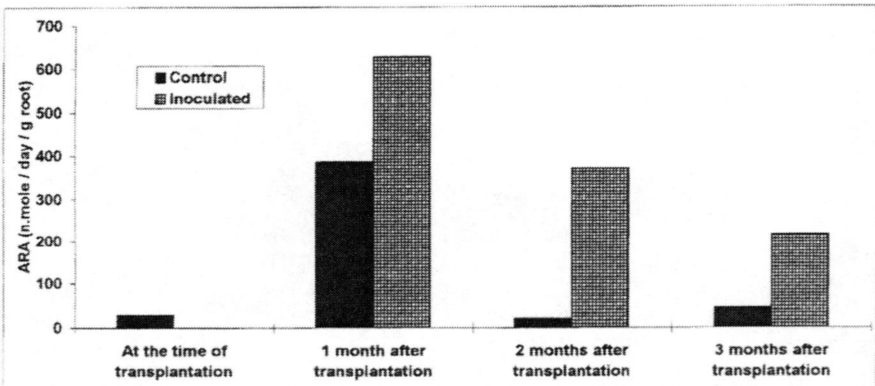

Figure 5. Nitrogenase activity estimated in rice roots.

and then on the basis of their cell morphology, ARA and by using specific fluorescent antibodies. One month after inoculation, approximately 1.4×10^5 cells of *Azospirillum* strain N-4 per gram of root were found. In a study carried out by Yanni *et al.* (1996), 1.1×10^6 cells of *Rhizobium leguminosarum* bv. *trifolii* were detected when used to inoculate rice plants. It was estimated that among the bacteria associated with roots (both originating from inoculations and the indigenous population), strain N-4 comprised about 0.4% of the total bacterial population and about 2% of the total nitrogen-fixing population when peak nitrogenase activity was recorded in roots (Figure 6). In a study carried out by Schloter *et al.* (1993), in which *Azospirillum brasilense* was used as inoculum for wheat, the inoculated strain comprised about 0.1% of the total bacterial community.

A beneficial effect of bacterial inoculation was observed on plant biomass and grain yield (Table 1). In inoculated plots an approximately 12% increase in both grain yield and straw weight was noted (Table 1). The beneficial effect of bacterial inoculation on total plant biomass and root weight was significant as a

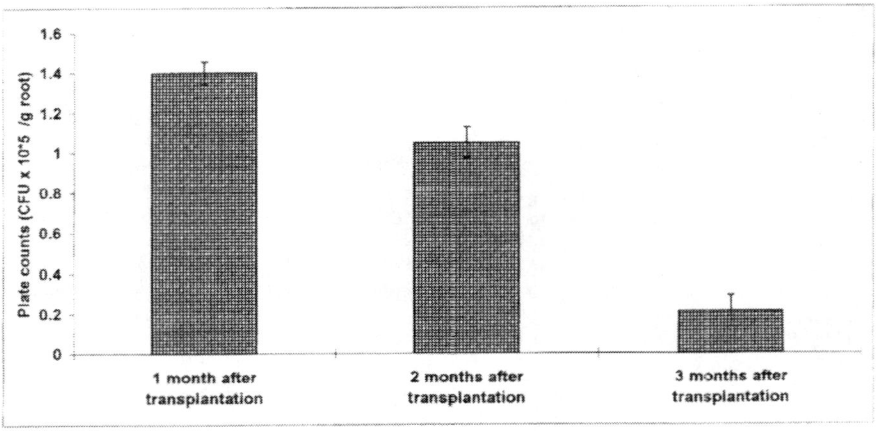

Figure 6. Detection of *Azospirillum* strain N-4 used as inoculum for rice.

41% increase in root weight and a 18.7% increase in plant biomass was recorded. Similar beneficial effects of PGPR inoculations on rice have been reported (Ali *et al*., 1995) where use of inoculum along with a low input of chemical fertilizer-N was found useful for increasing rice biomass, N uptake and fertilizer-N recovery in rice grown in pots containing saline soils.

The results of the present investigation clearly showed that the inoculated bacterial strains survived in the rhizosphere especially during early stages of plant growth and, as a result of these inoculations, plant growth and yield was improved. On the basis of these results this mixture of bacterial strains may be recommended to form an excellent active ingredient of a commercial biofertilizer for rice.

Acknowledgement

Financial support for this work was partially provided by PSF through PSF project P-NIBGE/Agr(153).

Table 1. Effect of PGPR inoculations on plant biomass of rice

Treatments	Straw weight (g/m^2)	Root weight (g/m^2)	Grain yield (g/m^2)	Total plant biomass (g/m^2)
30 kg N/ha	296	206	266	767
30 kg n/ha+inoculum	332$^{n.s.}$	291*	297$^{n.s.}$	911*
	(12)	(41)	(11.7)	(18.7)

* Significant difference between control and inoculated at the 5% level, n.s. = non significant. Figures in parentheses show percentage increase over control.

References

Alexander M 1965. Most probable number method for microbial population. In: Black C A, Evans D D, Ensuinger L E, White J K, Clark F F, eds. Methods of Soil Analysis. Part 2. Am. Soc. (*Sec.*) Agronomy, Madison, WI, 1467–1472.

Ali S, Hamid N, Rasul G, Malik K A 1995. Use of biofertilizers to enhance rice yield. Nitrogen uptake and fertilizer-N use efficiency in saline soils. Pak. J. Bot., 27(2), 275–281.

Baldani J J, Baldani V L D, Seldin L, Dobereiner J 1986. Characterization of *Herbaspirillum seropedicae* gen. nov., a root-associated nitrogen-fixing bacterium. Int. J. Syst. Bacteriol., 36, 86–93.

Bashan Y, Levanony H 1990. Current status of *Azospirillum* inoculation technology: *Azospirillum* as a challenge for agriculture. Can. J. Microbiol., 36, 591–608.

Berge O, Heulin T, Achouak W, Richard C, Bally R, Balandreau J 1991. *Rahnella aquatitis*–a nitrogen fixing enteric bacterium associated with the rhizosphere of wheat and maize. Can. J. Microbiol., 37, 195–203.

Boddey R M, Baldani V L D, Baldani J I, Dobereiner J 1986. Effect of inoculation of *Azospirillum* spp. on nitrogen accumulation by field grown wheat. Plant Soil, 95, 109–121.

Boddey R M, Alves B J R, Urquiaga S 1996. Evaluation of biological nitrogen fixation associated with non-legumes. Proceedings, Seventh International Symposium on BNF with Non-legumes, 16–21 October, Faisalabad, Pakistan. Eds. Malik K A, Mirza M S, Ladha J K, Kluwer Academic Publishers, The Netherlands.

Cavalcante V A, Dobereiner J. 1988. A new acid tolerant nitrogen fixing bacterium associated with sugar cane. Plant Soil, 108, 23–31.

De Troch P, Vanderleyden J 1996. Surface properties and motility of *Rhizobium* and *Azospirillum* in relation to plant root attachment. Microb. Ecol., 32, 149–169.

Haahtela K, Konkko R, Laakso T, Williams P H, Korhonen T K 1990. Root associated *Enterobacter* and *Klebsiella* in *Poa pratensis*: characterization of an iron-scavenging system and a substance stimulating root hair production. Mol. Plant Microb. Interact. 3, 358–365.

Hartmann A, Singh M, Klingmuller W 1983. Isolation and characterization of *Azospirillum* mutants excreting high amounts of indole acetic acid. Can. J. Microbiol., 29, 916–923.

Hartmann A, Assmus B, Lawrence J R, Reis V, Kirchhof G 1993. Development and application of 23S rRNA-directed oligonucleotide probes for *Azospirillum* spp., *Acetobacter diazotrophicus* and *Herbaspirillum*. In: Hegazi N A, Fayez M, Monib M, eds. Nitrogen Fixation with Non-legumes. Cairo University Press, Giza, Egypt. 283–289.

Kapulnik Y, Kigel J, Okon Y, Nur I, Henis Y 1981. Effects of *Azospirillum* inoculation on some growth parameters and N-content of wheat, sorghum and panicum. Plant Soil, 61, 65–70.

Kennedy I R, Tchan Y T 1992. Biological nitrogen fixation in non-leguminous field crops: recent advances. Plant Soil, 141, 93–118.

Kloepper J W, Lifshitz R, Zablotowicz R M 1989. Free living bacterial inocula for enhancing crop productivity. TIBTECH, 7, 39–44.

Levanony H, Bashan Y 1989. Localization of specific antigens of *Azospirillum brasilense* Cd in its exopolysaccharide by immuno-gold staining. Curr Microbiol., 18, 145–149.

Malik K A, Rasul G, Hassan U, Mehnaz S, Ashraf M 1994. Role of N_2-fixing and growth hormones producing bacteria in improving growth of wheat and rice. In: Hegazi N A, Fayez M, Monib M, eds, Nitrogen Fixation with Non-legumes. Cairo University Press, Giza, Egypt. 409–422.

Okon Y, Labandera-Gonzalez C A 1994. Agronomic applications of *Azospirillum*. in improving plant productivity with rhizosphere bacteria. In: Ryder M H, Stephens P M, Bowen G D, eds. Commonwealth Scientific and Industrial Research Organization, Adelaide, 274–278.

Pereira J A R, Cavalcante V A, Baldani J I, Dobereiner J 1988. Field inoculation of sorghum and rice with *Azospirillum* spp. and *Herbaspirillum seropedicae*. Plant Soil, 110, 269–274.

Rennie R J 1981. A single medium for the isolation of nitrogen fixing bacteria. Can. J. Microbiol., 27, 8–14.

Roger P A, Watanabe J 1986. Technologies for utilizing biological nitrogen fixation in wetland rice: potentialities, current usage and limiting factor. Fert. Res., 9, 39–77.

Schloter M, Kirchhof G, Heinzmann U, Dobereiner J, Hartmann A 1993. Immunological studies of the wheat root colonization by the *Azospirillum brasilense* strains Sp 7 and Sp 245 using strain specific monoclonal antibodies. In: Hegazi N A, Fayez M, Monib M, eds. Nitrogen Fixation with Non-legumes. Cairo University Press, Giza, Egypt. 291–297.

Tien T M, Gaskins M H, Hubbell D H 1979. Plant growth substances produced by *Azospirillum brasilense* and their effect on the growth of pearl millet (*Pennisetum americanum* L.) Appl. Environ. Microbiol., 37, 1016–1024.

Van Holm L H J, Nieuwenhove C V, Kumari I A D C T, Kilasooriya S A, Vlassak K 1993. Survival and effect of *Azorhizobium caulinodans* on rice root. In: Hegazi N A, Fayez M, Monib M, eds. Nitrogen Fixation with Non-legumes. Cairo University Press, Giza, Egypt. 327–332.

Vermeiren H, Hai W, Vanderleyden J 1996. Colonisation and *nif H* expression on rice roots by *Alcaligenes faecalis* A15. Proceedings, Seventh International Symposium on BNF with Non-legumes, 16–21 October, Faisalabad, Pakistan. Eds. Malik K A, Mirza M S, Ladha J K. Kluwer Academic Publishers, The Netherlands.

Yanni Y G, Rizk R Y, Corich V *et al.*, 1996. Natural endophytic association between *Rhizobium leguminosarum* bv. *trifolli* and rice roots and assessment of its potential to promote rice growth. Proceedings, Seventh International Symposium on BNF with Non-legumes, 16–21 October, Faisalabad, Pakistan. Eds. Malik K A, Mirza M S, Ladha J K, Kluwer Academic Publishers, The Netherlands.

You C B, Song W, Wang H X, Li J P, Lin M, Hai W L 1991. Association of *Alcaligenes faecalis* with wetland rice. Plant Soil, 137, 81–85.

Posters

9 Poster

Salt tolerance of diazotroph *Alaligenes faecalis* and its salt-tolerant association with host rice

S.Z. PING, M. LIN, Y.N. LIU, C.B. YOU AND KAUSER A. MALIK

There are about 20 million hectares of saline land in North and East China, which causes a loss of approximately 30 million tons of crop yield annually. Some studies conducted in China showed that biofertilizer containing salt-tolerant nitrogen-fixing bacteria could promote growth of plants and increase the yield of crops (Ping *et al.*, 1996). *Alcaligenes faecalis* A1501, which was isolated from paddy soils in South China in 1980, could grow well in solid or liquid medium containing 0.8 mol/L NaCl (Table 1). Growth of bacteria was repressed in solid medium containing 1.2 mol/L NaCl. A1501 could grow very slowly even if the NaCl concentration reached 1.6 mol/L in liquid medium. A1501 could produce IAA under salt stress and maintain N-fixing activity until the NaCl concentration was higher than

Table 1. Effect of various concentrations of NaCl on growth of *A. faecalis* grown in solid medium

Na^+ (mol/L)	0.2	0.4	0.6	0.8	1.0	1.2
Solid medium	+++	+++	++	++	+	–

+++, grow well; ++, grow normally; +, grow slowly; –, no growth.

Table 2. Identification of IAA produced by *A. faecalis* A1501 under salt stress (0.4 M NaCl)

Time of incubation (h)	A1501	*Klebsiella* sp. 101	*Klebsiella* NIAB–I
24	++	+	–
48	++	+	–
72	++	+	–
96	++	+	–
120	+++	++	–

+, IAA production; –, no IAA production.

K.A. Malik et al. (eds.), Nitrogen Fixation with Non-Legumes, 87–89.

Table 3. Associative nitrogen fixation between host rice and *A. faecalis* A1501

	Concentration of NaCl (%)			
	0.01	0.5	1.0	3.0
Rice roots inoculated with A1501	1045.3*	1565.4	956.7	799.1
Pure culture of A1501	681.0	755.1	803.6	456.5

* Activity unit is 'nmol C_2H_4 produced/h. g f.w. roots'.

Table 4. Effect of various NH_4^+ concentrations on expression of *nifL–lacZ* fusion in conjugant A1501(pUK121.2) under microaerobic conditions

		NH_4^+ concentration (mM)		
0.0	2.0	6.0	10.0	15.0
2380.7*	1966.8	1800.2	1655.5	970.0

* β-galactosidase activity (Miller units).

Table 5. Effect of various NH_4^+ concentrations on N_2-fixing activity of A1501 and conjugant A15L1 (pUK121.2) under microaerobic conditions

	NH_4^+ concentration (mM)				
	0.0	2.0	6.0	10.0	15.0
A1501	100	76	54	15	2
A15LI	10	8	3	0	0

* N_2-fixing activity (%).

Table 6. Effect of various concentrations of nalidixic acid on expressions of *nifL-lacZ* fusion in conjugant A15LI under microaerobic conditions

			Nalidixic acid (μg/ml)		
0.0	1.0	2.0	3.0	4.0	5.0
3120.5*	2966.8	2657.0	1655.5	970.0	0.0

* β-galactosidase activity (Miller units).

1.0 mol/L (Tables 2 and 3). Southern hybridization of total DNA from A1501 grown under condition of different NaCl concentration with *Klebsiella pneumoniae nifHDK* DNA probe show that homologous hybridization bands under the normal condition were identifical to those of the cells grown under the salt stress.

By conjugation, *Azotobacter vinelandii nifL* or *nifL-lacZ* gene fusion (kindly provided by Professor Das) was introduced into *Alcaligenes faecalis* wild-type A1501. This is confirmed by physical mapping of plasmid purified from conjugant. A study on salt tolerance, chemotaxis, nitrogen-fixing activity,

Table 7. Effect of NH_4^+ and NaCl on the colonization of A1501 and conjugant A15Ll (pUK121.2) to rice roots

Stains	Treatment		
	$-NH_4^+ - Na^+$	$+NH_4^+$ (15 mmol/L)	$+Na^+$ (0.5%)
A1501	2.7×10^9*	2.75×10^8	9.62×10^9
A15(pUK121.2)	4.37×10^9	1.8×10^8	2.4×10^9

* Bacterial cells/g fresh root of rice.

β-galactosidase activity of A1501 and the conjugant was performed (Tables 4 and 5). The conjugant possesses the same salt tolerance as A1501, and the poor ability of chemotaxis, nitrogen-fixing activity and colonization on rice roots compared to A1501. Expression of *A. vinelandii nifL* in *A. faecalis* was regulated by NH_4^+ and O_2 and was inhibited by nalidixic acid, which is an inhibitor of DNA gyrase (Table 6). This suggested that the product of heterogeneous *nifL* in *A. faecalis* may possess the regulatory function.

Inoculation with A1501 could enhance salt tolerance of host rice, promote growth of rice plants and stimulate formation of lateral roots and root hairs under salt stress (Table 7). Study of the colonization of rice roots by histological assay indicated that (1) *A. faecalis nifH, A. brasilense nifA* and *A. vinalandii nifL-lacZ* gene fusions (Zhang *et al.*, 1996) could express well on the surface or in the endorhizosphere of rice roots under salt stress; (2) the sites and activities of expression of three gene fusions were quite different in endorhizosphere of rice.

References

Ping S Z, Lin M, You C B, Malik K A 1996. Salt tolerance of diazotroph *Alaligenes faecalis* and its interaction with host rice under the salt stress. J. Agric. Biotechnol., 4, 87–92.

Zhang H Y, Lin M, Ping S Z, You C B, Zhu Y X 1996. Construction *nifH::gusA* of *Alcaligenes faecalis* and expression of fusion gene in association with rice roots. J. Agric. Biotechnol., 4, 93–100.

10

Poster

Interaction of *Azorhizobium caulinodans* with different rice cultivars for increased N_2 -fixation

SHAHIDA NASREEN KHOKHAR AND AZRA QURESHI

Introduction

Azorhizobium caulinodans is considered a highly promising microbe because of its ability to fix nitrogen in the free-living state as well as in symbiosis. Most of the estimates made for N_2 -fixation by *Azorhizobium*–rice associations are based on acetylene-reduction assay, a method that has been questioned for accuracy in absolute measurements. The present work aimed at finding the effecitveness of this microorganism to increase N-content in four Pakistani cultivars of rice by comparing total nitrogen per plant.

Materials and methods

Rice seedlings with young radicals (2–3 days old) were transferred to autoclaved potted soil (clay loam, N=8 ppm, pH = 7.8) at field capacity, under glasshouse conditions, at the rate of three biseedlings/pot. After one week of transplanting, the seedlings were inoculated with two test strains of *Azorhizobium caulinodans* (TAL 1926 and IBRG-42), obtained from IRRI, Philippines, and grown in TGY liquid medium for inoculum preparation. Two millilitres of 10^7 cells/ml were inoculated into the roots of each 1-week old rice seedlings. The pots were allowed to be flooded 1 week after inoculation and plants were grown normally for 2 months. The experimental design was split plot with 24 replicates/treatment. Dry biomass of shoot/plant, percentage nitrogen (%N) in shoot and total N-content of shoot/plant were recorded at the age of 5 weeks.

Data were analysed for difference in treatments and estimation of interaction of azorhizobia with different rice cultivars by using (MSTATC software) FACTOR & DMRT.

K.A. Malik et al. (eds.), Nitrogen Fixation with Non-Legumes, 91–93.

Table 1. Effect of inoculation of rice with *A. caulinodans* on shoot dry weight/plant

	Shoot dry weight/plant (g)		
Varieties	Control	TAL1926	IBRG-42
Basmati	1.18^{fg}	2.97^{bc}	1.76^{d}
Superbasmati	1.02^{g}	1.78^{d}	1.44^{def}
DR-92	1.25^{efg}	3.85^{a}	3.13^{b}
KS-282	1.63^{de}	2.58^{c}	1.77^{d}
Average	1.27	2.79	2.02

Average of 24 replicate plants. Mean separation in a column by DMRT at the 5% level.
$LSD^{0.05}$ (for rows) = 0.39.

Table 2. Effect of inoculation of rice with *A. caulinodans* on %N/plant

Varieties	Control	TAL1926	IBRG-42
Basmati-pak	1.7cd	2.2a	1.92bc
Superbasmati	1.45ef	1.92bc	1.31f
DR-92	1.33f	1.54def	2.03ab
KS-282	2.17a	2.01ab	1.63de
Average	1.66	1.91	1.72

Average of 24 replicate plants. Mean separation in a column by DMRT at the 5% level.
$LSD_{0.05}$ (for rows) = 0.25.

Results and discussion

Inoculation by Azorhizobia significantly increased shoot dry weight (F=111.1, LSD=0.39); %N (F=30.24, LSD=0.23) and total nitrogen (F=40.2, LSD=10.36)/ plant. TAL1926 caused a 120% overall increase in shoot dry weight (Table 1), a 28% increase in %N over the uninoculated control (Table 2) and a 140% increase in total N-content/plant (Table 3). IBRG-42 caused a 58% overall increase in shoot dry weight (Table 1), a 19% increase in %N over the un-inoculated control (Table 2) and a 68% increase in total N-content (Table 3). Previously nitrogen fixation by rice roots inoculated with Azorhizobia has been reported (Cocking

Table 3. Effect of inoculation of rice with *A. caulinodans* on total N-content/plant

Varieties	Control	TAL1926	IBRG-42
Basmati-pak	20.06^{de}	65.34^{a}	33.21^{c}
Superbasmati	14.79^{e}	34.18^{c}	18.86^{de}
DR-92	16.62^{e}	59.29^{ab}	63.54^{a}
KS-282	35.37^{c}	51.86^{b}	28.85^{cd}
Average	21.71	52.67	36.11

Average of 24 replicate plants. Mean separation in a column by DMRT at the 5% level.
$LSD^{0.05}$ (for rows) = 10.36.

et al., 1995). Highest N-content/plant recorded in the present study is 65.34 mg. Maximum N-content by *Klebsiella* inoculation of rice has been reported as 11 mg/plant (Fujii *et al.*, 1987).

Significant differences in shoot dry weight (F=47.8), %N-content (F=12.08) and total N-content (F=15.6) were also recorded because of cultivars. Differences in nitrogen-fixation activities between 14 cereal cultivars inoculated with *Azospirillum* have also been reported by Jagnow (1990). Interaction of *Azorhizobium* and rice cultivars affected shoot dry weight (F=13.03), %N content (F=11.43) and total N-content (F=7.41), significantly. The best overall response to *Azorhizobium* inoculation was shown by DR-92, in terms of shoot dry weight and total N-content. The best combination in terms of dry shoot weight was DR-92 × TAL 1926. Basmati Pak, which is unresponsive to high fertilizer dosage, when inoculated with TAL 1926, responded to give the highest total N-content/plant. DR-92 inoculated with IBRG-42 ranked equally in total N-content/plant (Table 3).

Acknowledgements

The authors are grateful to IRRI, Philippines, for providing the *Azorhizobium* cultures, and to Dr Ashraf, Coordinator, Rice Program, NARC for supplying the rice seeds. We also acknowledge Dr Ehsan Akhtar, Coordinator, MOP, for extending the facility of the autoanalyser for N-estimation.

References

Cocking E C, Kothari S L, Batchelor C A, Jain S, Webster G, Jones J, Joatham J, Davey M R 1995. Interactions of rhizobia with non-legume crops for symbiotic nitrogen fixation and nodulation. In: Fendrik *et al.*, eds. *Azospirillum* VI and related microorganisms. Springer-Verlag, Berlin, 197–205.

Fujii T, Huang Y D, Higashitani A, Nishimura Y, Iyama S, Hirota Y, Yoneyama T, Dixon R A 1987. Effect of inoculation with *Klebsiella oxytoca* and *Enterobacter cloacae* on dinitrigen fixation by rice-bacterium associations. Plant Soil, 103: 221–226.

Jagnow G 1990. Differences between cereal crop cultivars in root associated nitrogen fixation, possible cause of variable yield response to seed inoculation. Plant Soil, 123, 255–259.

11 Poster

Response of wheat to an associative diazotroph inoculum under different rates of nitrogen fertilizer

A. RASHID, M.R. SAJJAD, M.A. GILL, M.S. CHEEMA, M.S. SINDHU AND M.M. NAYYAR

Introduction

The effect of associative microorganisms in increasing crop yield and N_2-fixation has been reported by many research workers (Bala and Kundu, 1988, Hegazi, 1988, Markus, 1988, Rashid *et al.*, 1993).

Accumulated evidence in the past decade indicated that cereal crop responses to inoculation of heterotrophic N_2-fixing bacteria are not consistent (Boddey and Dobereiner, 1982) and increase in crop yield has been demonstrated due to the plant growth hormones produced by these bacteria (Brown and Burlingham, 1968, Barea and Brown, 1974, Brown, 1976).

In the present study we assessed the yield response of common wheat under field conditions to a mixed diazotroph inoculum of *Azotobacter chroococcum* and *Azospirillum lipoferum* with different levels of N-fertilizer application.

Materials and methods

Four field experiments were carried out during the winter season in the Punjab area of Pakistan; one in 1994–95 and three in 1995–96. The effects of inoculation with a mixed culture of *Azotobacter chroococcum* and *Azospirillum lipoferum* on the yield of wheat varieties 'Perwaz-94' and 'Inqilab-91' were tested using a factorial experiment in the randomized complete block design in four replicates of microplots. A mixture of two bacteria of *Azotobacter chroococcum* and *Azospirillum lipoferum*, in equal amounts, was used for inoculation each at the rate of 11×10^7 MPN (most probable number) bacteria g^{-1} of peat as carrier. Before sowing, peat-based inoculum was applied to the seed as a surface coating at the rate of 25 g kg^{-1} seed. The seed was used at the rate of 100 kg ha^{-1}.

K.A. Malik et al. (eds.), Nitrogen Fixation with Non-Legumes, 95–97.

Results and discussion

The grain yield of wheat inoculated with a mixture of diazotrophic microorganisms (*Azotobacter chroococcum* + *Azospirillum lipoferum*) was significantly greater than that of uninoculated treatment (Table 1). Wheat grain yield was also significantly affected by the addition of different N-fertilizer rates. The ability of the inoculum to produce more wheat grain by 13.3% and 17.9% became evident at 40 and 80 kg N ha^{-1} application (data not shown). At the higher level of 120 kg N ha^{-1} the inoculum response was decreased to a 7.9% increase in yield; on the other hand, without N addition, the inoculum also showed a small effect on wheat yield. The interaction effect of inoculation and N-fertilizer rates in increasing wheat grain weight was found to be non-significant (Table 1). The increase in total N and P uptake by wheat crop due to inoculation, N rates and their interaction effect was found to be significant (Table 1).

Differences in wheat grain yield of inoculated *vs*. (versus) non-inoculated treatments for the other three tested sites are presented in Table 2. A significant effect of inoculum was recorded at two sites, and at one site the effect was non-significant. The response of N-fertilizer application was found to be significant at all sites, while the interaction effect between inoculum and N-fertilizer was recorded only at one site.

This study points out that a mixed inoculum of *Azotobacter chroococcum* and *Azospirillum lipoferum* isolated from roots of maize has the potential to improve the yield of cultivated wheat and thus be agronomically valuable. The highly variable results obtained in the 1994–95 and 1995–96 experiments might reflect these variations in basic soil conditions, local environmental conditions and agronomic practices (Boddey and Dobereiner, 1982, Millet *et al.*, 1985).

Table 1. Effect of a diazotrophic (*Azotobacter chroococcum* + *Azospirillum lipoferum*) inoculation on wheat yield, nitrogen and phosphorus uptake under different rates of nitrogen fertilizer, 1994–95

	Grain weight (kg ha^{-1})	Straw weight (kg ha^{-1})	Total N uptake (kg ha^{-1})	Total P uptake (kg ha^{-1})
Inoculation[1]				
+	3283	6822	104.2	22.9
–	2957	6529	90.9	18.8
Nitrogen[2]				
0	1863	4122	50.1	14.0
40	2973	6420	88.9	21.3
80	3581	7901	114.5	24.0
120	4063	8258	136.6	24.1
SE[3]				
Inoculation	67**	76**	1.4**	0.3**
Nitrogen	95**	108*	2.0**	0.4**
Inoculation× nitrogen	NS	NS	2.9*	0.6**

[1]Inoculation: means for all nitrogen rates. [2]Nitrogen: means for both inoculation treatments. [3]SE = standard error for means group, *p = 0.05, ** p = 0.01. NS = No significant difference.

Table 2. Effect of a diazotrophic (*Azotobacter chroococcum* + *Azospirillum lipoferum*) inoculation on wheat grain yield under different rates of nitrogen fertilizer at different locations, 1995–96.

	Grain weight (kg ha^{-1})		
	At Karor	At Khanewal	At Bahawalpur
Inoculation[1]			
+	2976	4434	2927
–	1816	4319	2819
Nitrogen[2]			
0	1508	3403	1744
40	2128	4187	2513
80	2811	4638	3430
120	3137	5277	3803
SE[3]			
Inoculation	77**	32*	NS
Nitrogen	109**	45**	69**
Inoculation × nitrogen	154**	NS	NS

[1] Inoculation: means for all nitrogen rates. [2] Nitrogen: means for both inoculation treatments. [3]SE = standard error for means group. *$p = 0.05$, ** $p = 0.01$. NS = no significant difference.

Acknowledgements

Financial support for this work from the Punjab Government, Pakistan-Agricultural Research Service, is greatly appreciated.

References

Bala I, Kundu B S 1988. Interactions between *Azospirillum* species and wheat (*Triticum aestivum*) varieties. Ind J. Agric. Sci., 58(3), 227–229.

Barea J M, Brown M E 1974. Effects on plant growth produced by *Azotobacter paspali* related to synthesis of plant growth regulating substances. J. Appl. Bacteriol., 37, 583–593.

Boddey R M, Dobereiner J 1982. Association of *Azospirillum* and other diazotrophs with tropical Gramineae. Transactions of the 12th International Congress on Soil Science, Symposium Papers I, Non-Symbiotic Nitrogen Fixation and Organic Matter in the Tropics. New Delhi, India, 25–47.

Brown M E 1976. Role of *Azotobacter paspali* in association with *Paspalum notatum*. J. Appl. Bacteriol. 40, 341–348.

Brown M E, Burlingham S K 1968. Production of plant growth substances by *Azotobacter chroococcum*. J. Gen Microbiol. 53, 135–144.

Hegazi N A 1988. Modification of soil environment through straw application versus *Azospirillum* spp. inoculation. In: Klingmuller W, ed. Azospirillum. IV. Genetics, Physiology, Ecology. Proceedings of the Fourth Bayreuth German Federal Republic. Springer-Verlag, Berlin, 215–222.

Markus P 1988. Studies on the mode of N_2-fixing bacteria on the growth and yield of spring wheat in organic farming systems. Field Crop Abstr., 42, 4936.

Millet E, Avivi Y, Feldman M 1985. Effect of rhizospheric bacteria on wheat yield under field conditions. Plant Soil, 86, 347–355

Rashid A, Siddique A G, Gill S M, Aslam M 1993. Contribution of microbes in crop production: I. Effect of a diazotrophic and a commercial EM_4 inoculation on the yield and N economy of wheat. In: Proceedings, First National Seminar on Nature Farming, NFR Centre, University of Agriculture, Faisalabad, Pakistan, pp 54–66.

12 Poster

Effects of high temperature on nitrogen fixation in soybean cultivars

M.I. KEERIO AND J.M. WILSON

Introduction

The adverse effects of root temperatures on nitrogen fixation have been shown for numerous legumes (Gibson, 1977, Sprent and Minchin, 1984). Biological nitrogen fixation is often affected more by temperature than is the general growth and photosynthesis of the plant (Granhall, 1981). Steward (1966) has also suggested that nitrogen fixation is often especially inhibited by temperature extremes which have less effect on plant growth. Here the effect of temperature stresses on the ARA of two soybean varieties is discussed.

Material and methods

Seeds of two soybean cultivars (Williams-82 and Mago-80) were grown in (10 × 5 cm) pots filled with vermiculite. The seeds were soaked for 30 min in an *Brady rhizobium japonicum* RCR3407 inoculum. Two seeds were sown in each pot. Seven days after sowing, the seedlings were thinned to one per pot and inoculated again with 3 ml of *B. rhizobium japonicum* inoculum. All the pots were irrigated with half-strength Long Ashton nutrient (nitrogen-free) solution (Hewitt, 1966).

The plants were used for the heat stress treatments 5 weeks after inoculation. Heat stresses of 2, 3 or 4 h were applied at 35°C, 40°C or 45°C. After the heat stress treatments nitrogen fixation was measured by ARA using gas chromatography (Sveriges, Sweden) and the plants were then allowed to recover for 72 h at 27°C.

Heat stress was applied in a shaking water bath (Grant). For these treatments the pots were covered with a plastic lid which had two holes, one for the plant shoot and the other for injecting and withdrawing gas samples. Temperatures inside the pot were monitored by inserting a thermocouple probe into the vermiculite. After

K.A. Malik et al. (eds.), Nitrogen Fixation with Non-Legumes, 99–102.

heat treatment the pots were removed from the water bath and nitrogen fixation was measured.

Results and discussion

The data for cultivar Williams-82 (Figure 1) show that the control value was 14.1 μmol (100 mg f.w. t) $^{-1}$ h^{-1}. In the case of cultivar Williams-82 at 35°C, fixation was inhibited more by 3 h heat stress than by 2 or 4 h stress. Also, nitrogen fixation at 40°C was higher after 4 h stress than after 2 or 3 h stresses. This was

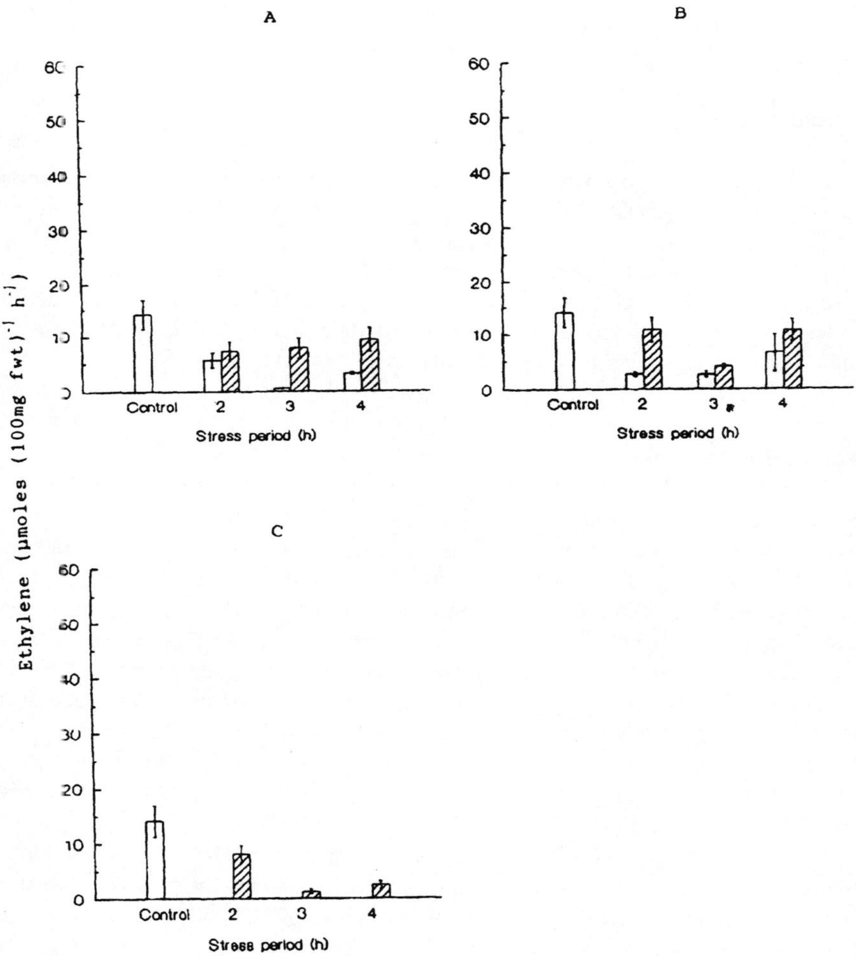

Figure 1. Effect of root heat stress on nitrogen fixation in cv. Williams-82. **A**: 35°C; **B**: 40°C: **C**: 45°C; ☐ at the end of heat stress period; ▨ following recovery for 72 h.

surprising, and may be due to unknown changes in nodule physiology affecting temperature sensitivity. No nitrogen fixation was detected afer 2, 3 or 4 h heat stress at 45°C, and the activity was not regained fully during subsequent recovery.

The data for cultivar Mago-80 (Figure 2) also contain some unexpected features. They indicate that nitrogen fixation was reduced by 2 and 3 h stresses at 35°C, but surprisingly no change was observed after 4 h stress. Also, no significant effect was found at 40°C. At 45°C, however, a decreasing trend, similar to that seen in cultivar Williams-82, was observed. After 72 h of recovery the acetylene reduction values were higher in most treatments compared with the controls, with the exception of the 2 h and 4 h stressed

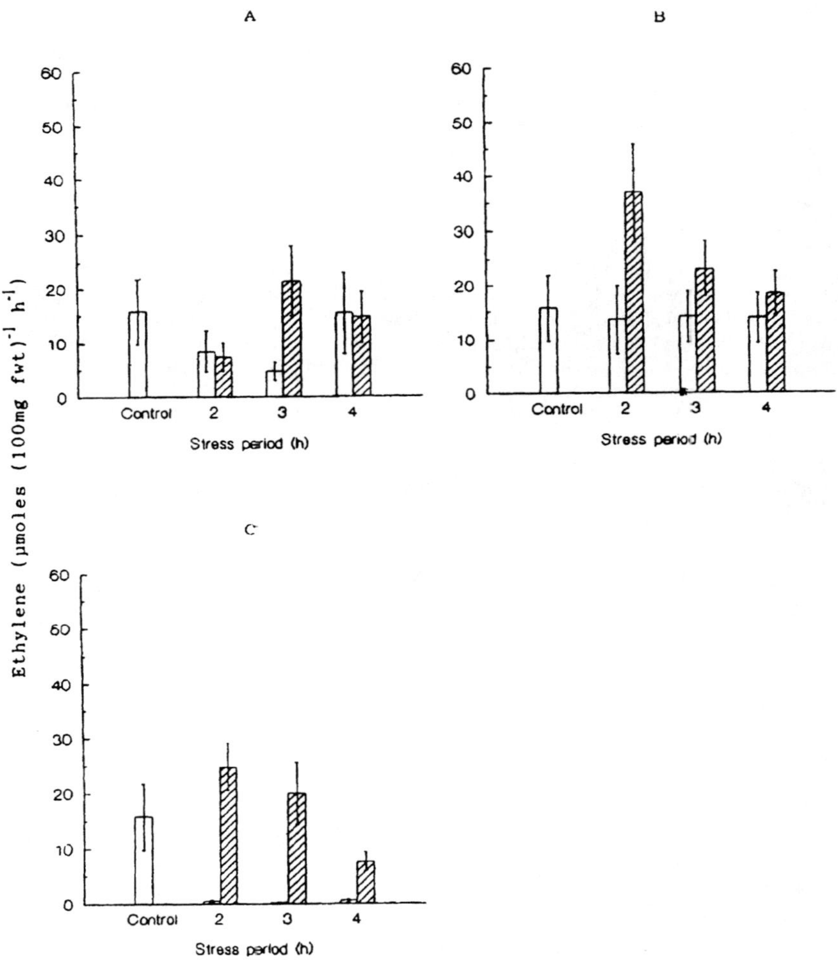

Figure 2. Effect of root heat stress on nitrogen fixation in cv. Mago-80. **A**: 35°C; **B**: 40°C; **C**: 45°C, □ at the end of heat stress period; ▨ following recovery for 72 h.

plants at 35°C and 45°C. Overall, cultivar Mago-80 appeared to recover better than the other cultivar tested here.

These results roughly agree with those of Dart and Day (1971) and Dart *et al.*, (1975), who reported that acetylene reduction by nitrogen-fixing organisms occurred over a wide temperature range with maximum activity between 24°C and 33°C, rapidly declining at higher temperatures. Sinclair and Weisz (1985) observed that acetylene reduction rates in soybean increased with soil temperatures up to 30°C. Rates then declined slightly up to 34°C, but above this temperature they were greatly reduced.

References

Dart P J, Day J M 1971. Effects of incubation temperature and oxygen tension on nitrogenase activity of legume root nodules. Plant Soil, spec. vol., 167–184.

Dart P J, Day J, Islam R, Dobereiner J 1975. Symbiosis in tropical grain legumes. Some effects of temperature and composition of the rooting medium. In: Nutman P S, ed. Symbiotic Nitrogen Fixation in Plants. Cambridge University Press. International Biology Programme, vol. 7, pp. 361–384.

Gibson A H 1977. The influence of the environment and managerial practices on the legume-rhizobium symbiosis. In: Hardy R W F, Gibson A H, eds. Treatise on Dinitrogen Fixation. IV. Agronomy and Ecology. John Wiley & Sons, New York, 393–450.

Granhall U 1981. Biological nitrogen fixation in relation to environmental factors and functioning of natural ecosystems. In: Clark F E, Rosswall T, eds. Terrestrial Nitrogen Cycles. John Wiley & Sons. New York; Ecol. Bull. Stockholm, 33, 131–144.

Hewitt E J 1966. Sand and water culture methods used in the study of plant nutrition. Technical Communication. No. 2 (revised 2nd ed) Commonwealth Bur. Hort. Plantn. Crops, Commonwealth Agricultural Bureaux, England.

Sinclair T R, Weisz P R 1985. Response to soil temperature of dinitrogen fixation (Acetylene reduction) rates by field grown soybean. Agron. J., **77**, 685–688.

Sprent J I, Minchin F R 1984. Environmental effects on the physiology of nodulation and nitrogen fixation. In: Jones D G, Davis D R, eds. Temperate Legumes: Physiology, Genetics and Nodulation. Pitman, London, pp. 269–317.

Steward W D 1966. Nitrogen Fixation in Plants. Athlone Press, University of London.

13 Effect of plant growth regulators and *Rhizobium* inoculum on N_2-fixation and yield of chickpea

Poster

ZARRIN FATIMA, A. BANO AND M. ASLAM

Introduction

Growth regulators are usually hormones regulating substances synthesized in one part of the plant and translocated to another part where they exert their effects. The first plant hormone identified was an auxin. The formation of nodules on legumes is associated with auxin-forming *Rhizobium* (Thimann, 1936). Extractable auxin levels per gram fresh weight of nodules were 40–60 times than the level in roots. Kinetin is a synthetic cytokinin which does not occur naturally in plants. Due to its property of actively promoting cell division (in conjunction with auxin) it was given the name of kinetin.

The role of ABA in root nodulation is not completely known. The development of root nodules was variously affected by ABA treatment; type and the magnitude of the effect depended on the developmental phase of the root nodules at the time of ABA treatment (Bano and Hillman, 1986).

Materials and methods

Growth conditions and plant materials

Surface-sterilized seeds of chickpea cv. CM-88, obtained from NARC, Islamabad, were soaked in distilled water (treated as control) and aqueous solution of IAA, ABA and kinetin (each at 10^{-5} M) for 6 h. These growth regulators were also used as foliar spray after 3 weeks of germination (each at 10^{-6} M concentration). Seeds were inoculated with two strain of *Rhizobium leguminoserum* TAL620 and TAL1148. Seeds were wetted with inoculum (16 g kg^{-1} seeds) and 48% sugar solution and put into a large Petri plate. Then powder form inoculant was dusted on this, and mixed thoroughly to a uniform coating of inoculum of the seeds.

K.A. Malik et al. (eds.), Nitrogen Fixation with Non-Legumes, 103–106.

Nitrogenase activity

The estimation of C_2H_4 produced by nodules was made following the procedure of Larue and Kurz (1973).

Soil nitrate and phosphorus content

Nitrate content and phosphorus content of soil was estimated following the procedure of Winkleman *et al.*, 1985.

Results and discussion

The maximum value of absorbance for C_2H_4, an indicator for nitrogenase, was recorded in kinetin treatment both in seed soaking as well as foliar spray, in combination with strain 1148 (Table 1). The IAA was least effective. The response of *Rhizobium* strain 1148 with regard to nodulation and nitrogen fixation was

Table 1. Plant physicochemical parameters as affected by growth regulators and *Rhizobium* inoculation on chickpea

Treatments	Nitrogenase activity of nodules (ppm)	Nodule weight plant^{-1} (g)	Grain weight plant $^{-1}$ (g)	Nitrate content of soil (ppm)	Phosphorus content of soil (ppm)
ABA+ss+A	0.84^{e}	1.12^{i}	2.94^{a}	6.06^{i}	158.66^{j}
ABA+ss+B	0.482^{h}	0.89^{p}	2.77^{c}	4.956^{l}	152.66^{k}
ABA+ss	0.467^{i}	1.01^{m}	1.103^{i}	4.196^{m}	199^{e}
ABA+sp+A	0.528^{g}	1.2033^{g}	2.87^{b}	3.52^{p}	187^{g}
ABA+sp+B	0.477^{b}	1.0266^{l}	2.43^{d}	3.026^{q}	227.86^{c}
ABA+sp	0.356^{m}	0.9033^{o}	1.66^{g}	2.02^{r}	175.3^{h}
IAA+ss+A	0.415^{j}	0.81^{q}	0.797^{k}	7.053^{g}	160.33^{g}
IAA+ss+B	0.303^{n}	1.323^{d}	0.721^{l}	2.2066^{h}	168^{i}
IAA+ss	0.06^{q}	0.7366^{r}	0.456^{m}	5.82^{j}	192.33^{f}
IAA+sp+A	0.371^{l}	1.1266^{h}	0.203^{o}	4.966^{l}	178.33^{h}
IAA+sp+B	0.064^{q}	0.9166^{n}	0.1466^{p}	4.116^{mn}	235.66^{b}
IAA+sp	0.1^{p}	1.69^{a}	0.11^{p}	4.03^{n}	218.33^{d}
Kinetin+ss+A	1.459^{a}	1.436^{b}	2.8766^{b}	17.95^{a}	48.66^{r}
Kinetin+ss+B	1.1436^{c}	1.3033^{e}	2.076^{e}	14.2^{b}	53.66^{q}
Kinetin+ss	0.674^{f}	1.2066^{g}	1.246^{h}	8.78^{e}	95^{n}
Kinetin+sp+A	1.376^{b}	1.39^{c}	2.09^{e}	10.133^{c}	68^{p}
Kinetin+sp+B	0.8616^{d}	1.263^{f}	1.843^{f}	9.14^{d}	87.33^{o}
Kinetin+sp	0.5316^{g}	1.1^{j}	0.94^{j}	8.1366^{f}	101.33^{m}
Control+A	0.396^{k}	1.063^{k}	0.4566^{m}	5.553^{k}	108.7^{l}
Control+B	0.2566^{o}	0.886^{j}	0.27^{n}	4.97^{l}	167.68^{i}
0.301p	0.67^{s}	0.41^{p}	3.773^{o}	248.66^{a}	

All such means which share a common superior letter are non-significantly different; otherwise they differ significantly.

ss = Seed soaking, sp = foliar spray, A = strain TAL 1148 and B = strain TAL620.

better than 620 when used either alone or in combination with growth regulators. The IAA, naphthalene acetic acid (NAA) and colchicine all reduced the rate of development and C_2H_2 reducing activity of paranodules (Kennedy *et al.*, 1991).

The maximum nodule weight was found in kinetin-treated seeds when used alone or in combination with strain 1148 as compared to control and other treatments. The minimum value was recorded in IAA treatment. The maximum weight was obtained due to kinetin treatment because of its property of actively promoting cell division. The cytokinin content of *Phaseolus vulgaris* nodules were higher than that of roots (Puppo *et al.*, 1974). The effect of ABA was less pronounced as compared to that of kinetin.

Kinetin showed maximum yield and larger seeds than that of ABA and IAA. The *Rhizobium* strain TAL1148 proved better than TAL620. Seed soaking treatment showed better results as compared to foliar spray. The maximum seed weight and large seed size due to kinetin tretment may be attributed to increased translocation of assimilates from vegetative to reproductive parts. It has been reported that kinetin mobilized the starch reserve and increased the flow of sugar in Chinese cabbage (Berridge and Ralph, 1971).

The soil nitrate content was higher when kinetin was applied through seed soaking or foliar spray. However, soil nitrate was lower than that of control when ABA was applied through foliar spray. The reasons for increased nitrate content of the soil during the growing season are reduced uptake rates of nitrate, or relative contribution of both N-sources, mainly influenced by the amount and availability of inorganic nitrogen in soil. Another possibility is that nitrogen fixation suppresses nitrate uptake of chickpea, and that is one of the reasons for high residual nitrate contents. In a previous report soybean genotypes have been identified which have more nodules in the presence of mineral N. (Betts and Herridge, 1987).

The residual phosphorus content was higher in soil in which seeds were inoculated with strain 620. Available soil phosphorus was less in the treatments for which seeds were soaked with growth regulators and inoculated with strain 1148. However, the available soil phosphorus was greater in the case of foliar spray of IAA and ABA plus strain 620, possibly due to reduced uptake or greater mobilization of phosphorus. It has been found that phosphorus plays an important role in dinitrogen fixation through an effective translocation of phosphorus to the leaf. Thus when the phosphorus supply is limited, efficient cultivars obtained a reasonably high yield through an effective translocation of the absorbed phosphorus to the leaf (Adu-Gyamfi *et al.*, 1989).

Acknowledgements

Thanks are extended to Mr Shahbaz Ahmed for providing chickpea seeds and strains of *Rhizobium*. Computer support was provided by my friend Miss Samia Kausar and Mr Aamir Nasir. Technical advice was provided by Mr Tariq Sultan and Mr Imdad. Support in statistical analysis was given by Miss Fareha Fyyaz.

References

Adu-Gyamfi-J J, Fujita K, Ogata S 1989. Phosphorus absorption and utilization efficiency of pigeon pea (*Cajnas cajan*) in relation to dry matter production and dinitrogen fixation. Plant Soil, 119, 315–324.

Bano A, Hillman J R 1986. Effect of abscisic acid on nodule morphology, nitrogenase activity and hydrogen evolution in *Faba vulgaris*. Ann. Bot., 58, 281–283.

Berridge M V, Ralph R 1971. Kinetin and carbohydrates metabolism in chinese cabbage. Plant Physiol., 47, 562–567.

Betts J H, Herridge D E 1987. Isolation of soybean lines capable of nodulation and nitrogen fixation under high levels of nitrate supply. Crop Sci., 27, 1156–1161.

Kennedy I R, Sriskandarajah S, Yu D, Nie Y F, Tchan Y T 1991. C_2H_2 reducing para-nodules in wheat: effect of growth regulators and colchicine. Proceedings, Ninth Australian Nitrogen Fixation Conference, vol. 9, 78–79.

Larue, T A, Kurz W G 1973. Estimation of nitrogenase using a calorimetric determination for ethylene. Plant Physiol., 51, 1074–1075.

Puppo A, Rigaud J, Barthe P 1974. Nodule development and senescence. C.R. Hebd Seanc. Acad. Sci. Paris, 1279–2029.

Thimann K V 1936. The Physiology of the formation of nodules on legume roots. Proc. Natl. Acad. Sci. USA 22, 511–514.

Winkleman G A, Rohul A, Tahir M B 1985. *Methods Manual, Soils Laboratory*. Pakistan Agricultural Research Council, Islamabad, pp. 60–68.

Section III

Paranodulation and forced association of diazotrophs with non-legumes

14

Prospects for facilitated evolution of effective N_2-fixing associations with cereals: comparative performance of *Azospirillum brasilense* Sp7-S with various free-living diazotrophs in *para*-nodulated wheat

IVAN R. KENNEDY, SUNIETHA KATUPITIYA, DAIGUAN YU, KATE GILCHRIST, ROSALIND DEAKER, LILY PEREG-GERK AND CRAIG WOOD

Abstract

By analogy with N_2-fixing sugar cane, achieving an effective N_2-fixing association between cereals and diazotrophs may require an endophytic mode of colonization, allowing better protection from oxygen and improved access to carbon substrates. Using *nifA-lacZ* as a genetic marker on a broad host-range plasmid pVK100, inserted into a range of free-living and associative N_2-fixing organisms, it has been possible to define their mode of colonization of the roots of wheat seedlings in addition to measuring the associated rates of nitrogen fixation. Of these bacterial strains, only *Herbaspirillum seropedicae, Azorhizobium caulinodans* and a mutant strain of *Azospirillum brasilense* (Sp7-S) displayed significant endophytic colonization of 2,4-D-treated wheat seedlings. By contrast, *Acetobacter diazotrophicus, Azotobacter vinelandii, Derxia gummosa* and other *Azospirillum* strains colonized at the rhizoplane. This chapter discusses these modes of colonization and the probable need for a stepwise process of facilitated evolution of selected diazatrophs and plants before an effective association can be achieved.

Introduction

There are several approaches to the study of the development of N_2-fixing symbioses with cereals (Cocking *et al.*, 1992; de Bruijn *et al.*, 1995). We have employed the *para*-nodule, induced with synthetic auxins such as 2,4-dichlorophenoxyacetic acid (2,4-D) as a laboratory model for our research on this topic, at least for preliminary stages. In 1992 we reviewed the status of biological N_2 fixation in non-leguminous field crops, drawing attention to facilitated colonization by crack entry of *Azospirillum brasilense* into structures derived from lateral roots induced by synthetic auxins such as 2,4-D applied at sub-lethal levels (Kennedy and Tchan,

K.A. Malik et al. (eds.), Nitrogen Fixation with Non-Legumes, 109–124.

1992); since then we have published several papers (see Sriskandarajah *et al.*, 1993; Katupitiya *et al.*, 1995a, b; Yu and Kennedy, 1995) describing our further development of this laboratory model of associative N_2 fixation. This approach was based on earlier observations in the 1980s of Yanfu Nie at Shandong University, that 2,4-D acting as an auxin could stimulate colonization by rhizobia of 'nodular structures' modified from lateral roots referred to as *para*-nodules. Typically, acetylene reduction rates under oxygen-controlled conditions of 200–2 000 nmol/g dry weight roots per hour are observed in young seedlings, several times greater than the rate observed without synthetic auxin treatment.

Recent research using this model has led to the recognition of several factors that facilitate improved N_2 fixation under our specific conditions of assay (Katupitiya *et al.*, 1995a). In this work the provision of genetic markers based on *lacZ* fusions to the promoters of *nif* genes of *Azospirillum brasilense* has allowed direct visual and enzymic detection of bacterial cells in association with wheat. This procedure enables individual bacterial cells attached to the root surface or occurring endophytically to be visualized by light microscopy, and provides a method for rapid screening of individual plants to reveal characteristic patterns of colonization (Arsène *et al.*, 1994; Katupitiya *et al.*, 1995b). For example, it showed that one mutant strain of *A. brasilense* Sp7 (Sp7-S), though deficient in factors allowing it to bind to the wheat root surface, obtained crack entry to the basal zone of *para*-nodules. Colonizing with a substantially endophytic character in the laboratory model of the *para*-nodule, Sp7-S produced much higher rates of acetylene reduction than those obtained with the Sp7 strain (Katupitiya *et al.*, 1995a).

Further development of this laboratory model requires that an essential set of desirable genetic characteristics be identified in both microbial and plant partners, and then applied in new combinations of microbial and plant germplasm or by deletions of genes that currently prevent desirable outcomes. A major challenge in this research is to find easily identified phenotypes indicative of effective associative systems.

Genetic tools such as the use of *lacZ* markers allow characteristics of colonization to be quantified. In this chapter, using the *nifA–lacZ* gene fusion as a marker, we compare a broader range of diazotrophs including species from *Azospirillum, Azorhizobium, Azotobacter, Acetobacter, Derxia* and *Herbaspirillum* to determine their patterns of colonization in association with wheat, and the effects of adding synthetic auxin on the success of each associative system. This comparison reinforces the relative uniqueness of *A. brasilense* Sp7-S in its endophytic character, but also reveals at least one new candidate as an endophyte for possible application in wheat.

Materials and methods

Bacterial strains and conjugation

Azospirillum brasilense Sp7 and Sp7-S were sourced as previously described (Katupitiya *et al.*, 1995a). *Azospirillum lipoferum* strains were obtained from a set previously described (New and Kennedy, 1989). *Azorhizobium caulinodans* ORS 571 was purchased from the LMG Culture-collection, Gent, Belgium, and the wild-type *Azotobacter vinelandii* strain (VW136) was a gift from C. Kennedy of the University of Texas, Houston. *Herbaspirillum seropedicae* and *Acetobacter diazotrophicus* were kindly supplied by Johanna Döbereiner, EMBRAPA, Brazil. The sources of all diazotrophs and related reagents are given in Table 1.

For all strains, conjugation was performed with *Escherichia coli* S17.1 carrying pAB576, a *nifA–lacZ* fusion of the *nifA* promoter of *A. brasilense* Sp7 (Liang *et al.*, 1991). In some cases *nodG–lacZ, ntrC–lacZ* and *nifH–lacZ* fusions were employed (see Arsène *et al.*, 1994, Katupitiya *et al.*, 1995a). Transconjugants were isolated on selective media containing Tc (5μg/ml) according to Katupitiya *et*

Table 1. Bacterial strains and plasmids

Strain	Source or description	Reference
Azospirillum brasilense		
Sp7	Institut Pasteur (C. Elmerich)	Tarrand *et al*, 1978
Sp7-S	Referred as Sp7 in previous studies (22, 45, 47, 53)	Katupitiya *et al.*, 1995a
Cd	(P.B. New)	Eskew *et al.*, 1977
Sp245	(P.B. New)	Baldani *et al.*, 1983
SpBr14	(P.B. New)	Tarrand *et al*, 1978
Azospirillum lipoferum		
Sp59b	(P.B. New)	Tarrand *et al.*, 1978
SpBr17	(P.B. New)	Tarrand *et al.*, 1978
596	Wheat soil	New and Kennedy, 1989
Azorhizobium caulinodans		
ORS571	Ghent, Belgium	Dreyfus *et al.*, 1983
Derxia gummosa		
SUT1	Tropical soil, University of Sydney	Tchan and Zeman, 1995
Escherichia coli		
S17.1	Institut Pasteur (C. Elmerich) pro. thi, hsdR, recA	
Azotobacter vinelandii		
VW136	Wild-type (C. Kennedy)	Bali *et al.*, 1992
Herbaspirillum seropedicae		
ATCC 35892	EMBRAPA, Brazil (J. Döbereiner)	Baldani *et al.*, 1986
Acetobacter diazotrophicus		
ATCC 49037	EMBRAPA, Brazil (J. Döbereiner)	Cavalcante and Döbereiner, 1988
Plasmid	Relevant characteristics	
pAB576	*nifA–lacZ* transcriptional fusion, Tc^R, Km^R, pVK100 derivative	

al., (1995a,b). *A. caulinodans* was selected in the same media as used for *A. brasilense* to which 0.1% yeast extract was added to stimulate growth.

Transconjugants were examined for nitrogen fixation activity in pure culture under reduced oxygen and the *lacZ* expression as ß-galactosidase confirmed (see Kayupitya *et al.*, 1995a,b). In some cases the presence of the plasmid in transconjugants was confirmed by gel electrophoresis (Dreyfus *et al.*, 1983).

Growth of wheat seedlings, acetylene reduction assay (ARA) and $^{15}N_2$ *exposures*

Wheat seedlings were grown hydroponically and treated with 2,4-D as previously described (Sriskandarajah *et al.*, 1993). The ARA was conducted in 30 ml McCartney bottles, generally with 0.025 atm oxygen in nitrogen (Zeman *et al.*, 1993). $^{15}N_2$ exposures were conducted as described by Yu *et al.*, (1993), with those involving *A. brasilense* Sp7-S being carried out in an especially constructed Perspex chamber in which test and control seedlings were incubated simultaneously, exposed to the same atmosphere.

ß-Galactosidase assays and studies on colonization patterns

Assays for measuring *lacZ* expression in pure culture or on wheat seedlings were conducted either using *o*-nitrophenol-ß-D-galactoside (ONPG) for enzyme activity or 5-bromo-4-chloro-3-indolyl-ß-D-galactoside (X-gal) for *in-situ* visualization by light microscopy as described as Arsène *et al.*, (1994). Background plant ß-galactosidase activity was eliminated by heating plant tissue to 50°C or using glutaraldehyde treatment (Arsène *et al.*, 1994). Conclusions regarding modes of colonization were made only after consistent patterns were obtained in experiments using replicate (five or greater) seedlings. Photographs were taken on automatic exposures using an Olympus BHA light microscope and Kodak Ektachrome film. Images were also stored using a digital camera or by scanning of photographs, and reproduced for this publication (Figure 1) using computer software (Microsoft Word and Adobe PhotoShop).

Results

Colonization patterns observed for free-living diazotrophs

The application of *lacZ* marker genes derived from *A. brasilense* Sp7 has enabled several free-living diazotrophs to be examined for their colonization patterns on wheat seedlings and their capacity to carry out N_2 fixation. With all the diazotrophs examined, *lacZ* transconjugants were formed with high frequency (about 10^{-3}), were stable and the fusions (e.g. *nifA–lacZ, nodG–lacZ, ntrC–lacZ, nifH–lacZ*) well expressed in pure culture and in association with plants. However, these

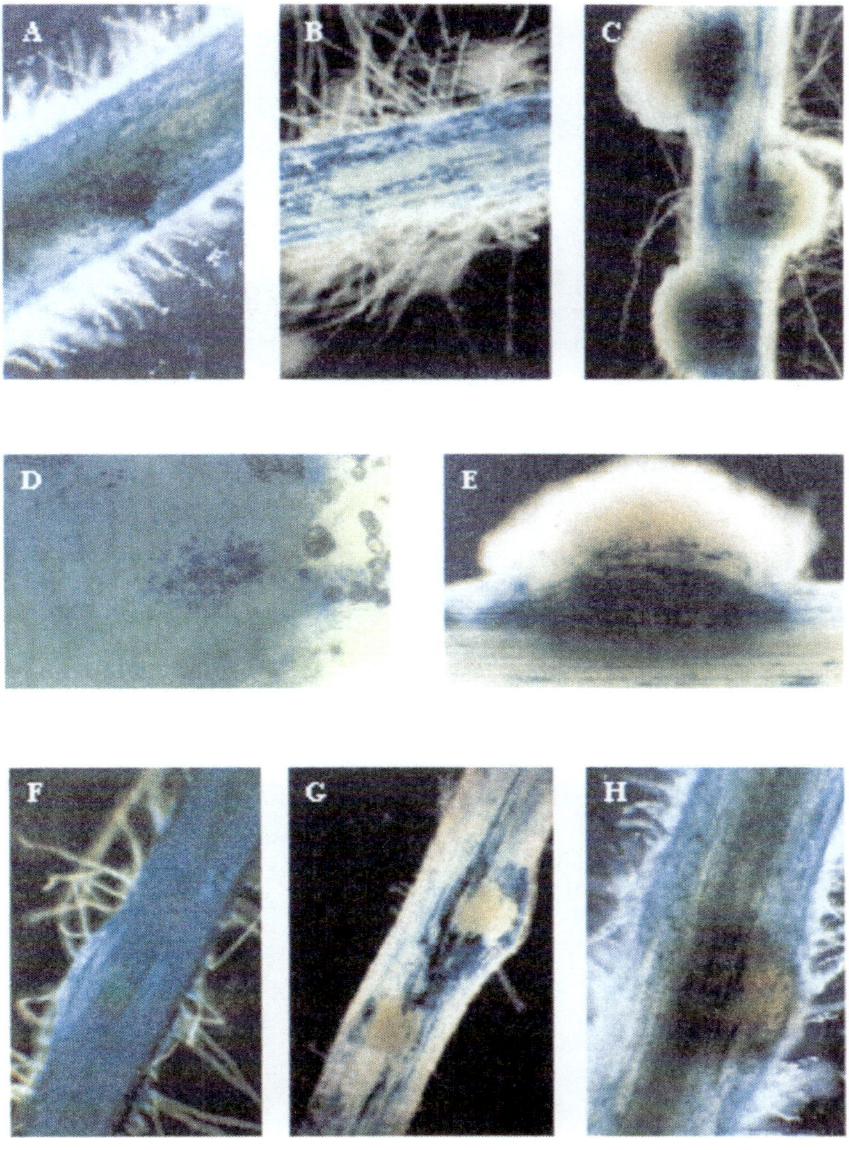

Figure 1. *Colonization patterns of free-living diazotrophs on para*-nodulated wheat. **A:** *Derxia gummosa*, rhizoplane, ×65; **B:** *Acetobacter diazotrophicus*, rhizoplane, ×60; **C:** *Herbaspirillum seropedicae*, endophytic, ×60; **D:** *Azotobacter vinelandii*, rhizoplane, ×900; E: *Azorhizobium caulinodans*, endophytic, ×110; **F:** *Azospirillum brasilense* Sp7, rhizoplane, ×60; **G:** *A. brasilense* Sp7-S, endophytic, ×60; **H:** *Derxia gummosa*, rhizoplane overlying uninfected *para*-nodule, ×70. Plants were grown hydroponically for 10 days after inoculation (Zeman *et al.*, 1992) and stained with X-gal overnight (Arsène *et al.*, 1994).

laboratory studies have shown that there are very distinct and reproducible species and strain differences in the patterns of colonization in hydroponic culture of wheat seedlings. As a consequence there are also differences in the effectiveness with which N_2 fixation can be carried out by inoculated seedlings. The results obtained with different diazotrophs, and the extent to which high activities are associated with an endophytic mode of colonization, are examined below.

Azospirillum brasilense *Sp7 versus* Sp7-S

As discussed in the Introduction, distinct differences in colonization pattern can be observed among different strains of *A. brasilense* and between some mutant and wild-type strains. In the case of most strains there is also a stimulation in the rate measured in the ARA when synthetic auxins are added, independently of colonization pattern. It has been observed that *A. brasilense* Sp7-S, a stable mutant of Sp7 with reduced expression of exopolysaccharides and superior nitrogen fixation rates in association with wheat (Katupitiya *et al.*, 1995a) colonize wheat roots as an endophyte, preferring crevices at the points of emergence of lateral roots, intercellular spaces in the basal zone of 2,4-D induced *para*-nodules for colonization and intercellular channels apparently extending from these points of crack entry (see Figure 1G). This confirms earlier data showing intercellular colonization of wheat roots (Zeman *et al.*, 1992), obtained with much less ease using electron microscopy and with laser confocal microscopy (Sriskandarajah *et al.*, 1993). Using *nifA–lacZ* staining of bacteria with X-gal, the stain appears to be very well associated with the location of individual bacteria; bacterial cells located on the root surface can be focused under high magnification during light microscopy, whereas bacterial cells occurring internally cannot readily be focused without clearing of the plant tissue. An internal location was confirmed by sectioning of wheat roots, confirming their endophytic location, although a significant proportion of these Sp7-S cells lie close to the root surface in the crevices surrounding lateral roots or *para*-nodules.

This internal pattern of colonization by Sp7-S contrasts with the extensive colonization of the root surface exhibited by the wild-type Sp7 (see Figure 1F). Almost as many cells as Sp7-S as of Sp7 are observed on wheat seedlings (Katupitiya *et al.*, 1995a), stimulated in numbers significantly about 5–10-fold by 2,4-D, but their distribution contrasts strongly.

In addition, morphological evidence of encystment of Sp7 within days of colonization of the root surface, is shown by the ovoid shape of the cells, compared to the vibrioid shape of cells of Sp7-S that apparently remain vegetative (Katupitiya *et al.*, 1995a, Kennedy *et al.*, 1997 and Figure 1). Elsewhere (Kennedy *et al.*, 1997, Deaker and Kennedy, unpublished), we show (using enzymatic analysis of *nifH–lacZ* expression, which reports N_2-fixing conditions) that strain Sp7-S shows greater potential for nitrogenase activity over a higher range of oxygen concentration on wheat roots, consistent with similar results obtained previously using ARA (Kennedy and Tchan, 1992).

A previous study (Yu *et al.*, 1993) has verified that the increased rates of ethylene production from added acetylene observed with this system is due to nitrogen fixation, by a range of techniques including the use of inhibitors and $^{15}N_2$. Some more recent $^{15}N_2$ data obtained under better conditions of assay demonstrate the significance of the nitrogenase activity observed with Sp7-S in wheat seedlings. The ^{15}N enrichments observed in Table 2 indicate that a significant proportion of the cells of *A. brasilense* Sp7-S colonizing the wheat root system (*ca* 10^8 organisms) were fixing N_2.

Azospirillum lipoferum

Under the experimental conditions described, three strains of *A. lipoferum* (Sp59b), SpBr17 and 596 – the latter an isolate from Australian wheat soils, (New and Kennedy, 1989) were found to be more effective in ARA activity than several other strains of *A. brasilense* (Sp7, Cd, Sp245, Sp13, SpBr14, 576, Sp107) (Katupitiya *et al.*, 1995b). Some characteristic data are shown in Table 2. Note that, in

Table 2. Nitrogen fixation by diazotrophs with *para*-nodulated wheat

Treatment	ARA activity (nmol/h per seedling)		^{15}N-enrichment	
	−2,4-D	+2,4-D	−2,4-D	+2,4-D
Control seedlings (-bacteria)	0[1]	0[1]	0[1]	0[1]
Acetobacter diazotrophicus	0[1]	0[1]	nd	
Azorhizobium caulinodans ORS571 (ß-galactosidase activity ng[5]. per seedling Miller units/min per seedling)	0.43±0.14[5] (5.4±)[1]	7.88±1.03[5] (12.1±)[1]	11.3±6.5	168.0±27.9 ng[5]
Azospirillum brasilense				
Sp7-S	0–5[1,2]	5–70[1,2]	–	0.487 atom%xs[1]
Sp7	0–3[2]	2–10[2]		
Sp245	3.5±1.5[3]	13.6±5.1[3]	nd	
Cd	4.4±1.2[3]	5.0±1.5[3]	nd	
Azospirillum lipoferum			nd	
Sp59b	14.2±5.2[3]	19.8±6.5[3]		
SpBr17	11.5±4.1[3]	38.5±9.7[3]		
596	3.7±1.4[3]	37.9±9.6[3]		
Azotobacter vinelandii	8.8±5.1[1]	22.7±9.3[1]		
Derxia gummosa	1.0±0.1[4]	18.8±8.5[4]		
Herbaspirillum seropedicae	0.4±0.2[1]	31.7±15.1[1]	–	
Rhizobium spp.	0[6]	0[6]		

ARA rates are means of five to eight replicate seedlings for root systems of about 100 mg fresh weight, with standard errors usually up to half the mean values as shown. For $^{15}N_2$ incorporation, seedlings were incubated separately in a gas mixture of 76% Ar, 10% $^{15}N_2$ (*ca.* 50 atom% ^{15}N) and 2.5% O_2 at 30°C for 36 h; some data shown are taken from previous publications from this laboratory ([1]this publication, [2]Sriskandarajah *et al.*, 1993, [3]Katupitiya *et al.*, 1995b, [4]Tchan and Zeman, 1995, [5]Yu and Kennedy, 1995, [6]Nie *et al.*, 1992).

contrast to the case with *A. brasilense* Sp7-S, stimulation by treating seedlings with 2,4-D was much less marked and probably not statistically significant except for strain 596. The most successful strains of *A. lipoferum* in ARA activity examined here were heavy colonizers of the root surface rather than endophytes. Apparently, the high activities resulted from oxygen restriction in possibly gummy surface layers (see Katupitiya *et al.*, 1995b). *NifA-lacZ* transconjugants of these strains (SpBr17, Sp59b and 596) were found to be very heavy colonizers of the root surface, even more prolific than *A. brasilense* Sp7. SpBr17 is naturally Congo red binding and flocculates in ß-hydroxybutyrate medium (Bleakley *et al.*, 1989) but not in fructrose. Both this strain of *A. lipoferum* and *A. brasilense* Sp7-S can be complemented for flocculation in fructose medium with the same DNA fragment originated from a gene bank of strain Sp7 (Pereg-Gerk *et al.*, 1997).

Azorhizobium caulinodans ORS 571

This diazotroph is the only example among those studied here capable of effectively nodulating a legume species (*Sesbania rostrata*). However, unlike most rhizobia, *A. caulinodans* is also capable of fixing N_2 independently of nodules. There are previous reports of its capacity to be incorporated in *para*-nodules and to express nitrogenase activity there (Chen *et al.*, 1992; Yu *et al.*, 1993). In fact, the nitrogenase activity by this species associated with wheat, even with added 2,4-D, is significantly less than with *A. brasilense* Sp7-S.

The plasmid pAB576 was estimated to be transferred to *A. caulinodans* at a frequency of about 10^{-3}, similar to that observed with all the other strains. The transconjugants were stable, as culture of transconjugants on media lacking Tc does not lead to a loss of expression of ß-galactosidase activity, even after many transfers. This probably results from the stability of the original plasmid pVK100 in a broad range of hosts.

There was a good expression of ß-galactosidase in *A. caulinodans* transconjugants (2267 Miller units min^{-1} mg $protein^{-1}$), similar to the activity of NifA-lacZ observed in *A. brasilense* (Arsène *et al.*, 1994; Katupitiya *et al.*, 1995a,b). The application of these *nifA–lacZ* fusions to demonstrate the localization of individual colonies and cells of *A. caulinodans* stained with X-gal showed that the most important reason for a lower ARA activity in the case of *A. caulinodans* (typically mean values of only 3–5 nmol $seedling^{-1}$ h^{-1}; Table 2) was probably the low rate of colonization of *para*-nodules compared to that by *A. brasilense* Sp7-S (typically mean values of 10–50 nmol C_4H_2 $seedling^{-1}$ h^{-1}).

Despite being an endophytic colonizer rather than a root surface colonizer (Figure 1E), *A. caulinodans* apparently has more difficulty in obtaining crack entry or multiplication in the wheat root interior. Only 20% of the lateral root structures induced by 2,4-D treatment were found to be colonized, compared to almost 100% in the case of *A. brasilense* Sp7-S. The nitrogenase activity of *para*-nodulated wheat and *nifA–lacZ*-induced ß-galactosidase activity were shown to be positively correlated, and both were stimulated by the addition of 2,4-D to inoculated seelings. Judging from the greater stimulation of ARA than ß-galactosidase activity by 2,4-D shown in Table 2, factors other than improved colonization may be involved, such as protection from oxygen or availability of carbon substrates.

As is the case with *A. brasilense* (Katupitiya *et al.*, 1994), blue-coloured cells stained with X-gal were sometimes observed within root hair cells. In these cases the infection is obviously intracellular, and intense staining of the basal root hair cell is observed, indicating that the normal cell contents have been disrupted. It seems unlikely that prior infection of root hairs is associated with endophytic colonization of the root, as is the case with many rhizobia, since no association between root hair infection and successful colonization of the basal cell zone of *para*-nodules was observed. Such root hair infection was also observed with *Azotobacter vinelandii*, suggesting that microbial infection is not species-specific, but the occurrence of root hair infection appears to be increased by 2,4-D treatment.

Azotobacter vinelandii

A. vinelandii, the well-known free-living soil diazotroph, is not considered to be a plant associative bacterium. The sole member of the genus isolated primarily from the roots of grasses is *A. paspali* (Döbereiner, 1974). The reported tolerance of the nitrogenase activity of *Azotobacter* to oxygen as a result of respiratory protection (Yates, 1988) was of particular interest for a possible association with wheat.

Tests showed that *A. vinelandii* had peak nitrogenase activities in pure culture (*ca.* 70 nmol min^{-1} mg $protein^{-1}$) with an atmosphere of 21% oxygen, although the activity at 5% oxygen was 60–80% as high (data not shown). 2,4-D (0–1.0 ppm) was shown to have no effect on either growth or nitrogenase activities of both strains in pure culture.

When *Azotobacter* was inoculated onto wheat seedlings, significant colonization of the root surface was observed using *nifA–lacZ* transconjugant strains and X-gal staining of individual cells (Figure 1D). As noted above, root hair cells were also occasionally infected, probably leading to the death of the underlying epidermal plant cell. In contrast to inoculation with *A. brasilense* Sp7-S, there was no noticeable endophytic colonization associated with *para*-nodules by *A. vinelandii* strains. Nor was there a consistent pattern of stimulation of nitrogenase activity from adding 2,4-D to the seedling culture solution, although some stimulation appears in Table 2, similar to the results obtained with *Azospirillum lipoferum*. This result is consistent with the failure of both species to colonize the endophytic niche produed by 2,4-D treatment and occupied by *A. brasilense* Sp7-S.

Significant nitrogenase activity on wheat plants inoculated with *A. vinelandii* was also observed (Table 2), similar to the maximum activities observed with *A. brasilense* Sp7-S and equivalent to about 1 000–2 000 nmol g^{-1} d.w. roots h^{-1}. It was of interest that the peak activity of nitrogenase activity was in the range 2–5% oxygen rather than reaching a maximum in air (data not shown) as observed with broth cultures well supplied with sugar as a carbon substrate. This suggests that the carbon-limited *Azotobacter* colonizing the root surface of wheat seedlings was unable to effectively operate respiratory protection of nitrogenase.

Derxia gummosa

Although previous work from this laboratory (Tchan and Zeman, 1995) had suggested that *D. gummosa* might colonize *para*-nodules of wheat, we were unable to confirm this using *nifA–lacZ* staining (see Figures 1A, 1H). On the contrary, the bacterial cells were found located exclusively on the root surface and clearly not colonizing *para*-nodules (Figure 1H), although colonies sometimes formed membranous sheet-like clumps overlying these structures (data not shown). Apparently the tolerance to oxygen shown by *D. gummosa* in association with wheat (Tchan and Zeman, 1995) results from the aggregation of cells in these surface colonies (Figure 1A), probably assisted by the strong gum production associated with this diazotroph (Hill and Postgate, 1969; Ravi Shankar *et al.*, 1986).

Herbaspirillum seropedicae

When tested in the same system employed for azospirilla, *H. seropedicae* was found to perform in a very similar fashion to *A. brasilense* Sp7-S. According to J. Döbereiner (personal communication), this organism is an obligate endophyte. Using *nifA–lacZ* as a marker we confirmed this conclusion (Figure 1C). No X-gal stained bacterial cells were observed in focus on the root surface of wheat seedlings and all the stained cells observed occurred endophytically. As mentioned above, unless root tissues are cleared by further treatment such as hypochlorite after *lacZ* staining with X-gal, it is relatively simple to decide if bacterial cells are endophytic since only surface cells can be focused. *Herbaspirillum* was a prolific endophytic colonizer, particularly in the basal zone of *para*-nodules similar to *A. brasilense* Sp7-S (Figure 1C) and the ARA activity observed, based on a limited number of observations, was probably comparable to that observed with Sp7-S (Table 2). *H. seropedicae* should be considered as a good candidate for further studies with this system.

Acetobacter diazotrophicus

Unlike *H. seropedicae, A. diazotrophicus* was only observed clumped on the wheat root surface following inoculation of seedlings (Figure 1B), without any evidence of endophytic colonization. The very high sugar requirement of this organism may prevent significant colonization, and it is possible that many of the bacterial cells observed on the root surface were provided from the inoculum. In the limited number of treated wheat seedlings tested, no ARA activity was observed (Table 2), consistent with the superficial location of bacteria on the root surface more exposed to oxygen and the restricted levels of sugars available.

Discussion

Key factors in colonization of para-nodules by A. brasilense Sp7-S

A number of factors have been recognized as of significance in the *para*-nodule and in adjacent cortical tissues. These include a fortunate spontaneous mutation of *A. brasilense* Sp7, resulting in the selection of a new strain (Sp7-S) showing inability of colonies to bind Congo red and the loss of flocculating ability (Katupitiya *et al.*, 1995a) from lowered exopolysaccaride (EPS) production (*flcA*$^-$) and some loss of motility (Pereg-Gerk *et al.*, 1996, 1997). This loss of EPS production was accompanied by loss of the ability of the bacterial cells to adhere to the wheat root surface when seedlings were inoculated. Recent work on other effects of synthetic auxin treatment on wheat seedlings (Feng, Copeland and Kennedy, unpublished) indicates that there is improved availability of carbon substrates. Our studies have shown that fructose, glucose and sucrose are the main soluble carbon substrates in wheat roots, with some organic acids. Sugars but not organic acids were increased by a factor of two in 2,4-D-treated seedlings. Preliminary results with a glucose-utilizing mutant of Sp7-S (Sp7-Sg) suggested that colonization of channels between cortical cells of wheat roots was more extensive with this mutant (Feng *et al.*, unpublished).

Accompanying the loss of EPS production there was an increased ability of Sp7-S to colonize wheat roots internally (endophytically), in an intercellular fashion, particularly in crevices at the point of emergence of lateral roots enabling crack entry, between cells in the basal region of the *para*-nodules and penetrating interstitially in channels between cortical cells. A simple interpretation of this is that the absence of EPS production provided a smooth-surfaced bacterial cell with lowered power of adhesion to the cell surface of plants. The stickiness of the wild-type Sp7 may tend to trap these cells on root surfaces, inhibiting crack-entry of cells into crevices at the root surface and near the points of emergence of lateral roots or *para*-nodules. The presence of such points of crack entry in the more loosely structured monocots such as wheat was pointed out previously (Kennedy and Tchan, 1992).

The absence of encystment and retention of the vegetative phenotype at all times by Sp7-S both in pure culture and on wheat roots is of importance. *A. brasilense* Sp7 cells adhering to the surface of wheat roots encyst within several days. This may be detrimental to the achievement of significant N_2 fixation, compared to the extended period of vegetative growth observed with Sp7-S. Encystment in pure culture occurs in aged cultures, after the phase of rapid cell division, and coincides with lessened nitrogenase expression (Papen and Werner, 1982). The *flcA* gene and EPS formation is associated with encystment (*flcA*$^-$ cells do not encyst) (Pereg-Gerk *et al.*, 1998). By contrast, Sp7-S in wheat roots may continue to colonize endophytically indefinitely if a suitable niche for its growth exists. Whether the ability to fix N_2 is also an advantage in endophytic colonization has not been established, and *nif*$^-$ mutants of Sp7-S could be instructive regarding an answer to this question.

Tn5-*flcA*$^-$ mutants also have reduced surface colonization of wheat roots, similar to Sp7-S, but it was observed that Sp7-S lacked the ability to swarm, probably as a result of the absence of lateral flagellae (Pereg-Gerk *et al.*, 1996, 1998). Our current interpretation of these findings is that the wild-type *A. brasilense* Sp7 displays features in this system primarily adapted to enhancing its capacity for survival. Binding to the root surface using both polar flagellae and exopolysaccarides (Bashan *et al.*, 1986; Michiels *et al.*, 1991) provides a strategic localization for obtaining some substrates for growth and maintenance, but not for protection of cells from oxygen to allow full expression of nitrogenase activity, since *nifH–lacZ* is incompletely expressed on the root surface. Encystment follows within several days, possibly responding to the relatively poor conditions for growth and consistent with cellular survival; one might conclude that azospirilla do not fix N_2 altruistically.

By contrast, Sp7-S is unable to firmly attach to the root surface and to produce sticky exopolysaccarides (Pereg *et al.*, 1995). Inability to adhere to the root surface may allow cells of this mutant to enter crevices and to penetrate more readily into internal spaces where cell division can continue in response to elevated levels of carbon (Feng *et al.*, unpublished) and nitrogenous substrates (Yu and Kennedy, unpublished) following 2,4-D treatment. Colonization of the root surface in the rhizosphere may give Sp7 an advantage in access to energy substrates by close anchoring to the root surface.

Microscopical observations also suggest that the synthetic auxin treatment loosens the adhesions between plant cells, possibly allowing Sp7-S to divide more freely and reducing the physical work of cell division in colonies established in the intercellular spaces. All these features may contribute to the establishment of a protected endophytic niche for Sp7-S, in which oxygen is restricted because respiratory consumption by colonizing bacteria exceeds diffusion and where energy substrates are more freely available than at the root surface. As a consequence of these factors favouring endophytic colonization and the absence of encystment, rates of nitrogen fixation (ARA) an order of magnitude higher are observed with Sp7-S, even though the total number of viable cells (colony-forming units, cfu) of Sp7 in the association with wheat seedlings treated with 2,4-D (10^7-10^8 cells) are slightly greater (Katupitiya *et al.*, 1995a).

Comparisons between diazotrophs as endosymbionts

The possible advantages of the endophytic mode of root colonization were discussed by Quispel (1991). However, most of the strains of free-living diazotrophs tested for colonization pattern in this study were found to occur at the rhizoplane (see Table 3 for summary). Only *A. brasilense* Sp7-S, *H. seropedicae* and *A. caulinodans* were found to colonize predominantly in an endophytic pattern in *para*-nodulated wheat seedlings. For a model of effective N_2-fixing endosymbiosis these strains therefore appear to be the best candidates for application in a practical system, although *A. caulinodans* colonized relatively poorly. Recent findings that flavonoids such as naringenin can stimulate the extent of crack entry by plant roots by *A. caulinodans* (Gough *et al.*, 1996) may indicate a means of overcoming this poorer colonization.

In terms of rates of nitrogenase activity the response of *A. brasilense* Sp7-S and *H. seropedicae* was strongest, although some strains of *A. lipoferum*, *A. vinelandii* and *D. gummosa* also produced significant activities. The greater insensitivity of the latter two diazotrophs to effects of oxygen on the rate of nitrogenase activity was also of interest. However, all of these latter strains colonized at the rhizoplane. Thus, in terms of the model criteria for effective endosymbiosis, that of an endophytic character, these diazotrophs fail the test. It is anticipated that the effects of competition for energy substrates with other rhizosphere organisms, greater variability in exposure to oxygen conditions and inefficiency in nitrogen transfer to the host plant will reduce the effectiveness of diazotrophs colonizing at the rhizoplane.

Despite this, some of the strains colonizing at the rhizoplane could still prove useful in a future effective symbiosis by using mixed inocula, acting as a back-up to an endophyte playing the main role in the symbiosis, augmenting the benefits obtained from the endophyte.

Facilitating evolution of a N_2-fixing symbiosis between a diazotroph and cereals

In our research programme the laboratory model afforded by the 2,4-D induced *para*-nodule is directed to a step-wide process of *facilitated evolution* in which

Table 3. Colonization pattern by different strains of diazotrophs in *para*-nodulated wheat

Strain	Predominant colonization mode	Oxygen sensitivity of nitrogenase in *para*-nodulated wheat
Azospirillum brasilense Sp7	Rhizoplane	High
A. brasilense Sp7-S	Endophytic	Medium
A. lipoferum	Rhizoplane	Medium
Azorhizobium caulinodans	Endophytic	Medium
Acetobacter diazotrophicus	Rhizoplane	–
Azotobacter vinelandii	Rhizoplane	Low
Derxia gummosa	Rhizoplane	Low
Herbaspirillum seropedicae	Endophytic	Medium
Rhizobium spp.	Endophytic	–

Wheat seedlings were treated with 2,4-D and inoculated with diazotrophs as previously described (Zeman *et al.*, 1993). Colonization mode was determined by light microscopy using *nifA–lacZ* transconjugants stained with X-gal (Arsène *et al.*, 1994).

the end point is an effective endosymbiotic system for N_2 fixation in cereals. In joint research work with the Institut Pasteur, considerable progress has now been made in defining the colonization phenotype (endophyte versus rhizoplane) and in detailed genetic information related to control of colonization (Pereg-Gerk *et al.*, 1997) by azospirilla (*flcA*). Using *A. brasilense* in the research programme has been attractive because of the strong background of genetic information available for this organism, and the availability of specific genetic tools to visualize and quantitate the results of modifications in the system and to study the mode of colonization. This is obviously essential for success in this stepwise process, that has the ultimate goal of providing a thermodynamically spontaneous system for conditions of monoculture of cereals such as wheat. Ideally, such monoculture of cereals would also have lower input costs than current systems using chemical fertilizers.

Although the use of 2,4-D, even in a concentration range selected to ensure it acts as a synthetic auxin without serious damage to wheat seedlings, is artificial, an effective procedure was needed to allow sufficient improvement in colonization by crack entry and N_2 fixation rates for key factors to be recognized and quantified. We envisage it will prove possible later to simulate 2,4-D treatment biologically (e.g. by localized indoleacetic acid production). However, it should be recognized that some systems of agricultural broadacre monoculture already employ very stringent chemical treatments in farming practice (e.g. cotton production using phytohormones to alter plant architecture, defoliants and many pesticides). Obtaining effective N_2 fixation in associations with wheat is an ambitious task that we suggest now demands this stepwise process of *facilitated evolution*, aiming to yield a spontaneous symbiosis effective in N_2 fixation. In principle the success so far in apparently obtaining an endophytic strain of *A. brasilense* (Sp7-S) as one of the key steps in this process and the subsequent development of new genetic tools (using transposon mutagenesis) possibly enabling control of the mode of colonization (see Pereg-Gerk *et al.*, 1997) may open the way to allowing other desirable genetic traits to be recombined with the endophytic mode of colonization. Ideally, selection of the most desirable strains should be carried out using a laboratory model such as the *para*-nodule, where effectiveness may be assessed in the plant system under conditions approaching endosymbiosis.

Acknowledgments

This work was supported by grants from the Australian Research Council and the Australian Grains Research and Development Corporation. Daiguan Yu would also like to acknowledge support from AusAID. We are grateful to Johanna Döbereiner and Christina Kennedy for providing microbial cultures.

References

Arsène F, Katupitiya S, Kennedy I R, Elmerich C 1994. Use of *lacZ* fusions to study the expression of *nif* genes of *Azospirillum brasilense* in association with plants. Mol. Plant-Microbe Int., 7, 748–757.

Baldani V, Baldani J, Döbereiner J 1983. Effects of *Azospirillum* inoculation on root infection and nitrogen incorporation in wheat. Can. J. Microbiol. 29, 924–929.

Baldani J I, Baldani V L D, Seldin L, Döbereiner J 1986. Characterization of *Herbaspirillum seropedicae* gen. nov., sp. nov., a root-associated nitrogen-fixing bacterium. Int. J. Syst. Bacteriol. 36, 86–93.

Bali A, Blanco G, Hill S, Kennedy C 1992. Excretion of ammonium by a *nifL* mutant of *Azotobacter vinelandii* fixing nitrogen. Appl. Environ. Microbiol., 58, 1711–1718.

Bashan Y, Levanony H, Klein E 1986. Evidence for a weak active external adsoption of *Azospirillum brasilense* Cd to wheat roots. J. Gen. Microbiol. 132, 3069–3073.

Bleakley B H, Gaskins M H, Hubbel D H, Zam S G 1988. Floc formation by *Azospirillum lipoferum* grown on poly-ß-hydroxybutyrate. Appl. Environ. Microbiol. 54, 2986–2995.

Cavalcante V A, Döbereiner J 1988. A new acid-tolerant nitrogen-fixing bacterium associated with sugarcane. Plant Soil, 108, 23–31.

Chen T W, Scherer S, and Böger P 1992. Nitrogen fixation of *Azorhizobium* in artificially induced root para-nodules in wheat. Sci. China (B), 35, 1463–1470.

Cocking E C, Davey M R, Kothari S L, Srivastava J S, Jing Y, Ridge R W, Rolfe B G 1992. Altering the specificity control of the interaction between rhizobia and plants. Symbiosis, 14, 123–130.

de Bruijn F J, Jing Y, Dazzo F B 1995. Potential and pitfalls of trying to extend symbiotic interactions of nitrogen-fixing organisms to presently non-nodulated plants, such as rice. Plant Soil, 174, 225–240.

Döbereiner J 1974. Nitrogen fixing bacteria in the rhizosphere. In: Quispel A, ed., The Biology of Nitrogen Fixation. North-Holland, Amsterdam, 86–120.

Dreyfus, B L, Elmerich C, Dommergues Y R 1983. Free-living *Rhizobium* strain able to grow under nitrogen as sole nitrogen source. Appl. Environ. Microbiol. 45, 711–713.

Eskew D L, Focht D D, Ting I P 1977. Nitrogen fixation, denitrification and pleomorphic growth in a highly pigmented *Spirillum lipoferum*. Appl. Environ. Microbiol. 34, 582–585.

Gough C, Webster G, Vasse J, Galera C, Batchelor C, O'Callaghan K, Davey M, Kothari S, Denarie J, Cocking E C 1996. Specific flavonoids stimulate intercellular colonization and non-legumes by *Azorhizobium caulinodans*. In: Stacey G, Mullin B, Gresshoff P M, eds. Biology of Plant-Microbial Interactions International Society for Molecular Microbe Interactions, Minnesota, 409–415.

Hill S, Postgate J R 1969. Failure of putative nitrogen fixing bacteria to fix nitrogen. J. Gen. Microbiol. 58, 277–285.

Katupitiya S, Liu X M, Elmerich C, Kennedy I R 1994. A genetic tool to study colonization of wheat roots by *Azospirillum*. In: Nie Y F, Kennedy J R, Chen T W, eds. International Symposium on Nitrogen Fixation with Non-leguminous Crops. Qingdao Ocean University Press, Qingdao, China, 77–81.

Katupitiya S, Millet J, Vesk M, Viccars L, Zeman A, Zhao L, Elmerich C, Kennedy I R 1995a. A mutant of *Azospirillum brasilense* Sp7 impaired in flocculation with modified colonization and superior nitrogen fixation in association with wheat. Appl. Environ. Microbiol. 61, 1987–1995.

Katupitiya S, New P B, Elmerich C, Kennedy I R 1995b. Improved nitrogen fixation in 2,4-D treated wheat roots associated with *Azospirillum lipoferum*: colonization using reporter genes. Soil Biol. Biochem., 27, 447–452.

Kennedy I R, Tchan Y T 1992. Biological nitrogen fixation in non-leguminous field crops: recent advances. Plant Soil, 141, 93–118.

Kennedy I R, Pereg-Gerk L, Wood C, Deaker R, Gilchrist K, Katupitiya S 1997. Biological nitrogen fixation in non-leguminous field crops: facilitating the evolution of an effective association between *Azospirillum* and wheat. Plant Soil, 194, 65–79.

Liang Y Y, Kaminski P A, Elmerich C 1991. Identification of a *nifA*-like regulatory gene of *Azospirillum brasilense* Sp7 expressed under conditions of nitrogen fixation and in air and ammonia. Mol. Microbiol. 5, 2735–2744.

Michiels K, Croes C L, Vanderleyden J 1991. Two different modes of attachment of *Azospirillum brasilense* Sp7 to wheat roots. J. Gen. Microbiol. 137, 2241–2246.

New P B, Kennedy I R 1989. Regional distribution and pH sensitivity of *Azospirillum* associated with wheat roots in eastern Australia. Microb. Ecol., 17, 299–309.

Nie Y F, Vesk M, Kennedy I R, Sriskandarajah S, Tchan Y T 1992. Structure of 2,4-dichlorophenoxyacetic acid induced *para*-nodules on wheat roots. Phytochem. Life Sci. Adv., 11, 67–73.

Papen M, Werner D 1982. Organic acid utilization, succinate excretion, encystation and oscillating nitrogenase activity in *Azospirillum brasilense* under microaerobic conditions. Arch. Microbiol. 132, 57–61.

Pereg L L, Millet J, Katupitiya S, Kennedy I R, Elmerich C 1995. Genetic analyis of an *Azospirillum brasilense* Sp7 mutant impaired in flocculation. In: Tikhhanovich I A, Romanov V I, Newton W E, eds. Nitrogen Fixation: Fundamentals and Applications. Kluwer, Dordrecht, 345.

Pereg-Gerk L L, Kennedy I R, Elmerich C 1996. Genetic factors controlling colonization of wheat roots by *Azospirillum brasilense* Sp7. Proceedings, 11th Australian Nitrogen Fixation Conference, Perth, WA, 122–123.

Pereg-Gerk L, Paquelin A, Gounon P, Kennedy I R, Elmerich C 1998. A transcriptional regulator of the LuxR-UhpA family, FlcA, controls flocculation and wheat root surface colonization by *Azospirillum brasilense* Sp7. Mol. Plant-Microbe Int. (In press).

Quispel A 1991. A critical evaluation of the prospects for nitrogen fixation with non-legumes. Plant Soil, 137, 1–11.

Ravi Shankar H N, Kennedy I R, New P B 1986. Autotrophic growth and nitrogen fixation in *Derxia gummosa*. J. Gen. Microbiol. 132, 1797–1803.

Sriskandarajah S, Kennedy I R, Yu D, Tchan Y T 1993. Effects of plant growth regulators on acetylene-reducing associations between *Azospirillum brasilense* and wheat. Plant Soil, 153, 165–178.

Tarrand J J, Krieg N R, Döbereiner J 1978. A taxonomic study of the *Spirillum lipoferum* group, with description of a new genus, *Azospirillum* gen. nov. and two species. *Azospirillum lipoferum* (Beijerinck) comb. nov. and *Azospirillum brasilense* sp. nov. Can. J. Microbiol. 24, 967–980.

Tchan Y T, Zeman A M M 1995. N_2 fixation (C_2H_2 reduction) in 2,4-diochlorophenoxyacetic acid (2,4-D) treated wheat inoculated with free-living diazotrophs. Soil Biol. Biochem. 27, 453–457.

Yates M G 1988. The role of oxygen and hydrogen in nitrogen fixation. In: Cole J A, Ferguson S J, eds. The Nitrogen and Sulphur Cycles, Proceedings of the 42nd Symposium of the Society for General Microbiology, Southampton, UK, January 1988, 383–416.

Yu D, Kennedy I R 1995. Nitrogenase activity (C_2H_2 reduction) of *Azorhizobium* in 2,4-D-induced root structures of wheat. Soil Biol. Biochem., 27, 459–462.

Yu D, Kennedy I R, Tchan Y T 1993. Verification of nitrogenase activity (C_2H_2 reduction) in *Azospirillum* populated 2,4-dichloracetic acid induced root structures of wheat. Austral. J. Plant Physiol. 20, 187–195.

Zeman A M M, Tchan Y T, Elmerich C, Kennedy I R 1992. Nitrogenase active wheat-root *para*-nodules formed by 2,4-dichlorophenoxyacetic acid (2,4-D)/*Azospirillum*. Res. Microb. 143, 847–855.

15

Transgenic rice plants with *Parasponia* haemoglobin gene and its expression

WEI GUO, SHAN XUEQIN, DING QUNXING, LIU BOLIN AND JING YUXIANG

Abstract

We transformed rice with *Parasponia* haemoglobin gene in order to create a low oxygen environment for protecting nitrogenase from oxygen damage if diazotrophs including *Rhizobium* strains could be penetrated into rice. Pollen-tube pathway and particle gun methods were used to transform the genomic gene of plant haemoglobin into rice. A total of 1512 florets were treated with pL305 DNA by using the pollen-tube pathway method and 707 seeds were obtained. After the germination, 206 out of 230 seeds became plantlets. DNA dot blotting showed that six out of 120 plantlets had positive results. Southern blotting was carried out with four of the six plantlets which showed that three plants had positive results. So the final transformation rate was 2.5% of 120 plantlets. Five hundred and five florets were treated with pLX412-Hb DNA containing T-DNA border sequence using the same method as above and 340 seeds were obtained. Southern blotting confirmed that the final transformation rate is 6.7%. Transformed rice immature embryos or calli with pLX412-Hb DNA by the particle-gun method were selected under hygromycin condition and 15 regeneration plantlets were obtained. DNA dot blotting showed that 14 plantlets gave a positive reaction. Seven of the plantlets were confirmed by Southern blotting and all of them had positive results. So the final transformation rate was 93% of the 15 plantlets.

Total RNA was extracted from roots and leaves of the plants which were confirmed to contain the *Parasponia* haemoglobin gene by Southern blotting. The positive dots were shown in RNA dot-blot of root RNA probed with the *Parasponia* haemoglobin gene cDNA, while the negative result was gained in RNA dot-blots of leaf RNA. This result preliminarily indicates that the *Parasponia* haemoglobin gene could be expressed in an organ-specific manner in the transgenic rice plants and similar regulation factor(s) for the haemoglobin gene expression would be present in rice.

K.A. Malik et al. (eds.), Nitrogen Fixation with Non-Legumes, 125–131.

Introduction

Leghaemoglobins in nitrogen-fixing nodules of leguminous plants are present to protect nitrogenase from oxygen damage. *Parasponia* is the only known non-legume genus which could be nodulated by *Rhizobium* strains and fix nitrogen in nature.

Haemoglobin in *Parasponia* nodules plays the same role in nitrogen fixation as in that of the leguminous nodules. Leghaemoglobin genes could be expressed only in the nodules of non-transgenic and transgenic legumes. However, the *Parasponia* haemoglobin gene could be transferred into tobacco, in which it could be expressed. We chose the *Parasponia* haemoglobin gene to perform transformation of rice. Its purpose is that, if the rice could be nodulated, the expression of the *Parasponia* haemoglobin gene at translational level may make an environment of low oxygen concentration beneficial to nitrogen fixation in rice nodules.

The pollen-tube pathway method and the particle-gun method were used to transform the *Parasponia* haemoglobin gene into rice, and transgenic rice plants were obtained. Southern blotting showed that the two methods were able to transform the exogenous gene into rice. RNA dot-blotting gave a preliminary demonstration that the *Parasponia* haemoglobin gene could be expressed especially in roots, not in leaves, of the transgenic rice plants.

Materials and methods

Plasmids and rice variety

pL305 containing *Parasponia andersonii* haemoglobin gene, pLX412-Hb, which is a binary transformation vector, and pL209 containing the haemoglobin gene cDNA were presented by Dr W.J. Peacock. *Oryza sativa* L. is a japonicum variety Zhonghua No. 8, obtained by the haploid breeding method and widely used in production, which was presented by Professor M.F. Li, Institute of Crop Breeding and Cultivation, Chinese Academy of Agriculture, Beijing 100080, PR China.

Tissue culture media

The NBD medium for callus generation was the NB medium supplemented with 2,4-D 2 mg/L and the NBBN medium for plant regeneration also was the NB medium supplemented with BAP 3 mg/L and NAA 0.5 mg/L.

Extraction and purification of plasmid DNA

Plasmid DNA was extracted as described by Sambrook *et al.* (1989), and purified according to the *Technical Bulletin* of the Wizard™ Miniprep DNA Purification System. *Parasponia* haemoglobin gene as probe was obtained after pL305 DNA was cut by BamH1 enzyme.

Table 1. Numbers of florets manipulated and seeds obtained from each ear via the pollen-tube pathway method to transform rice with the pL305 DNA

Florets/ear	29	31	17	30	41	26	20	34	41	30	46	23
Seeds/ear	24	19	10	5	26	25	12	11	33	28	26	10
Florets/ear	59	34	47	42	49	38	45	20	20	19	50	32
Seeds/ear	33	15	32	21	34	14	27	10	14	19	43	31
Florets/ear	46	80	42	36	35	60	34	23	28	19	31	
Seeds/ear	21	8	15	3	14	27	12	21	21	14	27	

Plant transformation

The most palea of a floret was diagonally cut away from outside to inside of the palea together with stigma and two-thirds to one-half of the style and the whole lemma was maintained. The particle-gun method was also used.

Extraction of DNA and RNA from transgenic rice plants

For extraction and purification of DNA and RNA, dot-blotting and Southern blotting, methods described by Sambrook *et al.*, (1989), were followed. For labelling and detection of DNA the instruction of the DIG DNA Labeling and Detection Kit were followed.

Results and discussion

Transformation of rice by using the pollen-tube pathway method

Altogether, 1267 florets of 35 rice ears were manipulated by dropping a 2–5 μl solution with the pL305 DNA of 0.5–1.0 μg/μl and 707 seeds were finally obtained. Seed set percentage was 56.3%; 36 florets per ear averaged 20 seeds. The numbers of florets and seeds obtained from the individual ear are listed in Table 1.

Numbers of rice florets manipulated and seeds obtained

Five hundred and eight florets of 17 rice ears were totally treated by dropping 2–5 μl of solution containing the pLX412-Hb DNA 0.5–1.0μg/μl and 340 seeds were obtained with the seed set percentage 58%. Each ear averaged 34 florets manipulated and 20 seeds obtained. The numbers of florets and seeds gained from the individual ear are listed in Table 2.

Rice seed germination test

Two hundred and thirty randomly selected seeds obtained from the treatment with the pL305 DNA via the pollen-tube pathway method were used for germination. The germinated seeds totalled 220, with a frequency of 96%; among them

Table 2. Numbers of florets manipulated and seeds obtained from each ear via the pollen-tube pathway method to transform rice with the pLX412-Hb DNA

Florets/ear	28	28	31	42	30	34	40	41	55	16	50	48
Seeds/ear	23	19	17	24	26	33	17	21	32	2	27	16
Florets/ear	21	33	27	45	17							
Seeds/ear	17	22	16	23	5							

206 seedlings with a frequency of 90% were gained. One hundred and twenty seeds from treatment with the pLX412-Hb DNA using the same method were also used for the germination; 117 seeds were germinated, with a frequency of 97.5%, and 111 seedlings were obtained with a frequency of 92.5%. Here it should be pointed out that a few albino rice plantlets appeared after germination of the seeds obtained by the pollen-tube pathway method. This phenomenon was not reported before, although other researchers have used the same method in transformation of rice and wheat.

Rice transformation via the particle-gun method

Immature rice embryos were derived from the immature seeds after 10–15 days pollenation. Particle-gun treatment of the embryos or calli from the immature embryos cultured on the NBD medium was performed. Then, they were transferred to NBD medium with the hygromycin, which was successively increased until the optimum concentration (50 mg/L) was achieved. In this process the fresh calli were continuously selected until no brown calli appeared. These fresh calli were transferred to the NBBN medium containing the hygromycin 50μg/L with light 16 h a day at 26°C. Several months later, green buds appeared on the calli, which gradually differentiated into plantlets. The plantlets which were regenerated from the calli with no exogenous DNA transformed were kept as the control. Both the regenerated plantlets above were cultured on NB or MS medium containing hygromycin, the control plantlets died and those with exogenous DNA grew into healthy plants.

DNA hybridization identification of the transgenic rice plants

DNA dot-blotting

DNA from each regenerated plant (1 μg) was hybridized with the *Parasponia* haemoglobin gene probe from pL305. Results are shown in Table 3 and Figure 1.

Southern blotting

Ten micrograms of total DNA cut with BamH1 from the plants, and positively confirmed by dot-blotting, was transferred to nylon membrane and hybridized with the *Parasponia* haemoglobin gene as probe (Table 4; Figure 2).

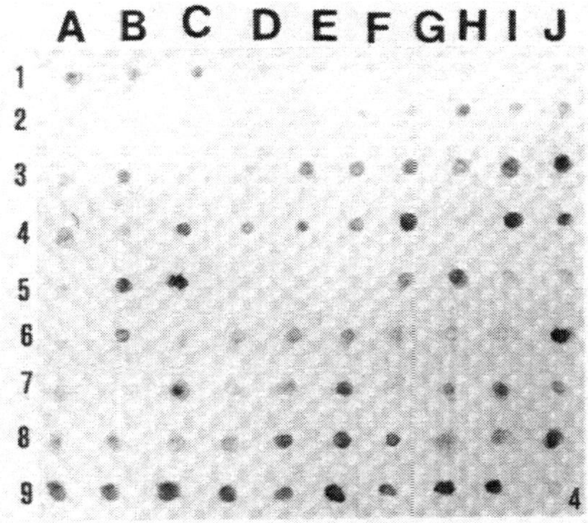

Figure 1. 1A-1H, 2A-7F (except 4C, 5C, 6B, 6J) belong to the plants transformed with the pL305 DNA using the pollen-tube pathway method. 7G-8H are those plants transformed with the PLX412-Hb DNA using the same method. 81–91 are plants transformed with the pLX412–HL DNA using the particle gun method. 1I, 1J are control plants and 4C, 5C, 6B, 6J are the pL305 DNA (10^4–10^5μg) as positive control.

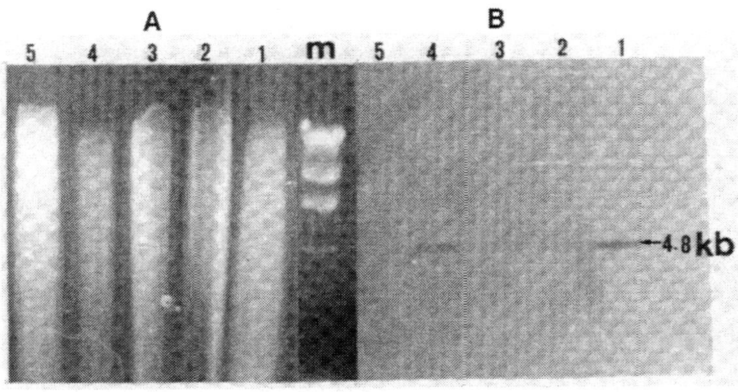

Figure 2. All the plants DNA cut with BamH1 enzyme. A. DNA-agarose gel electrophoresis. B. Southern hybridization. Lane 1, pL305 DNA as positive control; lane 2, DNA from the plant transformed with pL305 by the pollen-tube pathway method; lane 3, DNA from the plant transformed with pLX412-Hb by the pollen-tube pathway method, lane 4, DNA from the plant transformed with pLX412-Hb by the particle-gun method; lane 5, DNA from the non-transgenic plant as negative control.

The results above indicate that the *Parasponia* haemoglobin gene had been transformed into rice and integrated into its nucleus genome. Transformation frequency by the pollen-tube pathway method was 2.5%. If the transformed plasmid were a binary transformation vector, the frequency would be increased to 6.7%. It seems that the transformation rate by using the plasmid containing

Table 3. Identification of transgenic rice plants via dot blotting

Method	Plasmid DNA	No. of plants identified	No. of plants with positive reaction	Percentage
Pollen-tube pathway	pL305	120	6	5
Pollen-tube pathway	pLX412-Hb	30	4	13.3
Particle gun	pLX412-Hb	15	14	93

Table 4. Southern hybridization of DNA from the positive plants identified by dot-blotting

Method	Plasmid DNA	No. of positive plants by dot-blotting	No. of positive plants by Southern blotting	Percentage	Percentage of all plants identified positively
Pollen-tube pathway	pL305	4	3	75	2.5(3/120)
Pollen-tube pathway	pLX412-Hb	4	2	50	6.7 (2/30)
Particle gun	pLX412-Hb	7	7	100	93 (14/15)

T-DNA border sequence could be largely increased by 1.68-fold as compared to that by using the plasmid with no T-DNA border sequence. The transformation rate of the particle-gun method is very high at 93% (Table 3) in our experiment. Although the pollen-tube pathway method is a laborious one, it should still be a good method to use in agricultural practice, if we consider the absence of safety of the transgenic plants with antibiotic resistance reporter genes. We could construct the plasmid with no resistance gene for transformation in practice by using the pollen-tube pathway method.

Preliminary identification of transgenic rice plants at transcriptional level

Total RNA was extracted from roots and leaves of the plants which were positively confirmed by Southern blotting. Dot-blot hybridization with the *Parasponia* haemoglobin gene was performed. This indicated that the leaf RNA showed a negative result, and the roots RNA a positive result (Figure 3).

This also illustrates that the transformed *Parasponia* haemoglobin gene, could be expressed in the manner of organ specificity at the transcriptional level. This result implies that the rice may have the machinery for the expression of the haemoglobin gene, and its own haemoglobin gene.

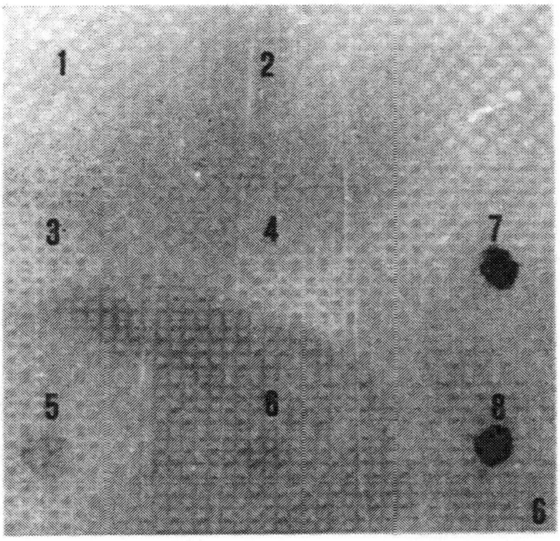

Figure 3. RNA dot-blot. 1, 2, RNA from the roots of non-transgenic rice plants as the negative control; 3, 4, RNA from the leaves of the transgenic rice plants with the *Parasponia* haemoglobin gene; 5, 6, RNA from the roots of the transgenic rice plants with the *Parasponia* haemoglobin gene; 7, DNA from the transgenic rice plant as the positive control; 8, pL305 DNA as the positive control.

Acknowledgements

This project was supported by grants from the National Natural Foundation of China and the Rockefeller Foundation, USA. We thank Mr Meng Zhouyi for technical assistance.

Reference

Sambrook J, Fritsch E F, Maniatis T 1989. Molecular Cloning; A Laboratory Manual, (2nd edn.), Cold Spring Harbor Laboratory Press, Cold Spring Harbor, New York.

16
Cell structure of pseudonodules formed on the roots of barley and rice

SHAN XUEQIN AND JING YUXIANG

Abstract

Two methods were used making *Rhizobium astragali* Huakui and *Rhizobium sesbania* sp. invade the roots of barley and rice, and form pseudonodules on their roots. One was to treat rhizobia and plants by a certain intensity of magnetic field and to inoculate rhizobia to plants. Secondly, a medium containing the extract of rice seedling roots was used to culture rhizobia, which was then inoculated to rice. The cell structures of pseudonodules on barley roots were very fine, with organelles and infection threads from which the rhizobial bacteria were released to cytoplasm. The bacteria enclosed in peribacteroid membranes become the bacteroids. These morphological structures are similar to those of the legume nodules with seemingly symbiotic characteristics, but the bacteroids were vesiculated. In the pseudonodules on rice roots formed by *Rhizobium sesbania* sp. cultured in the medium containing rice extract, the bacteria were distributed in intercellular spaces and cells themselves. The structures of the infected cells were coarse with naked bacteria not enclosed by the peribacteroid membrane, which are completely different from those of the legume nodules. However, no nitrogenase activity was detected, but plant growth was better than the control.

Introduction

Nodulation and nitrogen fixation of *Rhizobium* on roots of non-legumes, especially cereals, is one of the frontiers in the research of biological nitrogen fixation and a hard nut to crack. So far a unique example in nature is that the nodules on the roots of a non-legume genus *Parasponia* have been confirmed to contain rhizobia and fix nitrogen (Trinick, 1973, 1976). Since then a number of

K.A. Malik et al. (eds.), Nitrogen Fixation with Non-Legumes, 133–142.

scientists have made attempts to nodulate non-legumes and fix nitrogen by using *Rhizobium* strains (Al-Mallah *et al.*, 1989, 1990; Chen *et al.*, 1992; Jing *et al.*, 1990), but a final conclusion has not yet been reached. We have used two methods to work on the problem, and found that the rhizobia were able to invade barley and rice, and form pseudonodules on their roots. The fine cell structure of the pseudonodules on barley seems to be similar to those of legumes with seemingly symbiotic characteristics, but those of rice pseudonodules with asymbiotic association are completely different from legume nodules.

Materials and methods

Rhizobial strains and plants

Rhizobium astragali Huakai 38D was presented by the Institute of Plant and Fertilization, Chinese Academy of Agriculture, and *Rhizobium sesbania* sp. was isolated from the nodules of *Sesbania cannabina.* Barley is a local variety and the rice IR36 was presented by Dr E.C. Cocking.

Treatment of seeds and seedlings of barley with rhizobia

Barley seeds were surface-sterilized using conventional methods, and were germinated in sterile Patri dishes at 28°C. Rhizobia were cultured on AMS solid medium. Both the seeds and rhizobia were treated under a magnetic field with 1000 GS intensity at 28°C for 2–4 days (Jing *et al.*, 1990, 1992).

Rice treatments and rhizobia inoculation

After surface sterilization the rice seeds were soaked in sterile water for 3 days with change of water once a day. *Rhizobium sesbania* sp. was cultured on the modified AMS medium supplemented with rice extract from rice seedlings until the logarithm phase, then inoculated onto rice seedlings grown in pots in sterile vermiculite at 28°C with light 14 h and dark 10 h. One month later, observations of the pseudonodules were made.

Results

Cell structures of barley pseudonodules

We performed the experiments with soybean, and found that nodule number and weight, and efficiency of nitrogen fixation, could be increased by 2–3 and 4–5 times, respectively, under a certain magnetic intensity (Jing *et al.*, 1992).

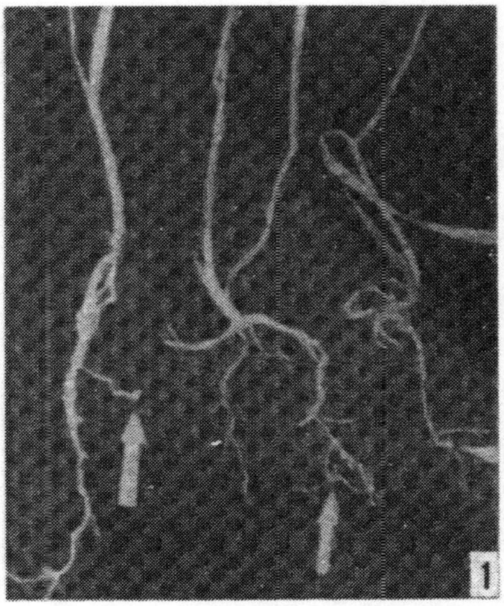

Figure 1. The pseudonodules (arrows) on the roots of barley seedling.

After the inoculation of rhizobia, the white pseudonodules appeared under the magnetic field in about 4–5 weeks (Figure 1). TEM observation indicated the fine structures of infected host cells with clearly seen organelles, such as nuclei and their nucleoli, nuclear membrane, mitochondria, Golgi bodies, tonoplasts, amyloplasts and a large number of ribosomes (Figure 2). A substantial number of rhizobia, each of them surrounded by the peribacteroid membrane, were

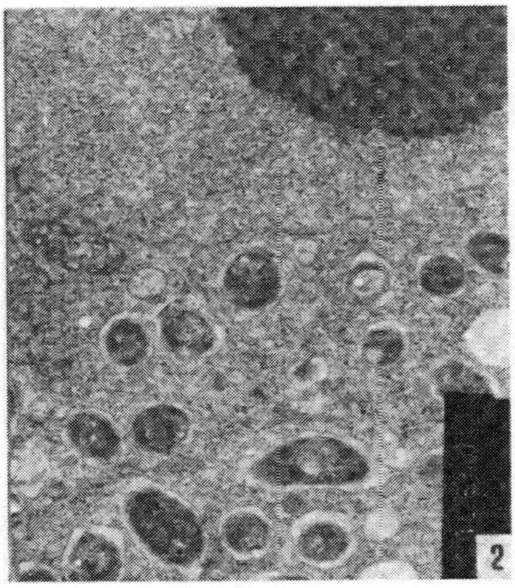

Figure 2. Fine structure of the infected host cell of barley pseudonodule. The nucleus and its nucleolus, nuclear membrane, mitochondria, Golgi body and the bacteroids, are clearly seen (× 10 000).

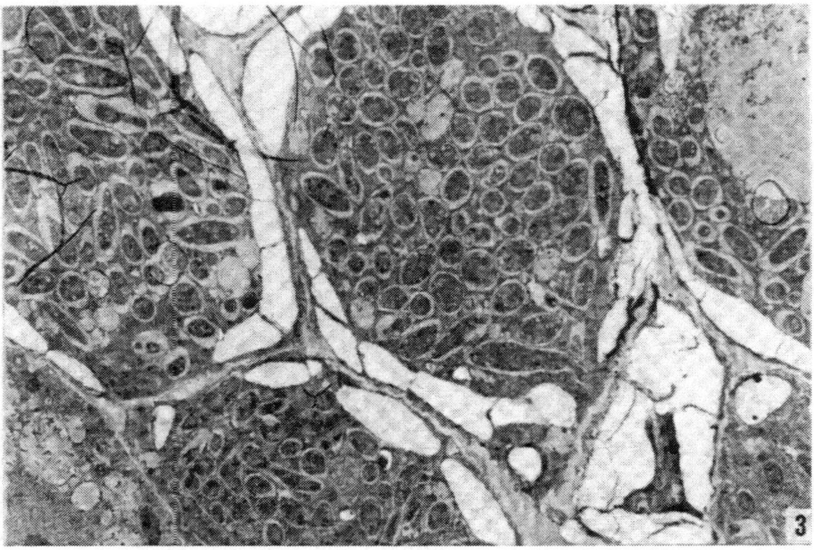

Figure 3. A large number of rhizobia, each of them enclosed in peribacteroid membrane, are distributed in the adjacent host cells of barley pseudonodules (× 3 400).

distributed in the infected cells, showing their similarity to bacteroids (Figure 3). Invasion of the host cells by rhizobia is conducted in the manner of infection threads (Figure 4). So all these structures of the infected cells are similar to those

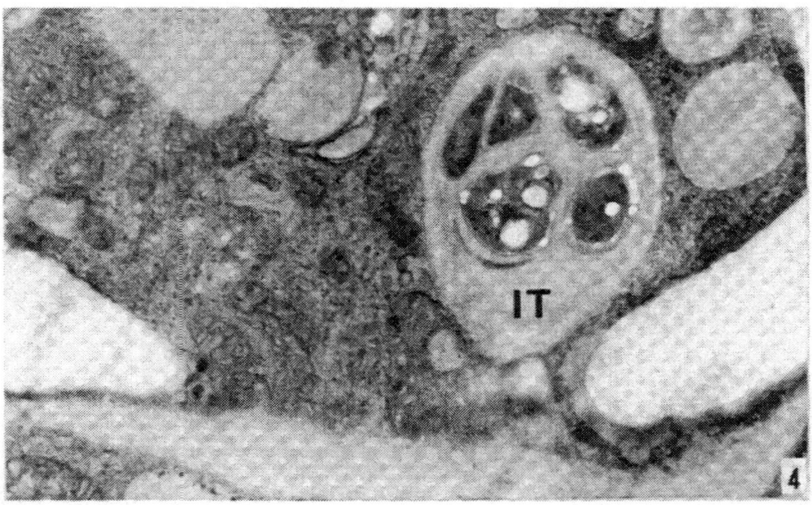

Figure 4. Rhizobial bacteria invading the host cell in the manner of infection thread (IT), and mitochondria, vacuole and electron transparent granules are clearly shown (× 20 000).

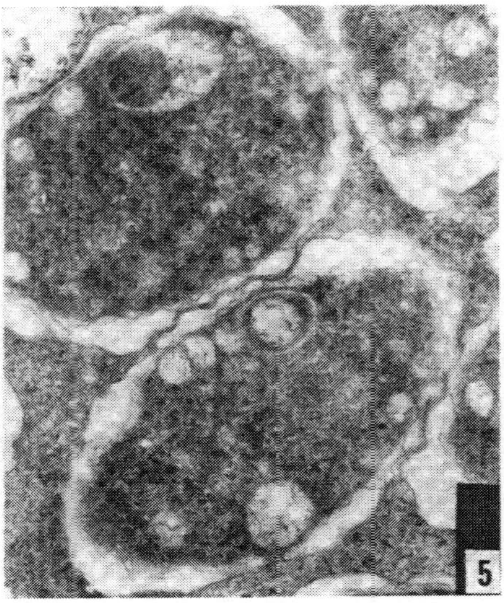

Figure 5. The bacteroids with a number of vesicles (× 10 000).

of legume nodules, but they are also different, e.g. in vesicular structure (Figure 5) and osmiophilic granules aggregated together in some of the bacteroids (Figure 6). Although no nitrogenase activity appeared in our experiments, plant growth was much better than the control (Figure 7).

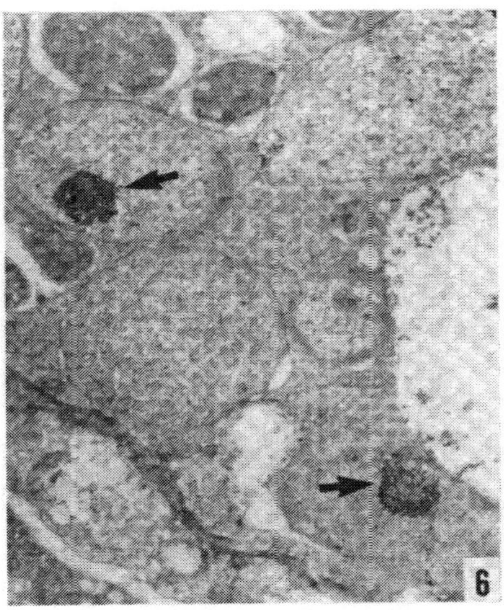

Figure 6. The osmiophilic granules aggregated together (arrows) in two bacteroids (× 10 000).

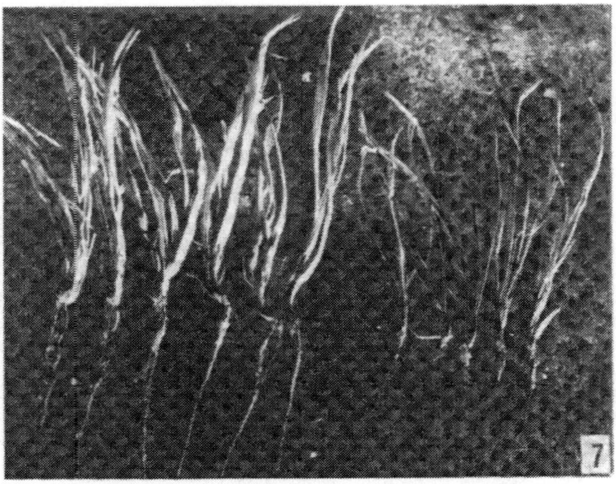

Figure 7. The plants on the left are much better than the control on the right.

Cell structures of rice pseudonodules

Three small protrusions close together on the root of rice could be seen (Figure 8) after the seedlings of rice variety IR36 were inoculated by rhizobia. Sections of the protrusions under the transmission electron microscope indicated that rhizobia existed densely in the intercellular space (Figure 9) and cell (Figure 10) respectively. The infected cell showed its coarse structure and the nuclear

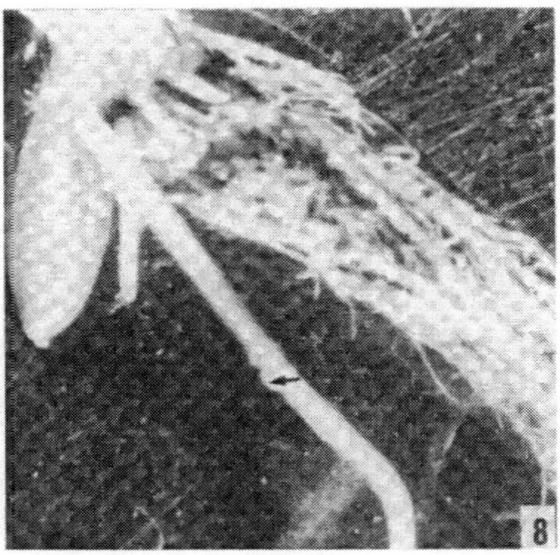

Figure 8. Three small protrusions (arrow) close together on a root of rice (× 10).

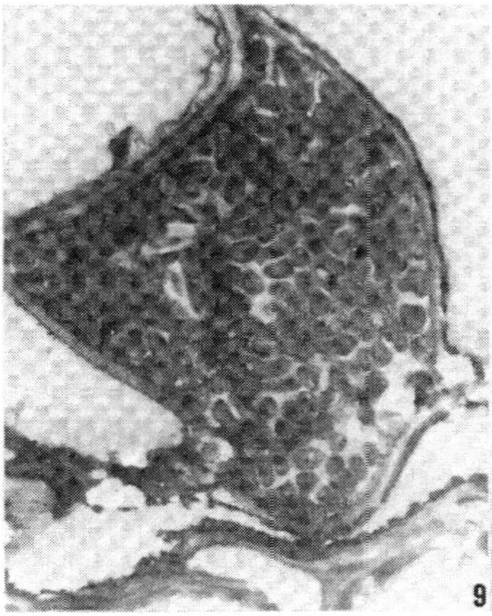

Figure 9. A large number of bacteria in the intercellular space (× 3 500).

membrane, ribosomes and other organelles wee not clearly seen. The rhizobia were naked, without surrounding membrane (Figure 10). Therefore, these structures are completely different from those of leguminous plants. No nitrogenase activity was present, but the rice plants grew much better than the control (Figure 11).

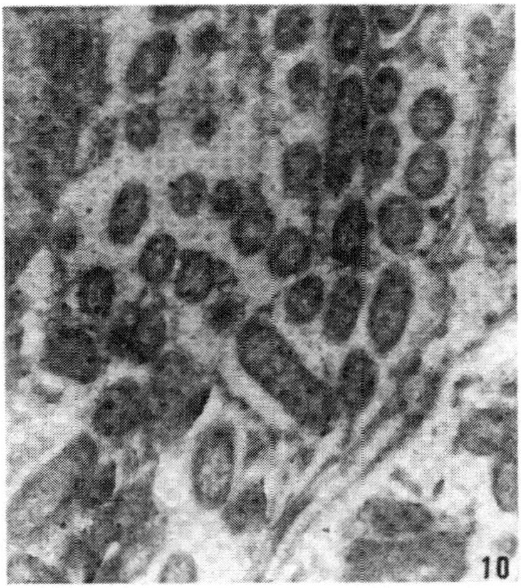

Figure 10. A large number of naked bacteria in the infected host cells (× 14 000).

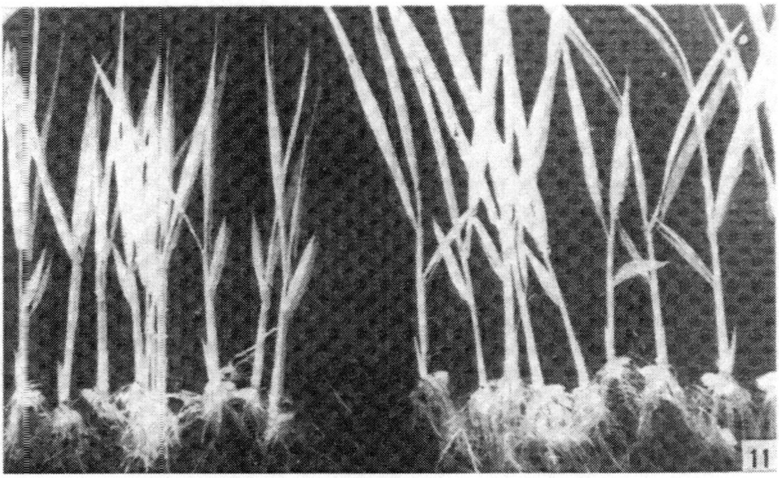

Figure 11. The rice plants inoculated with *Rhizobium astragalii* strain on the right are growing much better than those on the left (control).

Discussion

Role of magnetism in establishing the relationship between rhizobia and plants

What is the effect of magnetism on rhizobial bacteria? Nobody knows now. Probably the osmiophilic granules in the bacteria (Figure 6) would be the material containing iron. If it could be confirmed that the bacteria contain a kind of magnetic material, then this is what enables them to be magnetized, and possibly, is related to playing the role in the infection of barley roots. In fact, the diazotrophs contain abundant iron atoms, some of them existing in nitrogenase, possibly leading to the origin of magnetic field. If the organisms could be given a Lorentz force influencing the condition of motional electrons, it results in both producing a biomagnetic field, and affecting the force of the magnetic field. It would also be the same what influences could be given to the plants under the magnetic field. Therefore, the magnetic field may affect plant-microbe interaction, and perhaps is beneficial in establishing coexisting or symbiotic relationships between them.

Forms of rhizobia present in pseudonodules

It is indicated that there are two forms of rhizobial bacteria present in the infected cells of nodules. One is the bacteroids which are enclosed in the peribacteroid membranes originating from the infected host cells. The other is the rhizobia as vegetative cells present in the infected host cells. Even in leguminous nodules the bacteria also exist in two forms in the host cells, but the bacteroids occupy most of the infected host cells (Shan *et al.*, 1995). The unique example in nature of *Parasponia* nodules is evidence that shows all bacteria exist in the infection threads and are not released from them. These rhizobial bacteria without peribacteroid membranes enclosed also fix nitrogen (Trinick, 1976). So the rhizobial bacteria in the peribacteroid membrane are not a unique existing form in the nodules of legumes, or nodule-like structure of non-legumes.

Nitrogenase activity and effect on plant growth after invasion of the rhizobial bacteria into barley and rice

No nitrogenase activity of the pseudonodules on the roots of barley and rice was detected. This may be due to the fact that the essential requirements, such as anaerobic condition, etc., for nitrogenase activity could not be satisfied. However, plant growth was not influenced. Recently, *Azorhizobium caulinodans* ORS571 bacteria, which fix nitrogen in free-living condition, have been induced into wheat, and a small amount of nitrogenase activity has been indicated (Chen *et al.*, 1992; Cocking *et al.*, 1995), and there are a number of reports showing that free-living diazotrophs could be isolated from the vascular system and cortex of stems and roots (Dobereiner *et al.*, 1992) and could fix nitrogen beneficial to plant growth, but no data showing the relationship between bacteria and plants at the ultrastructural level has been published.

Acknowledgements

This work was supported by grants from the National Natural Science Foundation of China, and the Rockefeller Foundation of the USA. We thank Meng Zhaoyi for technical assistance.

References

Al-Mallah M K, Davey M R, Cocking E C 1989. Formation of nodular structures on rice seedling by rhizobia. J. Exp. Bot. 40, 473–478.

Al-Mallah M K, Davey M R, Cocking E C 1990. Nodulation of oilseed rape (*Brassica napus*) by rhizobia. J. Exp. Bot. 41, 1567–1572.

Chen T W, Schere S, Boger P 1992. Nitrogen fixation of paranodules on wheat roots artificially induced by *Azorhizobium* bacteria. Sci. China, B, 7, 712–716.

Cocking E C, Sabry R S, Saleh S A 1995. Symbiotic nitrogen fixation in cereals. In: International Workshop on Associative Interaction of Nitrogen-Fixing Bacteria with Plants. Saratov, Russia, 29–31.

Dobereiner J, Reis V M, Paula M A, Olivares F 1992. Endophytic diazotrophs in sugar cane, cereals and tuber plants. In: Palacios R, Mora J, Newton W E, eds. New Horizon in Nitrogen Fixation. Kluwer, Dordrecht, 671–676.

Jing Y, Zhang B T, Shan X Q 1990. Pseudonodules formation on barley roots induced by *Rhizobium astragali*. FEMS Microbiol. Lett., 69, 123–128.

Jing Y, Zhang B T, Wang Y P, Liu X 1992. Effect of magnetic field on symbiotic nitrogen fixation of soybean nodules. Acta Bot. Sin., 34, 364–368.

Shan X Q, Hou S Q, Jing Y 1995. Forms and fates of rhizobial bacteria present in the infected host cells of nodules. Acta Bot. Sin., 37, 522–526.
Trinick M J 1973. Symbiosis between *Rhizobium* and non-legumes–*Trema aspera.* Nature, 244, 459–460.
Trinick M J 1976. Structure of nitrogen-fixing nodules formed by *Rhizobium* on roots of *Parasponia andersonii* Planch. Can. J. Microbiol., 25, 565–578.

Posters

 Poster

Induction of paranodulation by 2,4-D, IAA and tryptophan in wheat inoculated with various diazotrophs

H. YOUSSEF, M. MONIB, M. FAYEZ AND N.A. HEGAZI

Introduction

The term *para*-nodule was introduced by Tchan and Kennedy (1989) to describe the chemically induced nodule since it differs from the naturally occurring legume nodule. Chemical treatment combined with inoculation with diazotrophs appears to be the most successful for inducing *para*-nodule formation (Zeman *et al.*, 1992). In this work, the phenomenon of *para*-nodulation in wheat cultivars inoculated with various diazotrophs was investigated in the presence of different concentrations of 2,4-D, IAA and tryptophan.

Materials and methods

Seeds of wheat (*Triticum aestivum* L.; cvs. Gemaza 1, Benisuif 1, Giza 155, 157 and 162) were surface-sterilized with 0.1% $HgCl_2$ (10 min) followed by five washes with sterile deionized water. Sterile seeds were then germinated on sterile semi-solid agar culture medium (0.6%) in dark. Uncontaminated 7-day-old seedlings were transferred into glass tubes (2.5 cm dia. and 14.5 cm length) containing 30 ml carbon- and nitrogen-deficient agar medium (Reinhold *et al.*, 1985). After 24 h, 2,4-dichlorophenoxyacetate (2,4-D), indoleacetic acid (IAA) and tryptophan were added at various concentrations. The chemicals were previously sterilized by membrane-filtration. Seedlings were inoculated with 1 ml suspension of each of tested diazotrophs; *Azospirillum* spp., *Klebsiella* spp. *Pseudomonas* spp. or composite of all in various inocula sizes of 10^5–10^{10} cells ml^{-1}. The tested diazotrophs used were broth-cultured to late log phase in selective culture media (Reinhold *et al.*, 1985 for *Azospirillum* spp.; Dixon *et al.*, 1980 for *Klebsiella* spp. and Haahtela *et al.*, 1983 for *Pseudomonas* spp.). Seedlings were kept in an environmentally controlled growth chamber having a day/night temperature of

K.A. Malik et al. (eds.), Nitrogen Fixation with Non-Legumes, 145–150.

21/19°C. After 45 days of transplanting, plants were harvested and *p*-nodules were examined by staining the root system with 10% methylene blue solution.

Results and discussion

Treatment with 2,4-D consistently induced structures that were round in shape. The outgrowths were distributed at irregular intervals along the primary roots. The concentrations of 1.0 and 2.0 ppm were found to produce relatively the largest structures (>140–290 μm) and higher concentrations slightly inhibited their development (Table 1). Such optimal concentration also corresponded to those giving the greatest number of root tumours. Root biomass decreased as 2,4-D concentration in the culture medium increased, although the differences were statistically insignificant. In a number of cases, *Azospirillum* inoculation intensified the effect of 2,4-D treatment on root development and *p*-nodulation was

Table 1. Root biomass and *p*-nodule formation in 45-day-old wheat cv. Gamaza 1 seedlings treated with different concentrations of 2,4-D, IAA and tryptophan and inoculated with *Azospirillum* spp.

Chemical compound (ppm)	*p*-Nodules		Root biomass (mg plant^{-1})
	No. per plant	Size (μm)	
2,4-D			
0.5	26(18–38)*	194 (172–220)	17.3(10.9–26.3)
1.0	29 (19–41)	208 (142–252)	17.9(12.8–21.1)
2.0	29 (19–48)	210 (162–290)	12.0(8.4–18.5)
4.0	24 (16–29)	182 (132–216)	10.6(7.8–11.7)
6.0	28 (20–43)	192 (136–226)	13.1(7.5–14.7)
Control†	29	184	15.9
LSD (0.05)	6	20	5.7
IAA			
2.0	32 (26–44)	230(146–38.4)	13.1(9.7–16.9)
4.0	32 (28–42)	216 (138–232)	11.0(9.1–14.4)
6.0	34 (29–45)	190 (144–206)	10.3(8.1–16.4)
8.0	33 (26–42)	236 (92–350)	10.2(6.3–13.4)
10.0	27 (16–34)	198 (146–230)	11.0(8.0–18.1)
Control	46	272	15.5
LSD (0.05)	6	40	4.1
Tryptophan			
25	22 (14–28)	134 (108–154)	14.3(10.3–21.4)
50	19 (18–22)	144 (116–174)	12.6(10.5–16.0)
100	17 (12–27)	144 (110–166)	9.9(8.8–11.4)
150	14 (10–19)	118 (112–166)	8.7(6.4–9.4)
200	15 (05–25)	130 (76–202)	9.8(7.2–13.0)
Control	32	194	15.2
LSD (0.05)	4	20	3.8

* Values in parenthesis indicate the range of numbers obtained for inocula sizes of *Azospirillum* (10^5–0^{10} ml^{-1}).

†Control plants are those supplemented with chemicals but receiving no inocula.

much better. The inoculum size had a unique influence on *p*-nodule formation. Numbers were highest with bacterial densities in the range of 10^6–10^8 cells ml^{-1}.

The relative increase of *p*-nodule formation through 2,4-D treatment, regardless of inoculation, supports the idea that *p*-nodulation is a physiological process independent of bacterial action (Kennedy and Tchan, 1992). However, our observation that 1.0 and 2.0 ppm 2,4-D treatment induced better *p*-nodulation in the presence of the inoculant indicates that *Azospirillum* might enhance 2,4-D-induced root tumours.

The effect of *Azospirillum* on IAA-induced wheat *p*-nodulation, in the form of *p*-nodule induction at 2.0–10.0 ppm, was considerably higher compared to 2,4-D (Table 1). This contradicts the findings of Yatazawa *et al.* (1967) that the effect of *p*-nodule-inducing compounds is not attributed simply to the auxin-like activity of such chemicals. Inoculation with *Azospirillum* had no significant influence on *p*-nodule formation in IAA-treated seedlings. Positive interaction of *Azospirillum* inoculation with 2,4-D, but not IAA, might indicate the interaction of auxin release by this particular diazotroph. Within inoculated treatments, considerable variations in numbers of *p*-nodules were attributed to inoculum size, being highest with 10^7 cells ml^{-1}.

Among the tested chemicals, tryptophan induced a lower degree of paranodulation compared to others. Low numbers of *p*-nodules (5–28 $plant^{-1}$) of small sizes (>76–202 μm) were formed on wheat cv. Gemaza 1 seedlings. Significant differences in *p*-nodule numbers were scored due to the chemical level in the culture medium. Lower concentrations (25 and 50 ppm) supported better *p*-nodulation than higher ones. *Azospirillum* inoculation did not significantly increase root dry weight of tryptophan-treated seedlings.

The interaction of *Klebsiella* spp. with *p*-nodulation was investigated in another set of experiments. Wheat cv. Benisuif 1 seedlings treated with 2,4-D, IAA and tryptophan developed tumour-knots all along primary roots. The highest *p*-nodule numbers were recorded at concentrations of 1.0, 6.0 and 50.0 ppm of 2,4-D, IAA and tryptophan, respectively (Table 2). Inoculation with *Klebsiella* spp. enhanced *p*-nodule formation for 2,4-D and IAA-treated seedlings. Slight fluctuations in *p*-nodule size and root biomass among the various treatments were also recorded. As mentioned by Bauer *et al.* (1985), a number of diazotrophs synthesize phytohormones such as auxin and cytokinine which, together with the endogenous hormones of the host plant, are involved in nodule formation.

No available information exists on cultivar effect in respect to *p*-nodulation. Therefore, a part of the present study dealt with the influence of the lowest concentrations of *p*-nodule inducers with or without diazotroph inoculation on the production of such structures by a number of wheat cultivars. Results presented in Table 3 demonstrate, irrespective of chemicals, that cv. Giza 157 responded well to all inocula applied compared to other cultivars. Superior cultivar-diazotroph interactions were Giza 157–*Pseudomonas* spp., Giza 162–*Klebsiella* spp. and Giza 155–*Pseudomonas* spp.

As to the lowest possible concentrations of *p*-nodule inducers, Table 4 shows that treatment with 0.5 ppm 2,4-D produced 45–68 *p*-nodules on wheat roots,

Table 2. Root biomass and *p*-nodule formation in 45-day-old wheat cv. Benisuif 1 seedlings treated with different concentrations of 2,4-D, IAA and tryptophan, and inoculated with *Klebsiella* in inoculum size of 10^6 cells ml^{-1}

Chemical compound (ppm)	*p*-Nodules		Root biomass (mg $plant^{-1}$)
	No. per plant	Size (µm)	
2,4-D			
0.5	9	136	4.8
1.0	14	172	4.9
2.0	3	180	4.5
Inoculation	18	128	4.4
IAA			
5.0	2	240	4.3
4.0	13	384	4.5
6.0	13	332	3.5
Inoculation	18	156	4.4
Tryptophan			
50.0	31	300	4.0
75.0	22	228	5.4
100.0	17	192	3.5
Inoculation	18	152	4.4
LSD (0.05)	4	70	1.4

Table 3. The most responsive cultivars to the various inocula irrespective of chemical inducers of *p*-nodulation

Cultivar	Inoculum	No. of paranodules $plant^{-1}$
Giza 157	*Azospirillum* spp.	49±3.7
	Klebsiella spp.	51±3.7
	Pseudomonas spp.	53±3.3
	Multi-strain	48±2.9
Giza 162	*Azospirillum* spp.	11±1.7
	Klebsiella spp	17±3.6
	Pseudomonas spp.	11±1.7
	Multi-strain	ND
Giza 155	*Azospirillum* spp	7±1.4
	Klebsiella spp.	6±1.3
	Pseudomonas spp.	7±0.9
	Multi-strain	5±0.7

ND, not determined.

being highest with *Pseudomonas* inoculation and lowest with uninoculated plants. As to IAA, *Pseudomonas* supported good *p*-nodulation (50 nodules per plant) on 2.0 ppm-treated plants followed by multi-strain inoculum (43 nodules $plant^{-1}$), and *Klebsiella* (40). In the presence of 50 ppm tryptophan, *Pseudomonas* and *Azospirillum* were superior and produced 52 and 50 *p*-nodules per root system,

Table 4. Paranodulation of wheat inoculated with various diazotrophs and treated with the lowest concentrations of *p*-nodule-inducing chemicals

Diazotrophs	2,4-D (0.5 ppm)	IAA (2.0 ppm)	Tryptophan (50.0 ppm)
Control	45±5.4	28±8.3	24±5.1
Azospirillum spp.	55±3.8	39±6.3	50±3.3
Klebsiella spp.	55±9.6	40±6.5	40±5.1
Pseudomonas spp.	68±9.3	50±5.2	52±6.1
Mixed	51±1.4	43±5.1	43±7.1

Table 5. Number of *p*-nodules of wheat cultivars due to the best diazotroph-chemical inducer interactions

Interactions	Giza 157	Giza 162	Giza 155
Azospirillum + 1 ppm 2,4-D	67.0±8.4	6.0±2.5	6.0±0.0
Klebsiella + 75 ppm Tryp.	67.8±5.3	37.0±17.7	3.0±0.0
Azospirillum + 4 ppm IAA	48.8±5.7	19.0±6.2	15.8±2.5

respectively. Contrary to other investigators (Kennedy and Tchan, 1992), the above results prove the necessity of bacterial introduction to the *p*-nodulation environment.

Table 5 summarizes the behaviour of various wheat cultivars to interactions of diazotrophs with lowest possible and stimulative concentrations of chemical inducers of paranodulation. Cultivar Giza 157 was the most promiscuous to various treatments yielding the highest number of nodules. The other two cultivars differed, as Giza 162 gave higher numbers with *Klebsiella* + 75 ppm tryptophan, while Giza 155 was more responsive to *Azospirillum* + 4 ppm IAA.

In conclusion, the mechanism of *p*-nodulation was proposed to be basically a chemical one. However, results present in this study demonstrate *p*-nodule formation in the presence of diazotrophs as such. They also positively interact with the lowest concentrations of chemical inducers recommended.

Acknowledgement

This study was supported by the national project on introducing agrobiotechnologies based on biological nitrogen for the development of Sinai, funded by the Egyptian Ministry of Agriculture and Land Reclamation and the French Food Aid Counterpart Fund (FFACF-BLAFE).

References

Bauer W D, Bhuvaneswari T V, Calvert H E, Law I J, Malik N S A, Vesper S J 1985. Recognition and infection by slow-growing rhizobia. In: Evans H, Bottomley P J, Newton W E, eds. Fixation Research Progress. Martinus Nijhoff, Dordrecht, 247–253.

Dixon R A, Eady R R, Espin G, Hill S, Laccarino M, Khan D, Merrich M 1980. Analysis of regulation of *K. pneumoniae* nitrogen fixation (nif.) gene cluster with gene. Nature (London), 286, 128–132.

Haahtela K, Helander I, Nurmiaho-Lassila E L, Sundman V 1983. Morphological and physiological characteristics and lipopolysaccharide composition of N_2-fixing (C_2H_2-reducing) root-associated *Pseudomonas* sp. Can. J. Microbiol., 29, 874–880.

Kennedy I R, Tchan Y T 1992. Biological nitrogen fixation in non-leguminous field crops: Recent advances. Plant Soil, 141, 93–118.

Reinhold B, Hurek T, Fendrik I 1985. Strain-specific chemotaxis of *Azospirillum* spp. J. Bacteriol., 162, 190–195.

Tchan Y T, Kennedy I R 1989. Possible N_2-fixing root nodule induced in non-legumes. Agric. Sci. (AIAS Melbourne), 2, 57–59.

Yatazawa M, Furuhashi A, Shimizu M 1967. Growth of callus tissue from rice root *in vitro*. Plant Cell Physiol., 8, 363–373.

Zeman A M M, Tchan Y T, Elmerich C, Kennedy I R 1992. Nitrogenase active wheat-root *para*-nodules formed by 2,4-dichlorophenoxyacetic acid (2,4-D)/*Azospirillum*. Res. Microbiol., 143, 847–855.

18 Poster

Optimizing zinc and phosphate statuses for enhancing field colonization and contribution of the symbiotic association *Azolla–Anabaena* to rice performance

Y.G. YANNI

Introduction

Green manuring with *Azolla* has been recommended for enhancing rice performance in the tropics and subtropics (Liu, 1979; Tuan and Thuyet, 1979). In addition to its capacity to contribute biologically fixed N, the *Azolla* canopy can inhibit growth of aquatic phototrophs and depress photodependent CO_2 uptake. Consequently, floodwater pH will not increase during the day, and this reduces N loss by ammonia volatilization, which can be high in the summer season, especially when large amounts of fertilizer-N are applied to rice fields (Singh, 1977; Watanabe *et al.*, 1977; Hamdi *et al.*, 1980; Varghese, 1990).

The aim of this study was to throw light on the nutritional requirements of *Azolla* in rice fields and the effect of application of the indigenous *Azolla* fern on rice performance when urea fertilizer, calcium superphosphate and $ZnSO_4$ were included in the rice fertilization schedule. Growth and rate of field colonization by *Azolla*, compared with its contribution to rice growth and performance, were taken as indicators of the interactions between N, P and Zn nutrition.

Materials and methods

A field experiment was conducted at the Sakha Agricultural Research Station, Kafr El-Sheikh, Egypt, using the Indica rice variety Giza 181. $ZnSO_4$ at 10 ppm Zn was broadcast over some sub-subplots after they were submerged with water and just before puddling. One-month-old rice seedlings were transplanted in split-splitplots, each of 4 × 5 m area, at 20 × 20 cm spacing so that each sub-subplot contained 500 seedlings with 456 spaces of 20 × 20 cm between them. The *Azolla caroliniana* at 2 kg fresh fern/10 m^2 was broadcast symmetrically on the water surface of some of the sub-subplots 5 days after the rice was transplanted.

K.A. Malik et al. (eds.), Nitrogen Fixation with Non-Legumes, 151–157.

Calcium superphosphate, at 54 kg, P_2O_5/ha, was broadcast on the water surface of some of the sub-subplots 5 days later, 10 days post-transplanting (DPT). The propagated *Azolla* was incorporated into the soil by foot during three hand-weedings at 20, 40 and 60 DPT. Urea at 0, 72 or 144 kg N/ha was applied in two equal doses: 25 DPT and at mid-tillering. Colonization of the field with *Azolla* was followed by eye, and started 5 DPT, i.e. just after *Azolla* distribution. The proportion of the (456) spaces between each four seedlings which contained *Azolla* was determined for each sub-subplot and given as tenths (Yanni, 1991). The data of growth and yield attributes were computed as a split-splitplot design experiment and the mean differences were compared to their corresponding least significant differences.

Results and discussion

Plant performance

Productive tillers/m^2, grain and straw yields and the harvest index were increased by application of *Azolla* and/or increasing the amount of urea fertilizer (Tables 1 and 2). Soil application of $ZnSO_4$ and/or calcium superphosphate actively enhanced plant performance. The maximum grain yield was obtained by application of *Azolla*, $ZnSO_4$, calcium superphosphate and urea at 72 kg N/ha. Urea at 144 kg N/ha enhanced vegetative growth and straw, rather than grain production, as the values for the harvest index (Table 2) were lowest when *Azolla* was applied with $ZnSO_4$ or calcium superphosphate under urea fertilizer at 144 kg N/ha. Whereas $ZnSO_4$ seemed more effective when no *Azolla* had been applied, application of phosphate was more effective with *Azolla* application. *Azolla* maintained its contribution to the rice even with urea fertilizer at 144 kg N/ha, may be due to split application of the fertilizer-N in two equal doses 1 month apart. This was found effective in minimizing urea's inhibitory effect on N_2-fixation systems in rice fields (Yanni and Hegazy, 1990; Yanni, 1991, Yanni and Sehly, 1991).

Rate and type of colonization with Azolla

Colonization decreased after incorporation of the fern at 20, 40 and 60 DPT and after addition of urea fertilizer. Calcium superphosphate enhanced colonization and frond size compared with addition of $ZnSO_4$ (Table 3). In most cases, maximum colonization was observed 20–40 DPT; it tended to decrease after 60 DPT due to rice vegetative growth and active tillering, which naturally increase plant canopy above the water surface and diminished light penetration to the ecosystem. Field observations revealed that colonization could be positively correlated with productive tillering, grain and straw yields and their N-contents, harvest index and the agronomic fertilizer N-use efficiency (Tables 1 and 2).

Table 1. Effect of application of *Azolla caroliniana*, urea fertilizer, calcium superphosphate and $ZnSO_4$ on performance of the Indica rice Giza 181

Treatment	Productive tillers/m²				Grain yield (t/ha)				Straw yield (t/ha)			
	–	Zn	P	Zn+P	–	Zn	P	Zn+P	–	Zn	P	Zn+P
0 N	350	468	405	509	4.42	5.40	5.08	6.10	4.63	6.99	6.87	7.87
0 N + *Azolla*	420	536	599	615	6.29	6.48	7.37	7.98	5.31	6.34	7.61	7.94
72 N	441	533	481	566	5.69	6.44	6.41	8.20	5.35	6.71	6.60	6.61
72 N + *Azolla*	515	571	669	655	7.29	7.72	8.96	9.24	6.52	7.88	9.41	7.93
144 N	536	599	530	631	6.50	8.02	7.81	8.14	6.25	7.63	7.58	9.03
144 N + *Azolla*	546	663	639	626	7.61	8.92	8.59	8.61	7.78	10.89	11.75	9.52
LSD		0.05*	0.01†			0.05	0.01			0.05	0.01	
Azolla (A)		37	n.s.			0.46	0.84			0.53	0.97	
N-levels (N)		29	52			0.21	0.30			0.36	0.51	
Elements (E)		24	32			0.22	0.29			0.47	0.63	
A × N		41	74			n.s.	n.s.			0.51	0.72	
A × E		33	45			0.30	0.41			0.67	0.89	
N × E		41	55			0.37	0.50			0.82	1.09	
A × N × E		58	77			n.s.	n.s.			1.16	1.54	

Azolla: inoculation with *Azolla caroliniana*, at 2 kg fresh fern/10 m², 5 DPT.
N: urea fertilizer at 0, 72 or 144 kg N/ha in two equal doses: 25 DPT and mid-tillering.
P: application of calcium superphosphate at 54 kg P_2O_5/ha just before submergence.
Zn: application of $ZnSO_4$ at 10 ppm Zn after submergence and just before puddling.
E: application of P and/or Zn.
LSD: least significant differences: n.s., not significant; * significant; †highly significant.

Table 2. Effect of application of *Azolla caroliniana*, urea fertilizer, calcium superphosphate and $ZnSO_4$ on harvest index and N-attributes of the Indica rice Giza 181

Treatment	Harvest index (%)[a]				Grain N-content (kg N/ha)				Straw N-content (kg N/ha)				Agronomic fertilizer N use efficiency[b]			
	–	Zn	P	Zn+P	–	Zn	P	Zn+P	–	Zn	P	Zn+P	–	Zn	P	Zn+P
0 N	48.8	43.6	42.5	43.7	34.6	36.9	40.8	49.0	16.5	21.2	18.1	29.3	–	–	–	–
0 N + *Azolla*	54.2	50.5	49.2	50.1	50.3	53.2	62.7	68.8	19.9	24.5	28.9	31.9	–	–	–	–
72 N	51.5	49.0	49.3	55.4	45.6	60.7	52.3	69.4	22.4	24.8	25.7	24.0	79	89	89	114
72 N + *Azolla*	52.8	49.5	48.8	53.8	60.1	65.6	68.9	81.3	25.8	31.4	37.2	32.4	101	107	124	128
144 N	51.0	51.2	50.7	47.4	53.2	67.2	56.9	69.2	23.8	26.3	29.8	33.2	45	56	54	57
144 N + *Azolla*	49.4	45.0	42.2	47.5	65.3	69.2	70.4	74.1	31.7	44.9	48.5	48.7	53	62	60	60
L.S.D.		0.05*	0.01			0.05	0.01			0.05	0.01			0.05	0.01	
Azolla (A)		1.1	n.s.			4.8	8.7			2.6	4.8			4	7	
N-levels (N)		1.2	1.7			2.0	2.8			1.2	1.7			3	4	
Elements (E)		1.4	1.9			1.8	2.4			2.2	2.9			2	3	
A × N		1.7	2.4			n.s.	n.s.			1.8	2.5			4	5	
A × E		2.0	2.7			2.5	3.4			3.1	4.1			3	4	
N × E		2.5	3.3			3.1	4.2			3.7	5.0			4	5	
A × N × E		3.5	4.6			n.s.	n.s.			5.3	7.1			5	n.s.	

Azolla: inoculation with *Azolla caroliniana*, at 2 kg fresh fern/10 m^2, 5 DPT.
N: urea fertilizer at 0, 72 or 144 kg N/ha in two equal doses: 25 DPT and mid-tillering.
P: application of calcium superphosphate at 54 kg P_2O_5/ha just before submergence.
Zn: application of $ZnSO_4$ at 10 ppm Zn after submergence and just before puddling.
E: application of P and/or Zn.
[a] Grain yield as a proportion of grain + straw yields.
[b] kg grain yield/kg fertilizer-N.
LSD: least significant differences: n.s., not significant; * significant; highly significant.

Table 3. Rate of colonization of rice sub-subplots by *Azolla caroliniana* as affected by time and fertilization with urea, calcium superphosphate and $ZnSO_4$

Days post-transplanting (DPT)	Without urea				72 kg N/ha				144 kg N/ha			
	–	Zn	P	Zn+P	–	Zn	P	Zn+P	–	Zn	P	Zn+P
5	2	2	2	2	2	2	2	2	2	2	2	2
10	5	5	6	5	4	5	6	5	4	5	5	6
15	7	6	9	9	6	6	8	8	7	5	8	8
20	7	8	10	10	7	8	10	9	7	8	10	9
25	4	5	5	5	5	3	4	5	3	3	4	4
30	7	8	9	8	5	5	6	6	4	4	6	5
35	8	8	10	10	7	8	7	8	6	6	7	8
40	8	9	10	10	7	8	9	9	7	7	8	8
45	5	4	5	5	4	4	5	5	3	3	4	3
50	7	7	8	9	5	5	6	6	3	3	5	5
55	7	8	9	9	7	6	8	7	6	6	7	7
60	8	8	10	10	7	7	8	8	6	7	8	7
65	4	4	5	5	3	3	4	4	2	3	4	4
Mean for Zn + P	6.08	6.31	7.54	7.46	5.31	5.38	6.38	6.31	4.62	4.77	6.00	5.85
Mean for N level	6.85				5.85				5.31			

Inc.: incorporation of *Azolla* in the soil by foot during handweeding 20, 40 and 60 DPT.
N: addition of urea fertilizer (in two equal doses, 25 DPT and at mid-tillering).
P: application of calcium superphosphate at 54 P_2O_5/ha just before submergence.
Zn: application of $ZnSO_4$ at 10 ppm Zn just after submergence and before puddling.
* The proportion of spaces between each four rice seedlings covered by *Azolla*, given to the nearest tenth (mean of four replicates).

Nitrogen attributes

Nitrogen contents of grain and straw were increased by addition of urea up to 144 kg N/ha. Further increases were obtained by green manuring with *Azolla* and application of $ZnSO_4$ and/or calcium superphosphate. Without application of *Azolla*, $ZnSO_4$ enhanced N-accumulation more efficiently than PO_4, which seemed more effective when *Azolla* was added (Table 2). Maximum N-content in grain developed with the sub-subtreatment of $ZnSO_4$ at 10 ppm, calcium superphosphate at 54 kg P_2O_5/ha and urea at 72 kg N/ha, i.e. the same one which gave the highest grain yield. Highest straw N-contents were observed when *Azolla*, $ZnSO_4$ or calcium superphosphate and urea at 144 kg N/ha were used, but these were at the expense of grain N-content and percentage. Application of *Azolla* increased grain N/ha by 45%, 32% and 23% when urea was added at 0, 72 and 144 kg N/ha, respectively. The corresponding increases when calcium superphosphate was applied instead of *Azolla* were 54%, 32% and 12%, respectively, indicating greater enhancement with phosphate under low levels of urea fertilizer. Regardless of grain yield, the data (Table 1) indicate that, without fertilizer-N, *Azolla* contributed 19.1 kg N/ha to grain plus straw yields, and this increased to 19.6, 32.7 and 22.4 kg N/ha when either $ZnSO_4$, calcium superphosphate or both were applied, respectively. Application of $ZnSO_4$, calcium superphosphate or both increased the agronomic benefit of urea (kg grain yield/kg fertilizer-N), probably due to more enhancement of vegetative growth and N absorption. Green manuring with *Azolla* especially when either or both of the two mineral fertilizers were applied increased the efficiencies, indicating additional indirect contribution of both elements to rice performance via activation of *Azolla* growth development and field colonization with the fern. Although efficiencies decreased with increasing amount of urea fertilizer, they maintained the same pattern of changes towards *Azolla*, $ZnSO_4$ and calcium superphosphate application.

Discussion

Field observations made in the present study on rice growth, crop yield and N-contents of grain and straw, along with development, the growth features and field colonization with *Azolla*, indicate the possibility that *Azolla* makes an effective contribution to rice performance, especially when the N, Zn and P statuses of the soil are optimized.

References

Hamdi Y A, Alaa El-Din M N, Shalan S N, Hassan M E 1980. Report on Methodology of Azolla Preservation and Transport. Ministry of Agriculture, Egypt, and Food and Agriculture Organization, Rome.

Liu C 1979. Use of *Azolla* in rice production in China. In: Nitrogen and Rice, pp. 375–394. IRRI, Los Banos, Philippines.

Singh P K 1977. Effects of *Azolla* on the yield of paddy with and without application of N-fertilizer. Current Science, 46, 642–644.

Tuan D T, Thuyet T R 1979. The use of *Azolla* in rice production in Vietnam. In: Nitrogen and Rice, pp. 395–504. IRRI, Los Banos, Philippines.

Varghese A 1990. Effect of organic manure on natural occurrence of *Azolla pinnata* and its effect on rice yield. International Rice Research Newsletter, 15, 16.

Watanabe I, Espinas C R, Berja N S, Alimango B V 1977. Utilization of the *Azolla–Anabaena* complex as a nitrogen fertilizer for rice. Research Paper Series No. 11. IRRI, Los Banos, Philippines.

Yanni Y G 1991. Potential of indigenous cyanobacteria to contribute to rice performance under different schedules of nitrogen application. World J. Microbiol. Biotechnol., 7, 48–52.

Yanni Y G, Hegazy M H 1990. Efficiency of algalization of rice in the Nile Delta with different amounts of nitrogen and potassium fertilization. World J. Microbiol. Biotechnol., 6, 395–399.

Yanni Y G, Sehly M R 1991. Rice performance and natural infection with blast (*Pyricularia oryzae* Cav.) under different algalization techniques and rates of fertilizer nitrogen. World J. Microbiol. Biotechnol., 7, 43–47.

19 Poster

Enhancement of rice performance and incidence of natural fungal and insect damage with cyanobacterization, microelements, combined nitrogen and certain pesticides

Y.G. YANNI, M.R. SEHLY AND F.E. ABDALLAH

Introduction

Cyanobacterization of rice fields is capable of meeting a considerable proportion of rice nitrogen demand (Subramanyan *et al.*, 1965; Yanni and Abd-El-Rahman 1993) and of regulating the N status of the rice ecosystem (Yanni *et al.*, 1988). This work is concerned with the contribution of cyanobacteria to the incidence of infection with some fungal and insect pests of rice under different schedules of agrochemical applications.

Materials and methods

The rice variety Giza-176 or IR-28 was used in two separate field experiments. One-month-old seedlings were transplanted 20 cm apart in 20 m^2 sub-subplots. For the rice variety Giza 176, each of the macro plus microelements combination Vitafort or the fungicide Sportake-45 (Table 1), or a combination of 50/50 of the two compounds was applied during the late panicle initiation stage. For the variety IR 28, the stem borer (*Chilo agamemnon* Bles.) controlling insecticides: Cyfen: 0,O-dimethyl O-(3-methyl-4-nitrophenyl) phosphorothioate, Dursban: 0,O-diethyl-O-(3,5,6-trichloro-2 pyridyl) phosphorothioate, or Fipronil: Phenyl pyrazole ($C_{12}H_4C_{12}F_6N_4OS$) were used. The statistical design comprised two main-plot treatments for inoculation with cyanobacteria, two subplot treatments for fertilization with combined-N and four sub-subplot treatments for application of the Vitafort and/or Sportake-45 in the first experiment or the stem borer insecticides in the second one (containing untreated controls).

Results and discussion

Rice performance, as indexed by productive tillering, grain yield, straw yield and their N-contents and the agronomic fertilizer N-use efficiency (kg grain yield/kg

K.A. Malik et al. (eds.), Nitrogen Fixation with Non-Legumes, 159–164.

Table 1a. Effect of cyanobacterization, N-fertilization and foliar application with Sportake-45 and Vitafort on growth, performance and agronomic N-use efficiency of the transplanted rice Giza 176

Treatments	Productive tillers/m^2		Grain yield (t/ha)		Straw yield (t/ha)		Grain N-content (kg N/ha)		Straw N-content (kg N/ha)		N-use efficiency*	
	72 N	144 N	72 N	144 N	72 N	144 N	72 N	144 N	72 N	144 N	72 N	144 N
S	505	572	7.28	8.20	11.74	13.78	59.3	67.8	35.6	43.4	101	57
S + Cyan	565	587	7.45	9.37	13.23	15.57	61.8	81.2	42.6	52.4	103	65
V	530	558	7.40	8.85	12.51	14.34	60.8	73.6	38.9	45.8	103	62
V + Cyan	542	593	8.85	10.54	14.48	16.58	73.9	89.0	48.4	57.4	123	73
S + V (50/50)	545	487	8.42	8.70	13.14	15.50	69.3	72.8	41.2	50.4	117	60
S + V (50/50) + Cyan	593	553	9.83	10.10	15.32	18.73	82.5	86.1	52.3	65.9	137	70
Control	518	482	6.77	7.80	9.15	11.62	54.4	63.4	27.5	35.8	94	54
Cyan	628	635	7.55	8.27	12.42	14.53	62.1	68.5	39.2	46.6	105	57
L.S.D.	0.05	0.01	0.05	0.01	0.05	0.01	0.05	0.01	0.05	0.01	0.05	0.01
Cyanobacteria (Cyan)	30	n.s.	0.14	0.31	0.39	0.90	0.6	1.4	1.4	3.2	1	3
Fertilizer-N (N)	n.s.	n.s.	0.17	0.28	0.26	0.44	0.5	0.9	0.9	1.4	1	2
Foliar application (F)	n.s.	n.s.	0.15	0.21	0.31	0.41	1.2	1.6	1.3	1.7	2	3
Cyan. × N	n.s.	n.s.	n.s.	n.s.	n.s.	n.s.	0.7	1.2	1.2	n.s.	2	3
Cyan. × F	38	51	0.22	0.29	0.43	0.59	1.7	2.3	1.3	1.7	2	3
N × F	38	51	0.22	0.29	0.43	n.s.	1.7	2.3	1.8	2.4	2	3
Cyan. × N × F	n.s.	n.s.	0.30	0.41	n.s.	n.s.	2.4	3.3	1.8	n.s.	3	5

S: the fungicide Sportake-45 (Prochloraz EC 45): CN-propyl-N(2,4–6-trichlorophenoxy methyl) ethyl carbamyl imidazle at 0.1 ml/m^2 for foliar application during the late panicle initiation stage.

V: the nutrient solution Vitafort (N,P,K,Mg,Fe,Mn,Zn,Cu,B,Mo,Co, Vitamin B, meta-indole-acetic acid, amino acids and growth promotion substances) at 0.25 g/m^2 foliar application during the late panicle initation stage. Cyanobacteria: inoculation with a soil-based inoculum of cyanobacteria containing: *Anabena cylindrica, Anabena oryzae, Aulosira fertilissima, Nostoc muscorum* and *Tolypothrix tenuis* applied at 100 kg fresh material (90 moisture) ha 5 days after transplanting (main-plot treatments).

N: fertilization with urea (46% N) at 72 or 144 kg N/ha in two equal doses: 25 days after transplanting and at the mid-tillering stage (subplot treatments).

* The agronomic fertilizer N-use efficiency (kg grain yield/kg fertilizer-N). LSD (0.05, 0.01): the least significant differences at 95% and 99% confidence levels.

Table 1b.

Treatments	Panicle blast				Brown spot				Leaf spot				Stem rot			
	Infection		Severity		Infection		Severity		Infection		Severity		Infection		Severity	
	72 N	144 N	72 N	144 N	72 N	144 N	72 N	144 N	72 N	144 N	72 N	144 N	72 N	144 N	72 N	144 N
S	10.7	15.7	7.7	11.1	20	20	14	14	20	20	4	4	20	20	4	4
S + Cyan	7.0	6.3	3.5	4.7	2	20	0	5	0	0	0	0	0	20	0	5
V	18.7	23.3	12.4	18.4	20	60	8	26	20	40	2	4	40	20	8	5
V + Cyan	18.0	17.3	10.6	12.8	0	40	0	6	0	0	0	0	20	0	2	0
S + V (50/50)	18.7	20.0	12.4	15.9	20	20	6	5	20	40	2	2	0	0	0	0
S + V (50/50) + Cyan	15.3	16.7	14.7	13.0	10	20	3	2	0	20	0	2	0	0	0	0
Control	27.7	22.7	23.5	17.4	40	60	8	13	40	20	2	2	0	40	0	3
Cyan	20.0	17.7	16.1	14.3	20	20	20	12	20	20	4	2	0	0	0	0

S: the fungicide Sportake-45 (Prochloraz EC 45): CN-propyl-N(2,4–6-trichlorophenoxy methyl) ethyl carbamyl imidazle at 0.1 ml/m^2 for foliar application during the late panicle initiation stage.

V: the nutrient solution Vitafort (N,P,K,Mg,Fe,Mn,Zn,Cu,B,Mo,Co, Vitamin B, meta-indole-acetic acid, amino acids and growth promotion substances) at 0.25 g/m^2 foliar application during the late panicle initation stage. Cyanobacteria: inoculation with a soil-based inoculum of cyanobacteria containing: *Anabena cylindrica, Anabena oryzae, Aulosira fertilissima, Nostoc muscorum* and *Tolypothrix tenuis* applied at 100 kg fresh material (90 moisture) ha 5 days after transplanting (main-plot treatments).

N: fertilization with urea (46% N) at 72 or 144 kg N/ha in two equal doses: 25 days after transplanting and at the mid-tillering stage (subplot treatments).

Panicle blast (*Pyricularia oryzae*), brown spot (*Helminthosporium oryzae*), leaf spot (*Alternaria* sp.) and stem rot (*Sclerotium oryzae*). Infection: Mean percentage of the infected panicles over a sample size of 100 panicles. Severity: Mean percentage of the panicle area show natural sporulation with the fungus.

Table 2a. Effect of cyanobacterization, nitrogen fertilization and some insecticides on growth, yield, N attributes and agronomic N-use efficiency of the rice IR-28

Treatments	Productive tillers/m^2		Grain yield (ton/ha)		Straw yield (ton/ha)		Grain N-content (kg N/ha)		Straw N-content (kg N/ha)		N-use efficiency*	
	No Cyan.	Cyan.	No. Cyan.	Cyan.	No. Cyan.	Cyan.	No. Cyan.	Cyan.	No. Cyan.	Cyan.	No. Cyan	Cyan.
72 N	390	462	7.41	8.63	9.67	13.44	58.8	69.8	29.2	44.3	103	120
144 N	460	530	8.65	8.93	13.62	14.53	70.0	72.4	44.0	48.0	60	62
72 N + Dursban	407	452	7.82	8.84	10.03	14.34	62.8	72.7	32.2	48.4	109	123
144 N + Dursban	462	552	8.58	9.12	14.51	17.46	70.6	76.3	48.2	58.6	60	63
72 N + Fipronil	430	490	7.54	9.29	10.39	16.79	60.8	77.5	33.9	57.5	105	129
144 N + Fipronil	468	502	8.44	9.34	16.70	18.59	69.9	79.7	56.1	65.6	59	65
72 N + Cyfen	382	493	8.65	9.56	11.63	15.83	69.4	78.9	38.6	56.0	120	133
144 N + Cyfen	513	548	9.17	9.62	17.62	18.24	75.7	83.1	60.0	65.6	64	67
L.S.D.	0.05	0.01	0.05	0.01	0.05	0.01	0.05	0.01	0.05	0.01	0.05	0.01
Cyano	19	43	0.17	0.40	0.30	0.70	1.4	3.1	0.9	2.0	2	5
N	16	27	0.06	0.10	0.12	0.21	0.5	0.9	0.5	0.9	1	2
Insec.	ns	ns	0.17	0.23	0.22	0.30	1.4	1.9	0.8	1.1	2	3
Cyano. × N	ns	ns	0.09	0.14	0.18	0.29	0.7	1.2	0.8	1.2	2	3
Cyano. × Insec.	ns	ns	0.24	0.33	0.31	0.42	1.9	2.6	1.2	1.6	3	4
N × Insec.	28	37	ns	ns	0.31	0.42	ns	ns	1.2	1.6	3	4
Cyano × N × Insec.	39	ns	ns	ns	0.44	0.60	ns	ns	1.4	1.9	4	ns

Cyanobacteria: inoculation with a soil-based inoculum of cyanobacteria containing: *Anabaena cylindrica, Anabaena oryzae, Aulosira fertilissima, Nostoc muscorum* and *Tolypothrix tenuis* applied at 100 kg fresh material (90 moisture)/ha 5 days after transplanting.
N: fertilization with urea (46%) at 72 or 144 kg N/ha in two equal doses: 25 days after transplanting and at the mid-tillering stage.
Insecticides: application of Dursban, Fipronil or Cyfen at 12, 24 or 28.8 kg/ha, respectively during the early panicle initiation stage.
LSD (0.05, 0.01): the least significant differences at 95% and 99% confidence levels, respectively.
n.s.: statistically not-significant.
* The agronomic fertilizer N-use efficiency (kg grain yield/kg fertilizer-N) as it is appeared in this table.

Table 2b. Effect of cyanobacterization, nitrogen fertilization and some insecticides on symptoms of infestation with stem borer *Chilo agamemnon* Bles. in the rice variety IR-28

Treatment	Dead heart/m^2		White heads/m^2	
	No. cyan.	Cyan.	No. cyan.	Cyan.
72 N	35.0	13.0	39	15
144 N	53.0	33.0	60	37
72 N + Dursban	20.0	5.0	22	6
144 N + Dursban	22.0	10.0	24	11
72 N + Fipronil	16.7	13.0	19	15
144 N + Fipronil	25.0	18.0	28	20
72 N + Cyfen	30.0	17.0	34	19
144 N + Cyfen	48.0	17.0	54	19

LSD	0.05	0.01	0.05	0.01
Cyano	1	2	1	3
N	4	6	4	7
Insec.	3	4	3	5
Cyano. × N	n.s.	n.s.	6	9
Cyano. × Insec.	4	6	5	6
N × Insec.	4	6	5	6
Cyano × N × Insec.	6	8	7	9

Cyanobacteria: inoculation with a soil-based inoculum of cyanobacteria containing: *Anabaena cylindrica, Anabaena oryzae, Aulosira fertilissima, Nostoc muscorum* and *Tolypothrix tenuis* applied at 100 kg fresh material (90 moisture)/ha 5 days after transplanting.

N: fertilization with urea (46%) at 72 or 144 kg N/ha in two equal doses: 25 days after transplanting and at the mid-tillering stage.

Insecticides: application of Dursban, Fipronil or Cyfen at 12, 24 or 28.8 kg/ha, respectively during the early panicle initiation stage.

Dead heart and white heads: stems showing dead heart and plants showing white heads due to infestation with the stem borer/m^2

LSD (0.05, 0.01): the least significant differences at 95% and 99% confidence levels, respectively.

n.s.: statistically not-significant.

fertilizer-N) (Table 1a), was enhanced by inoculation with the cyanobacteria along with 72 kg N/ha and 50/50 of the macro plus microelement combination Vitafort and the fungicide Sportake-45. Natural damage (infection and severity) due to *Pyricularia oryzae, Helminthosporium oryzae, Alernaria* sp. and *Sclerotium oryzae*, the causative agents of the rice blast, brown spot, leaf spot and stem rot diseases, respectively (Table 1b) was less in the inoculated sub-subplots compared to their non-inoculated counterparts and in case of application of 72 rather than 144 kg N/ha.

Inoculation of the rice variety IR-28 along with urea at 72 rather than 144 kg N/ha, and Cyfen at 28.8 kg/ha, registered the maximum grain yield and N-content and the highest agronomic fertilizer N-use efficiency (Table 2a). Natural infestation with the stem borer (*Chilo agamemnon*), as indexed by numbers of dead hearts or white heads/m^2 (Table 2b), increased with increasing urea fertilizer and was significantly less in the inoculated sub-subplots than the non-inoculated ones.

In this study, enhanced rice growth and performance was achieved by cyanobacterization along with 72 rather than 144 kg N/ha without inoculation. Concomitant increases in the fungal and insect disease symptoms with lapse of time were found to be less in the inoculated sub-subplots compared with their non-inoculated counterparts. Effects of the pesticides differed according to the amount of combined-N and inoculation or not with the dinitrogen fixing cyanobacteria.

The results of the present study suggest that a mechanism involving a sort of regulation of available-N in the rice ecosystem by the dinitrogen-fixing cyanobacteria played a role in diminishing excessive or sub-minimal N absorption by plant. This decreased rice susceptibility to fungal and insect pests which is normally associated with high or low nitrogen status in the plant material.

Reference

Subramanyan R, Relwani L L, Manna G B 1965. Fertility build-up of rice field soils by blue-green algae. Proceedings of the Indian Academy of Science, Series B 62, 252–277.

Yanni Y G, Abd El-Rahman A A M 1993. Assessing phosphorus fertilization of rice in the Nile Delta involving nitrogen and cyanobacteria. Soil Biology and Biochemistry, 25, 289–293.

Yanni Y G, Shalan S N, Mahrous F N 1988. Effect of algalization on rice plant growth, yield, N-content and efficiency of N-use under different nitrogen fertilization treatments. Proceedings the Second Conference of Agricultural Development Research, Ains Shams University, Cairo, Egypt, vol. II, 204–215.

Section IV

Molecular biology of plant–diazotroph interaction

20
Colonization and *nifH* expression on rice roots by *Alcaligenes faecalis* A15

HANS VERMEIREN, JOZEF VANDERLEYDEN AND WEILI HAI

Abstract

The strain under study, *Alcaligenes faecalis* A15, is a diazotroph isolated from the (endo)rhizosphere of paddy rice. To study the interaction of *Alcaligenes faecalis* A15 with rice we have applied the *gusA* reporter-gene technology. With a fusion constitutively expressing the *gusA* gene we have analysed the colonization patterns of rice roots and the environmental factors that affect colonization. With a transcriptional fusion of the *Alcaligenes faecalis* A15 *nifH* gene and *gusA* we were able to visualize the expression of the nitrogen fixation genes, both *in planta* and in free-living conditions. Finally, we have compared *nifH* expression in *Azospirillum brasilense, Azospirillum irakense,* and *Alcaligenes faecalis* A15 under plant root-associated conditions.

Introduction

When using bacteria to replenish the lack of nitrogen of plants, two different strategies can be chosen. One can either try to extend the so-called symbiotic nitrogen-fixing system to non-leguminosae (rice, wheat, maize), or one can try to improve the efficiency (in terms of N transfer) of diazotrophs that already live in close association with the plants of interest. Several groups have already reported on the nodulation of non-leguminosae (Bender *et al.*, 1990; Cocking *et al.*, 1990; Jing *et al.*, 1990) but so far the system seems to be too complex to be merely transferred to non-leguminosae (de Bruijn *et al.*, 1995). Therefore we adopted the latter strategy and chose to study *Alcaligenes faecalis* A15. This strain was isolated from surface-sterilized rice roots (You *et al.*, 1991). *Alcaligenes faecalis* A15 has a considerable potential to be used in a rice-beneficial association, because, besides being a diazotroph and one of the main rice rhizosphere inhabitants, it also appears to be a rice endophyte. Nitrogen-fixing endophytes have

K.A. Malik et al. (eds.), Nitrogen Fixation with Non-Legumes, 167–177.

recently gained attention, since high sugarcane or kallar grass yields on poor soils are possibly due to the presence of endophytes (*Acetobacter diazotrophicus* and *Azoarcus*; Döbereiner *et al.*, 1993 and Reinhold-Hurek *et al.*, 1993, respectively).

We are aware of the fact that not many reports mention the presence of *Alcaligenes faecalis* in the rice rhizosphere. Recently analysed sequence data of the 16S rDNA of this strain show that the strain under study might be related to *Pseudomonas* sp. Pseudomonads are recognized to be abundant in soil; however, the occurrence of N_2-fixing *Pseudomonas* spp. (Watanabe *et al.*, 1987; Haahtela *et al.*, 1983; Chan, 1985) is still controversial. A more extensive taxonomic study of *Alcaligenes faecalis* strain A15 is currently ongoing.

A problem in the study of associative diazotrophs is the lack of an easily detectable plant phenotype indicating effective nitrogen-fixing associations. Vande Broek *et al.*, (1993) successfully developed the *gusA* reporter gene technology to visualize the association of *Azospirillum* with wheat roots. The same approach was applied to study the colonization pattern of *Alcaligenes faecalis* A15 on rice and to give more conclusive evidence on the endophytic character of *Alcaligenes faecalis* A15. To monitor nitrogen fixation *in planta, a nifH::gusA* transcriptional fusion was constructed.

Materials and methods

Bacterial strains, plasmids and media

The bacterial strains and plasmids used in this study are listed in Table 1. *E. coli* strains were routinely grown at 37°C in LB medium (Miller, 1972). LB or minimal MMAB medium (Vanstockem *et al.*, 1987) was used for growing *Alcaligenes faecalis, Azospirillum brasilense* and *Azospirillum irakense* (30°C). When appropriate, the media were supplemented with tetracycline (Tc) at a final concentration of 10 μg/ml, or kanamycin (Km) at 25 μg/ml. Indicator plates for strains carrying *gusA* fusions contained 50 μg/ml 5-bromo-4-chloro-3-indolyl-β,D-glucuronide (X-Gluc) (Jefferson, 1987).

Isolation of recombinant plasmids expressing the gusA *gene in Alcaligenes faecalis A15*

Genomic DNA of *Alcaligenes faecalis* A15 was partially digested with *Sau3*A and ligated into the *Bam*HI site of pFAJ31 (Vande Broek *et al.*, 1993). The ligation mixture was packaged and used to transfect *Escherichia coli* HB101 (Maniatis *et al.*, 1982). Plasmids from approximately 2 000 *E. coli* clones, pooled in LB, were mobilized to A15, using the helper plasmid pRK2013. Recombinant plasmids expressing the *gusA* gene in *Alcaligenes faecalis* A15 were directly selected as blue transconjugants on MMAB medium containing Tc and X-Gluc. Fifteen

Table 1. Bacterial strains and plasmids used in this study

Strain or plasmid	Relevant characteristics	Reference
Strains		
Alcaligenes faecalis A15	Wild-type strain, isolated from rice rhizosphere soil	You *et al.*, 1991
Azospirillum brasilense Sp245	Wild-type strain, isolated from surface sterilised wheat roots (Brasil)	Baldani *et al.*, 1986
Azospirillum irakense KBC1	Wild-type strain, isolated from roots and rhizosphere of rice	Khammas *et al.*, 1989
Escherichia coli HB101	F^- *hsd*S20 ($r^-_R m^-_R$) *rec*A13 *ara*-14 *pro*A2 *lac*Y1 *gal*K2 *rps*L20 *xyl*-5 *mtl*-1 *sup*E44	Maniatis *et al.*, 1982
Plasmids		
pFAJ31	Tc^r, pLAFR3 derivative containing the promoterless *E. coli gusA* gene, cloned into the *Eco*RI–*Bam*HI sites	Vande Broek *et al.*, 1993
pFAJ21	Tc^r, pFAJ31 derivative containing an *Azospirillum brasilense nifH::gusA* fusion	Vande Broek *et al.*, 1993
pAHG	Tc^r, *Alcaligenes faecalis* A15 *nifH::gusA* fusion	This study
pFAJ701	Tc^r, pFAJ31 derivatives containing constitutively	This study
pFAJ707	expressed *Alcaligenes faecalis* promoter::*gusA* fusions	This study
pBI101	pBIN19 derivative, promoterless *gusA* cassette	Jefferson, 1987
pRK2013	Km^r, *rep* (ColE1) mob^+ tra^+	Figurski and Helinski, 1979

blue-staining clones were selected for further analysis. Two clones, pFAJ701 and pFAJ707, with the highest glucuronidase activity when growing aerobically in rich medium (analysed by spectrophotometry using *p*-nitrophenyl-glucuronide as a substrate), were tested under several physiological conditions (in rich and minimal medium, with and without N, in 21% O_2 and 1.5% O_2), and showed more or less equal activities under these growth conditions. Restriction analysis proved that the two recombinants tested did not contain the same DNA fragments, so we used these two constructs simultaneously in the plant experiments to rule out possible artefacts due to the use of one particular promoter. Since plant work does not permit the use of tetracycline, we checked also whether the plasmid (pLAFR3) remained stable in *Alcaligenes faecalis* without selection pressure. After 10 days of growth without selection pressure (in rich medium), 84% of the CFUs still contained the plasmid.

Construction of the nifH::gusA *fusion plasmid pAHG*

On the basis of the sequence of the *Alcaligenes faecalis* A15 *nifH* gene (data not shown), two oligonucleotides were designed to amplify the *nifH* upstream region using PCR 94°C: 50s, 50°C: 60s, 72°C : 120s, 35 cycles, *Taq*I polymerase). The primers used were:

5' AAGG AAGCTT GAT TCA GAT CAT CAG CCG3' and
5' GGAA GGATCC AAT TAA GTT CCT CAGG3'

each containing either an extra *Hind*III or *Bam*HI recognition site (underlined). Using these sites the amplified *nifH* promotor region was cloned in front of the *gusA* startcodon in PBI101. The 2.6 kbp. *nifH::gusA* fusion was subsequently cloned in pLAFR3 (using *Eco*RI and *Hind*III). The resulting construct (pAHG) was conjugated into *Alcaligenes faecalis* A15, using pRK2013 as a helper in a triparental conjugation. *Alcaligenes faecalis* A15 (pAHG) was tested in semi-solid medium for glucuronidase activity. Blue staining of the bacterial pellicle was observed only in nitrogen-fixing conditions (semi-solid medium, without fixed N).

Germination, inoculation, and growth of rice seedlings

Rice seeds (*Oryzae sativa indica* HD428 IRRI) were peeled and surface-sterilized by consecutive immersions in 70% ethanol (1 min) and 45 min in Domestos solution, containing 3% commercial bleach, 0.1% Na_2CO_3, 3% NaCl and a few drops of SDS (10%). The seeds were rinsed five times with sterile distilled water and further incubated overnight in the last change of water (room temperature). The treated seeds were germinated in the dark on sterile agar plates (0.7% agar). Three-day-old seedlings with radicles of approximately 1 cm were planted in aseptic test tubes (20 mm × 200 mm) containing 55 ml of rice medium (Yoshida *et al.*, 1971), with a

N-source (40 ppm NH_4NO_3) and a C-source (0.1% Na-malate), when indicated. The seeds were placed on a metal grid positioned at the surface of the nutrient solution (150 mm high). The seedlings were grown in a growth chamber (12 h day; 27°C day, 25°C night, 70% relative humidity). Exponential cultures of *Alcaligenes faecalis* A15, *Azospirillum brasilense* or *Azospirillum irakense* were harvested, washed with physiological water (0.85% NaCl) and resuspended to an OD_{600} of 0.2, corresponding with a cell density of approximately 10^7–10^8 CFU per ml. After 2–5 days of seed germination, 1 ml portions of bacterial suspension were added directly into the nutrient solution of the rice seedlings.

Preparation of rice roots for histochemical analysis

Roots were excised and gently rinsed in 0.85% NaCl. Washed roots were transferred to 0.1 M phosphate buffer (pH 7), containing 0.5 mg/ml X-Gluc, 0.5 mM $K_3Fe(CN)_6$ and 0.5 mM $K_4Fe(CN)_6$ as an oxidation catalyst, and 10 mM EDTA to mitigate the partial inhibition of the enzyme by the oxidation catalyst (Gallagher, 1992). Roots were stained overnight in this buffer (at 37°C), rinsed in 0.85% NaCl and examined for specific colonization sites. To prepare the roots for light microscopy, blue root segments were excised and fixed for 2 h at room temperature in 2.5% glutaraldehyde in phosphate buffer. After removal of the fixative and two rinses in distilled water, the segments were dehydrated in a graded ethanol series. Samples were embedded in LR White resin (Fisons, Loughborough, UK). Sections of 1.5–2 μm in thickness were cut (Microm HM360 microtome with glass knives) and examined using a Nikon Optiphot-2, under bright field, and phase contrast objectives.

Results and discussion

Surface colonization patterns of Alcaligenes faecalis *A15 on rice roots*

The two strains, *Alcaligenes faecalis* A15 (pFAJ701) and *Alcaligenes faecalis* A15 (pFAJ707), which we proved to be well marked with the *gusA* reporter gene, were used to inoculate rice seedlings (see Materials and Methods). Figure 1 shows the correlation between blue staining and the presence of bacteria. The negative controls, that is uninoculated rice roots (no staining, data not shown) and the absence of bacteria on non-staining roots (from the rice medium with nitrogen, without carbon) were also satisfactory, confirming the lack of endogenous β-glucuronidase activity in rice roots and proving the specificity of the visualization procedure.

In the first inoculation tests we used only two different culture solutions: rice medium with (40 ppm NH_4NO_3) and without a nitrogen source. Remarkably, only rice grown in medium without nitrogen showed a very specific colonization pattern within 3 days after inoculation (DAI), namely at the root caps and at the sites of emergence of lateral roots. A similar pattern was also observed during colonization studies of *Arabidopsis thaliana* by *Azorhizobium caulinodans* (Gough *et al.*, 1992). Figures 2 and 3 show the typical colonization of the base of a side root and of the sub-apical part of rice roots grown without nitrogen. Although Figure 3 suggests that the root cap is completely filled with bacteria, transverse sections showed only colonization of the outer cell layer (the epidermis).

In an attempt to explain the absence of staining (and hence colonization by *Alcaligenes faecalis* A15) of rice growing in N-supplemented medium, we investigated the influence of adding carbon source (0.1% malate). Malate is present in rice root exudates (You *et al.*, 1991) and can be metabolized by *Alcaligenes faecalis* (Qiu *et al.*, 1991). In rice medium containing both carbon and nitrogen a massive colonization of the root system was observed (see Figure 4). Hence we conclude that only in plants stressed by the lack of nitrogen is there sufficient exudation of C-compounds to support bacterial growth, explaining the specific pattern, since root caps are known to excrete mucilage.

Endophytic character of Alcaligenes faecalis *A15*

Alcaligenes faecalis A15 was isolated from surface-sterilized rice roots and TEM micrographs showed bacteria inside rice roots inoculated with *Alcaligenes faecalis* A15 (You *et al.*, 1991). Although not foolproof, these were the first indications that *Alcaligenes faecalis* A15 is able to infect the rice root. By inoculating axenic cultures of rice seedlings with a *gusA* marked *Alcaligenes faecalis* A15, we were able to prove that *Alcaligenes faecalis* A15 can penetrate rice root epidermis cells (Figure 5). To prove that both plant and bacterial cells are viable, we will have to do additional vital staining (DAPI, or other) but the difference between completely destroyed plant cells and colonized plant cells with intact vacuoles is evident on Figure 5.

Associative nifH *expression of* Azospirillum brasilense, Azospirillum irakense *and* Alcaligenes faecalis *on rice*

To analyse whether there is any difference in *nifH* expression in free-living and associative conditions, and to visualize the sites on the roots suitable for nitrogen fixation, rice seedlings were inoculated with *Alcaligenes faecalis* A15 (pAHG). Blue staining was detected only when no combined nitrogen was available in the growth medium. In case of rice grown in nitrogen- and carbon-supplemented medium, similar intensive colonization was observed microscopically as with the constitutive fusions (pFAJ701, pFAJ707), except that the bacteria did not stain and were difficult to detect (data not shown). In Figure 6 a transverse section of a blue-stained area on the main root is shown. Besides being colonized, rice root epidermal cells seem to offer an ideal environment to express the nitrogenase genes.

We also tested *Azospirillum brasilense* Sp245 and *Azospirillum irakense* containing an *Azospirillum nifH::gusA* fusion (pFAJ21, Vande Broek *et al.*, 1993) on rice in order to test the role of the bacterial partner in the association (Figures 7 and 8, respectively). The main difference observed was that

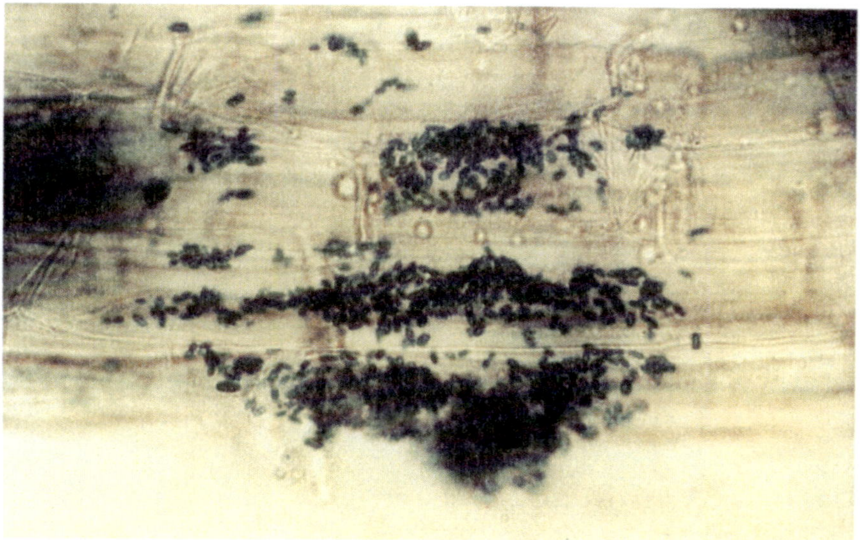

Figure 1. *Alcaligenes faecalis* A15 (pFAJ701) on a rice root surface.

Figure 2. *Alcaligenes faecalis* A15 (pFAJ707) showing colonization of emergence point of lateral root.

only *Alcaligenes faecalis* (Fig. 6) and *Azospirillum irakense* (Fig. 8) were able to penetrate plant cells. This coincides with the observation that *Azospirillum irakense* is the only *Azospirillum* species showing pectinolytic activities (Khammas *et al.*, 1989).

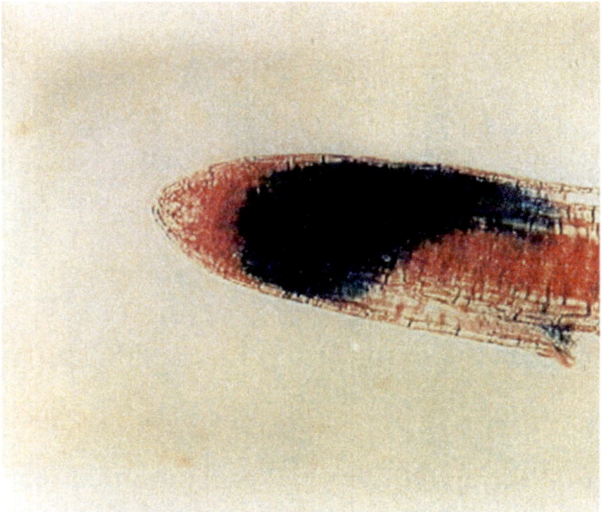

Figure 3. *Alcaligenes faecalis* A15 (pFAJ701) showing root cap colonization.

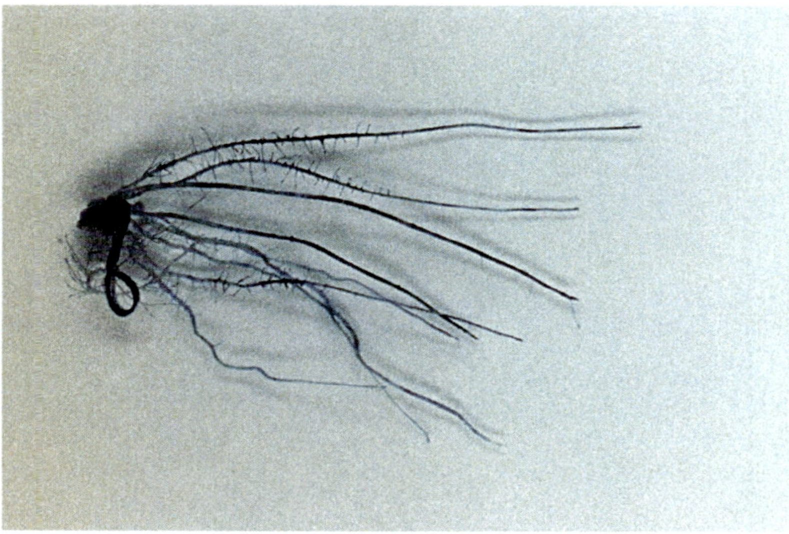

Figure 4. *Alcaligenes faecalis* A15 (pFAJ707) showing rice root colonization in medium with nitrogen and carbon.

For a full understanding of the *Alcaligenes faecalis* A15/rice association the use of the reporter gene constructs presented in this study will have to be combined with mutagenesis in order to determine which bacterial gene products are essential in this process.

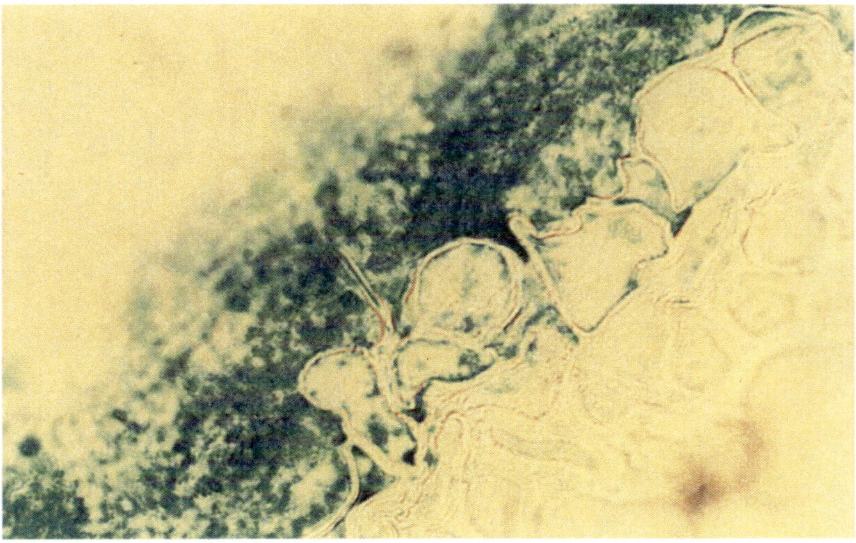

Figure 5. Cross-section of a rice root. *Alcaligenes faecalis* A15 (pFAJ707) showing intracellular epidermis colonization.

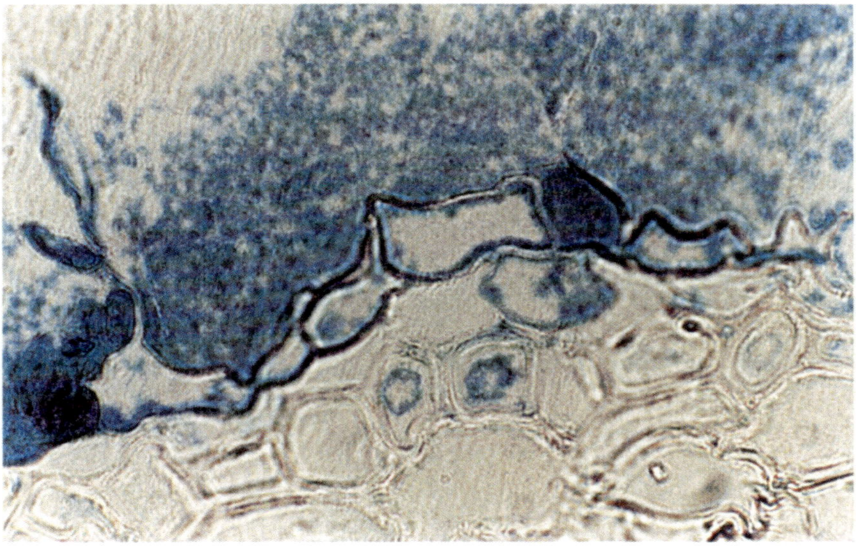

Figure 6. Cross-section of a rice root. *Alcaligenes faecalis* A15 (pAHG) showing intracellular *nifH* expression.

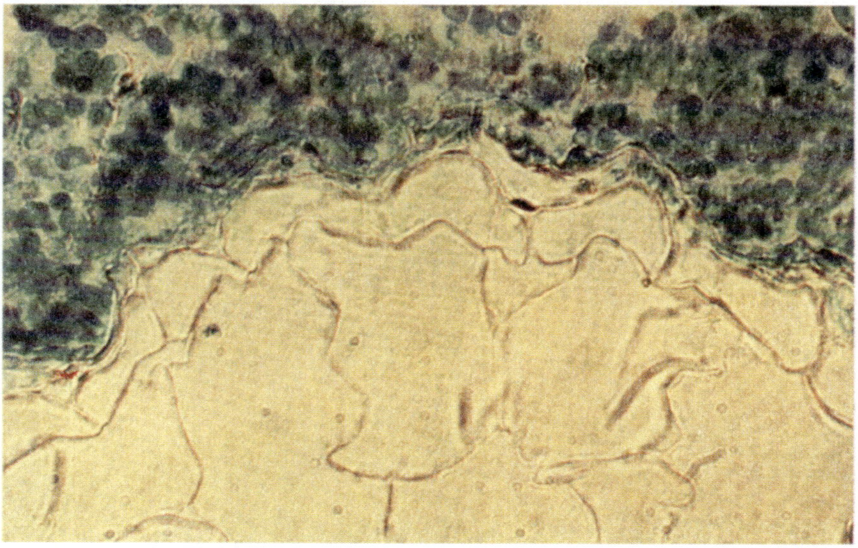

Figure 7. Cross-section of a rice root. *Azospirillum brasilense* Sp245 (pFAJ21) showing *nifH* expression on rice root surface.

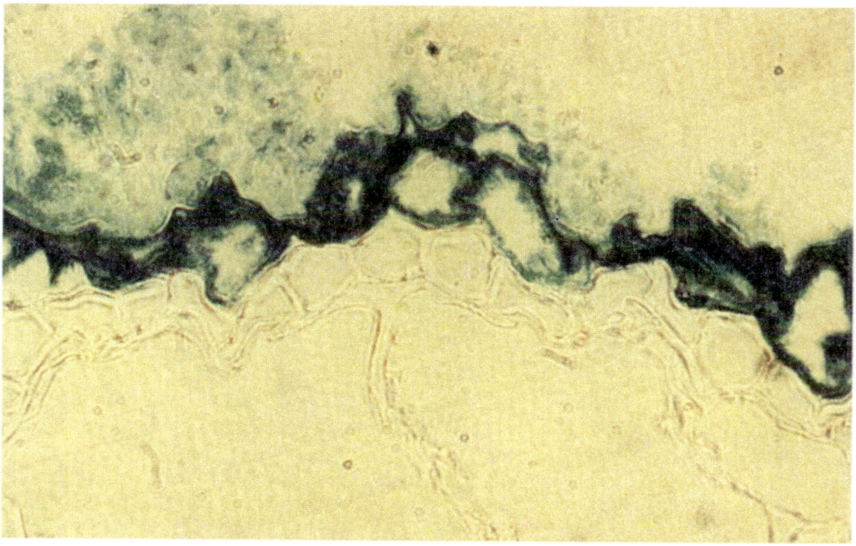

Figure 8. Cross-section of a rice root. *Azospirillum irakense* KBC1 (pFAJ21) showing intracellular *nifH* expression.

References

Baldini V L D, Alvarez M A, Baldini J I, Döbereiner J 1986. Establishment of inoculated *Azospirillum* spp in the rhizosphere and in roots of field grown wheat and sorghum. Plant Soil. 90, 35–46.

Bender G L, Preston L, Barnard D, Rolfe B G 1990. Formation of nodule-like structures on the roots of the non-legumes rice and wheat. In: Gresshoff P M, Roth L E, Stacey G, Newton W E, eds. Nitrogen Fixation: Achievements and Objectives. Chapman & Hall, London, 825.

Chan Y K 1985. Denitrification by a diazotrophic *Pseudomonas* species. Can. J. Microbiol., 31, 1136–1141.

Cocking E C, Al-Mallah M K, Benson E, Davey M R 1990. Nodulation of non-legumes by rhizobia. In: Gresshoff P M, Roth L E, Stacey G, Newton W E, eds. Nitrogen Fixation: Achievements and Objectives. Chapman & Hall, London, 813–823.

deBruijn F J, Jing Y, Dazzo F B 1995. Potential and pitfalls of trying to extend symbiotic interactions of nitrogen-fixing organisms to presently non-nodulated plants, such as rice. Plant Soil, 174, 225–240.

Döbereiner J, Reis V M, Paula M A, Olivares F 1993. Endophytic diazotrophs in sugar cane, cereals and tuber plants. In: Palacios J, Mora J, Newton W E, eds. New Horizons in Nitrogen Fixation. Proceedings of the 9th International Symposium on Nitrogen Fixation. Kluwer, Dordrecht, 671–676.

Figurski D H, Helinski D R 1979. Replication of an origin containing derivative of plasmid RK2 dependent on a plasmid function provided in trans. Proc. Natl. Acad. Sci. USA, 76, 1648–1652.

Gallagher S R (ed.) 1992. GUS Protocols: using the GUS gene as a reporter of gene expression. Academic Press, London.

Gough C, Vasse J, Galera C, Webster G, Cocking E, Denarie J 1997. Interactions between bacterial diazotrophs and non-legume dicots: Arabidopsis thaliana as a model plant. Plant Scil, 194, 123–130.

Haahtela K, Helander I, Nurmiaho-Lassila E L, Sundman V 1983. Morphological and physiological characteristics and lypopolysaccharide composition of nitrogen-fixing (C_2H_2 reducing) root-associated *Pseudomonas* sp. Can. J. Microbiol., 29, 874–880.

Jefferson R A 1987. Assaying chimeric genes in plants: the GUS gene fusion system. Plant Mol. Biol. Rep., 5, 387–405.

Jing Y, Li G, Jin G, Shan X, Zhang B, Guang C, Li J 1990. Rice root nodules with acetylene reduction activity. In: Gresshoff P M, Roth L E, Stacey G, Newton W E, eds. Nitrogen fixation: achievements and objectives. Chapman & Hall, London, 829.

Khammas K M, Ageron E, Grimont P A D, Kaiser P 1989. *Azospirillum irakense* sp. nov. a nitrogen fixing bacterium associated with rice roots and rhizosphere soil. Res. Microbiol., 140, 679–693.

Maniatis T, Fritsch E F, Sambrook J 1982. Molecular Cloning: A Laboratory Manual. Cold Spring Harbor, New York.

Miller J H 1972. Experiments in Molecular Genetics. Cold Spring Harbor, New York.

Qiu Y, Zhou S, Mo X, Wang D, Hong J 1991. Study of nitrogen fixing bacteria associated with rice root. Isolation and identification of organisms. In: You C B (ed.) The Associative Nitrogen Fixation in the Rice Rhizosphere. Agricultural Publishing House, Beijing, China, 13–18.

Reinhold-Hurek B, Hurek T, Gillis M, Hoste B, Vancanneyt M, Kersters K, DeLey J 1993. *Azoarcus* gen. nov., nitrogen fixing proteobacteria associated with the roots of Kallar grass (*Leptochloa fusca* (L.) Kunth), and description of two species, *Azoarcus indigens* sp. nov. and *Azoarcus communis* sp. nov. Int. J. Syst. Bacteriol., 43, 574–584.

Vande Broek A, Michiels J, Van Gool A, Vanderleyden J 1993. Spatial-temporal colonization patterns of *Azospirillum brasilense* on the wheat root surface and expression of the bacterial *nifH* gene during association. Mol. Plant Microb. Interact, 6, 592–600.

Vanstockem M, Michiels K, Vanderleyden J, Van Gool A 1987. Transposon mutagenesis of *Azospirillum brasilense* and *Azospirillum lipoferum*, physical analysis of *Tn*5 and *Tn*5-mob insertion mutants. Appl. Environ. Microbiol., 53, 410–415.

Watanabe I, So R, Ladha J K, Katajama-Fujimura Y, Kuraishi H 1987. A new nitrogen-fixing species of pseudomonad: *Pseudomonas diazotrophicus* sp. nov. isolated from the root of wetland rice. Can. J. Microbiol., 33, 670–678.

Yoshida S, Formo D A, Cock J H 1971. Laboratory Manual for Physiological Studies of Rice. IRRI, Los Banos, Philippines.

You C B, Song W, Wang H X, Li J P, Lin M, Hai W L 1991. Association of *Alcaligenes faecalis* with wetland rice. Plant Soil, 137, 81–85.

21

Responses of *Azospirillum brasilense* to salinity stress

A.K. TRIPATHI AND B.M. MISHRA

Abstract

Increase in sodium chloride concentrations inhibited growth, acetylene-reducing activity and indoleacetic acid production in *Azospirillum brasilense* Cd. Supplementation of glycine betaine did not relieve the salinity stress-mediated inhibition of growth; instead, it added to the growth inhibition. Inhibition of growth of *A. brasilense* Cd by 2,4-dehydroproline in the presence of 200 mM NaCl suggested the existence of a ProP- or ProU- like transporter which can transport both proline and glycine betaine under osmotic stress. Utilization of glycine betaine as the sole source of nitrogen explained the inability of *A. brasilense* Cd to use it as an osmoprotectant. Osmoregulatory properties of the two strains of *A. brasilense* (Sp7 and Cd) suggest that it might be strain-specific.

Introduction

Substantial areas of agricultural potential in the tropics and subtropics are affected by salinity which is known to have an adverse affect on plant productivity and on the activities of plant-associated bacteria. Understanding the mechanisms of osmoadaptation in plant growth-promoting bacteria is expected to contribute to the long-term goal of enhancing plant–microbe interaction for the improvement of crop productivity in salinity-affected fields (Gloux and Le Rudulier, 1989; Madkour *et al.*, 1990).

Bacteria of the genus *Azospirillum* colonize the roots of several cereals and grasses and are regarded as plant growth-promoting rhizobacteria due to their ability to produce phytohormones (indole acetic acid, gibberellins, cytokinins, etc.) and to fix atmospheric nitrogen (Bashan and Lavanony, 1990; Okon and Labandera-Gonzalez, 1994). Knowledge of the mechanisms of osmoregulation

K.A. Malik et al. (eds.), Nitrogen Fixation with Non-Legumes, 179–185.

in *Azospirillum* is rather limited. Osmotolerance depends on the ability of bacteria to scavenge potential osmoprotectants such as proline, betaine, choline, glutamate, etc., from the root exudates under salinity stress. In *Azospirillum,* osmotolerance has been reported to decrease in the order *A. halopraeferens, A. irakense, A. brasilense, A. lipoferum* and *A. amazonense* (Hartmann 1989; Hartmann *et al.*, 1991). The utilization pattern of different compounds such as choline, glycine betaine, proline, glutamate, etc., differs in different species of *Azospirillum. A. halopraeferens* and *A. brasilense* used glycine betaine, proline and glutamate as osmoprotectants whereas *A. lipoferum* and *A. amazonense* used them as sources of carbon, nitrogen and energy (Hartmann, 1988).

Azospirillum brasilense Sp7 has been shown to utilize glycine betaine as an osmoprotectant (Riou and Le Rudulier, 1990). The presence of glycine betaine binding protein (GBBP) was also demonstrated in this bacterium by immunological cross-reaction with anti-GBBP (*E. coli*) antibody (Riou *et al.*, 1991). Glycine betaine, however, failed to elicit a similar response in *A. brasilense* Cd (Tripathi *et al.*, 1995) which is considered to be very closely related to Sp7 (Schloter *et al.*, 1994). This report presents further observations on the response of *A. brasilense* Cd to sodium chloride.

Materials and methods

Azospirillum brasilense Sp7 (ATCC 29145) and Cd (ATCC 29710) were grown at 30°C in the minimal medium (Albrecht and Okon, 1980). Growth in liquid cultures was monitored as increase in absorbance at 560 nm. The acetylene-reducing activity of the cultures grown in semi-solid minimal medium was measured as described earlier (Singh and Klingmüller, 1986). For estimating indole acetic acid production, cultures were grown for 48 h in minimal medium supplemented with 100 μg/ml tryptophan followed by measuring absorbance at 530 nm after the Salkowski reaction (Hartmann *et al.*, 1983). Utilization of glycine betaine as the sole source of carbon and nitrogen for growth was tested by replacing malate and NH_4Cl with glycine betaine, respectively.

Results

Figure 1 shows the effect of increasing sodium chloride concentration on acetylene-reducing activity (Figure 1a) and indole acetic acid-producing activity (Figure 1b) of *A. brasilense* SpCd. Increase in salinity affected both activities adversely. Inhibition of both the activities was proportional to increase in salinity. Indole acetic acid production, however, was stimulated at low levels of salinity, i.e. 50 mM NaCl.

Figure 2 (a and b) shows the effect of NaCl on the growth of *A. brasilense* Cd and the role of proline and glycine betaine in osmoprotection. Increase in NaCl concentrations inhibited growth proportionally. Though proline supplementation partially relieved NaCl-mediated growth inhibition at all the NaCl

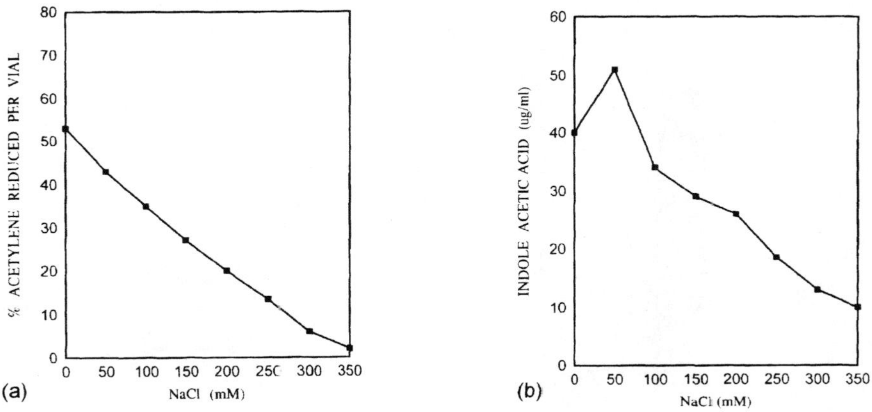

Figure 1. **(a)** Effect of sodium chloride on acetylene-reducing ability of a *A. brasilense* Sp Cd. Acetylene reduced in each vial was measured after 24 h by a gas chromatograph. **(b)** Effect of sodium chloride on indoleacetic acid (IAA) production by *A. brasilense* Sp Cd. IAA content was estimated with Salkowski reagent after growing the culture for 48 h.

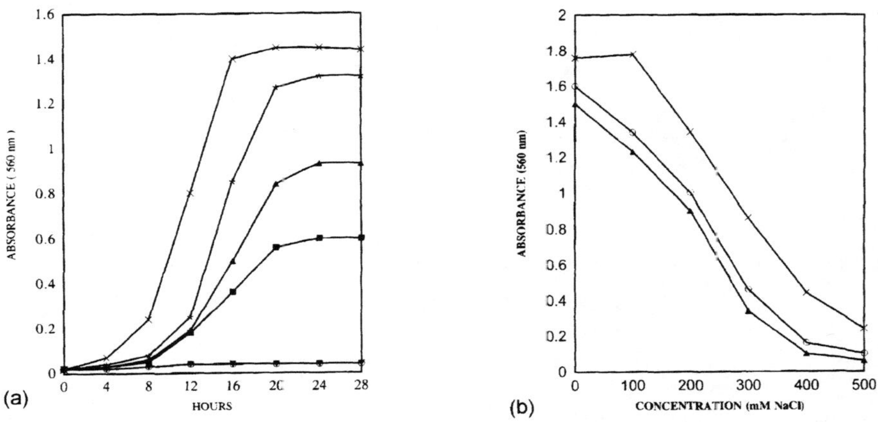

Figure 2. **(a)** Effect of glycine betaine on *A. brasilense* Sp Cd grown with sodium chloride. (×) Minimal medium (MM), (*) 200 mM NaCl, (▲) 300 mM NaCl, (■) 300 mM NaCl + 1 mM glycine betaine, (○) 500 mM NaCl, (▼) 500 mM NaCl + 1 mM glycine betaine. **(b)** Effect of glycine betaine and proline on sodium chloride-mediated growth inhibition in *A. brasilense* Sp Cd. (○) MM, (▲) MM + 10 mM glycine betaine, (×) MM + 10 mM proline. Growth was monitored by recording $O\Delta_{560}$ after 24 h.

concentrations (Figure 2b) glycine betaine failed to show a similar effect (Figures 2a and 2b). Figure 3 shows a comparison of the osmoprotective effect of proline and glycine betaine on *A. brasilense* Sp7 and *A. brasilense* SpCd strains. Both proline and glycine betaine displayed an osmoprotective effect in Sp7, but only proline showed some osmoprotective effect in the strain Cd.

Figure 4 shows the growth of *A. brasilense* with glycine betaine as the sole

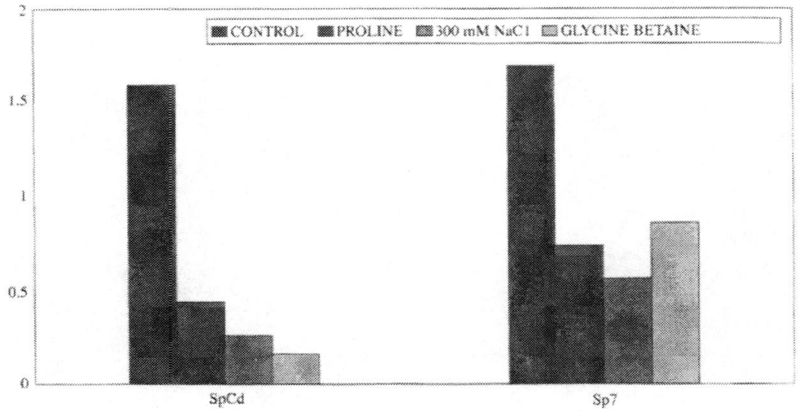

Figure 3. Comparison of the effect of proline and glycine betaine on relieving sodium chloride-mediated growth inhibition in *A. brasilense* Sp Cd. and *A. brasilense* Sp7. Growth was monitored by recording $O\Delta_{560}$ after 24 h.

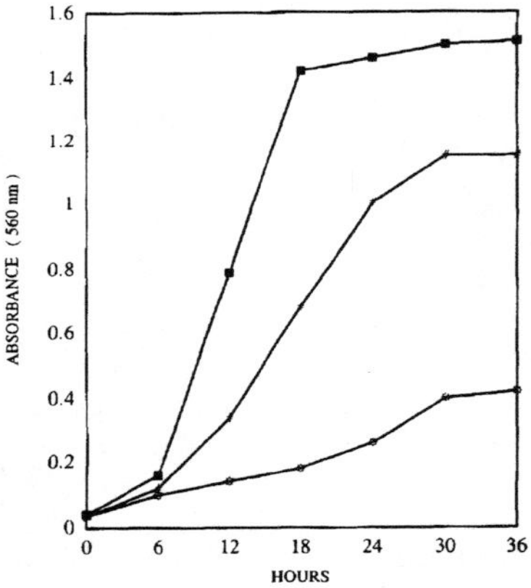

Figure 4. Utilization of glycine betaine as sole source of nitrogen (#) and carbon (-○-) by *A. brasilense* Sp. Cd. (■) growth of *A. brasilense* in MM containing NH_4Cl as N-source and malate as C-source.

source of carbon and nitrogen. It was observed that glycine betaine supported the growth of *A. brasilense* SpCd as sole nitrogen source. We have earlier observed that the acetylene-reducing ability of *A. brasilense* SpCd was inhibited by glycine betaine (Tripathi *et al.*, 1995).

Figure 5 shows a comparison of the colonies of *A. brasilense* Cd on plates with 250 mM NaCl and 0.3 mM 2,4-dehydroproline, supplemented together and separately. Sodium chloride (250 mM) or 2,4-dehydroproline permitted growth when supplemented separately, but inhibited growth completely when supplemented together.

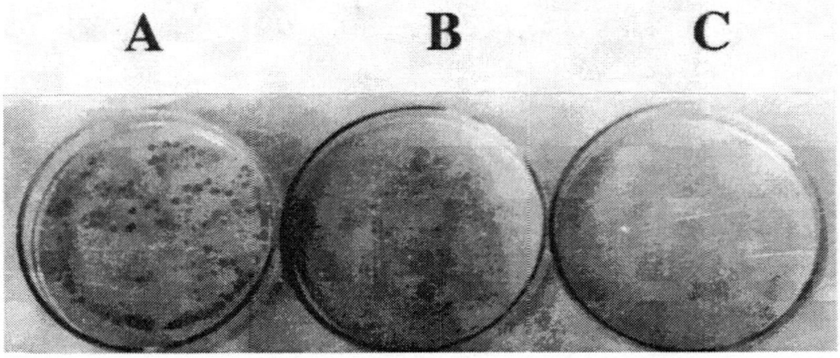

Figure 5. Effect of sodium chloride (300 mM) on the toxicity of 2,4-dehydroproline (DHP) on *A. brasilense* SpCd. **A**: MM + 0.2 mM DHP, **B**: MM + 300 mM NaCl **C**: 0.2 mM DHP + 300 mM NaCl.

Discussion

Sodium chloride inhibited growth and plant growth-promoting activities, e.g. nitrogen fixation and indoleacetic acid production, in *A. brasilense* Cd. Indole production was not inhibited as strongly as was acetylene-reducing activity. The slight increase in salinity was found to stimulate indoleacetic acid production. This type of response may result in the proliferation of roots, resulting in enhanced exudation of osmoprotectants such as proline or betaines by the plant partner. At the low level of salinity this might be one of the strategies for the better survival of both the plant and the associated rhizobacteria.

Under salinity stress proline, but not glycine betaine, was found to have an osmoprotective effect in *A. brasilense* Cd. The effect of 2,4-dehydroproline, a proline analogue, on the survival of *A. brasilense* Cd in the presence and absence of NaCl stress indicated that this bacterium possessed a salinity stress-induced proline transport system.

In contrast to *A. brasilense* Sp7, the strain Cd could not use glycine betaine as an osmoprotectant. Both strains, however, have been shown to possess glycine betaine-binding protein (GBBP). Our earlier study indicated that GBBP in *A. brasilense* SpCd is constitutively synthesized and is not regulated osmotically (Tripathi *et al.*, 1995). The ability of *A. brasilense* SpCd to grow on glycine betaine as the sole source of nitrogen was in complete contrast to *A. brasilense* Sp7, which could not use glycine betaine as nitrogen source but accumulated it as an osmoprotectant under salinity stress (Riou and Le Rudulier, 1990). Mutations in the adenylate kinase gene have been shown to be responsible for the inability of glycine betaine to act as an osmoprotectant in *Salmonella typhimurium* grown in media having elevated osmolarity (Gutierrez and Csonka, 1995). Although *A. brasilense* Sp7 and Cd are closely related (de Mot and Vanderleyden, 1989), they have been reported to differ at least with regard to their carotenoid content (Eskew *et al.*, 1977), chemotactic response to benzoic acid (Lopez-de-Victoria and Lovell,

1993), lipopolysaccharide and outer-membrane protein content (Schloter *et al.*, 1994). Osmoregulatory properties in *Azospirillum* thus appear to be strain-specific.

Glycine betaine, in *Rhizobium meliloti*, has been shown to act as an osmoprotectant under osmotic stress, but is utilized as a source of carbon, nitrogen and energy in the absence of osmotic stress (Smith *et al.*, 1988). Proline, on the other hand, is catabolized as a source of carbon, nitrogen and energy and, hence, is not used as an osmoprotectant in this bacterium (Maloy, 1989).

Acknowledgements

This work was supported by grants from the Board of Research in Nuclear Sciences, Bombay and the Deutsche Forschungsanstalt fur Luft-und Raumfahrt e. V. (DLR), Germany.

References

Albrecht S L, Okon Y 1980. Cultures of *Azospirillum*. In: San Pietro A, ed. Methods in Enzymology, vol. 69, Academic Press, New York, 740–749.

Bashan Y, Lavanony H 1990. Current status of *Azospirillum* inoculation technology: *Azospirillum* as a challenge to agriculture. Can. J. Microbiol., 36, 591–608.

de Mot R, Vanderleyden J 1989. Application of two-dimensional protein analysis for strain finger printing and mutant analysis of *Azospirillum* species. Can. J. Microbiol., 35, 960–967.

Eskew D L, Focht D D, Ting J P 1977. Nitrogen fixation, denitrification and pleomorphic growth in a highly pigmented *Spirillum lipoferum*, Appl. Environ. Microbiol., 34, 582–585.

Gloux K, Le Rudulier D 1989. Transport and catabolism of proline betaine in salt-stressed *Rhizobium meliloti*. Arch. Microbiol., 151, 143–148.

Gutierrez J A, Csonka L N 1995. Isolation and characterization of adenylate kinase (adk) mutations in *Salmonella typhimurium* which block the ability of glycine betaine to function as an osmoprotectant. J. Bacteriol., 177, 390–400.

Hartmann A, Singh M, Klingmuller W 1983. Isolation and characterization of *Azospirillum* mutants, excreting high amounts of indoleacetic acid. Can. J. Microbiol., 29, 916–923.

Hartmann A 1988. Influence of amino acids on nitrogen fixation ability and growth of *Azospirillum* spp. Appl. Environ Microbiol., 54, 87–93.

Hartmann A 1989. Ecophysiological aspects of growth and nitrogen fixation in *Azospirillum* spp. Plant Soil, 110, 225–238.

Hartmann A, Prabhu S R, Galinski E A 1991. Osmotolerance of diazotrophic rhizosphere bacteria. Plant Soil, 137, 105–109.

Lopez-de-Victoria G, Lovell C R 1993. Chemotaxis of *Azospirillum* species to aromatic compounds. J. Bacteriol., 59, 2951–2955.

Madkour M A, Smith L T, Smith G M 1990. Preferential osmolyte accumulation: a mechanism of osmotic stress adaptation in diazotrophic bacteria. Appl. Environ. Microbiol., 56, 2876–2881.

Maloy S R 1989. The proline utilization operon. In: Neidhardt F C, Ingraham J L, Brooks-Low K, Magasanik B, Schaechter M, Umbarger H E, eds. *Escherichia coli* and *Salmonella typhimurium: Cellular and Molecular Biology*. American Society of Microbiology, Washington, DC, 1513–1519.

Okon Y, Labandera-Gonzalez C A 1994. Agronomic applications of *Azospirillum*: An evaluation of 20 years world-wide field inoculation. Soil Biol. Biochem, 26, 1591–1601.

Riou N, Le Rudulier D 1990. Osmoregulation in *Azospirillum brasilense*: glycine betaine transport enhances growth and nitrogen fixation under salt stress. J. Gen. Microbiol., 136, 1455–1461.

Riou N, Poggi M C, Le Rudulier D 1991. Characterization of an osmoregulated periplasmic glycine betaine binding protein in *Azospirillum brasilense* Sp7. Biochimie, 73, 1187–1193.

Schloter M, Moens S, Croes C, *et al.*, 1994. Characterization of cell surface components of *Azospirillum brasilense* Sp7 as antigenic determinants for strain-specific monoclonal antibodies, Microbiology, 140, 823–828.

Singh M, Klingumuller W 1986. Transposon mutagenesis in *Azospirillum brasilense*: Isolation of auxotrophic and NIF⁻ mutants and molecular cloning of the mutagenized *nif* DNA. Mol. Gen. Genet., 202, 136–142.

Smith L T, Pocard J A, Bernard T, Le Rudulier D 1988. Osmotic control of glycine betaine biosynthesis and degradation in *Rhizobium meliloti*, J. Bacteriol., 170, 3142–3149.

Tripathi A K. Mishra B M, Schloter M 1995. Role of choline and glycine betaine in *Azospirillum brasilense* Cd under salinity stress. In: Fendrik I, Del Gallo M, Vanderleyden J, eds. NATO-ASI series, vol. G37, *Azospirillum* VI and related microorganisms, Springer-Verlag, Berlin, Heidelberg, 383–391.

22
Studies on the diploid indeterminate nodule-forming plant *Melilotus alba* Desr., white sweetclover

ANN M. HIRSCH, THOMAS A. LaRUE, JOANNA H. NORRIS, AND NORMAN F. WEEDEN

Abstract

Melilotus alba is a small-seeded, autogamous diploid (n = 8) with a relatively short life cycle. U389 and its dwarf derivative U390 are prolific seed producers considering their sizes. The genome size of *Melilotus* has been measured as 1000–1250 Mbp; a genomic library has been constructed. A map of morphological, isozyme, and RAPDs markers has been generated for *M. alba* and *M. polonica*. EMS, azide, and neutron irradiation mutants have been generated. A transformation system for white sweetclover is being developed.

Introduction

The study of *Melilotus alba* as a genetic system was begun by H.J. Gorz and F.A. Haskins, now Professors Emeriti, of the University of Nebraska. Gorz and Haskins obtained a number of USDA accessions of sweetclover. One of these, from India, was PI165554, a diploid (n = 8) sweetclover. Most sweetclovers, and probably all that most biologists have seen, are tall, branched, and bushy. These common weeds are annuals whereas commercial cultivars are usually biennials. Nearly all sweetclovers need to be 'tripped' to promote self-fertilization. However, PI165554 is autogamous, monopodial, and fast-flowering. Numerous flowers are borne in a raceme and one or two seeds are set per pod.

U389, a single seed line selected from PI165554 by Gorz *et al.*, 1975, is a small-seeded, dwarfed annual, having a life cycle of approximately 8–10 weeks, seed to seed. Because of its small size, many plants can be grown in a small space. Another advantage is that *M. alba* is easily mutable. It is second only to pea in the number of *sym* mutants obtained. In addition, a large number of morphological and chlorophyll mutants have been generated by the University

K.A. Malik et al. (eds.), Nitrogen Fixation with Non-Legumes, 187–194.

of Nebraska and Boyce Thompson Institute laboratories. Sweetclover is nodulated by *Rhizobium meliloti* and exhibits a close relationship to alfalfa (*Medicago sativa*) and fenugreek (*Trigonella* sp.) (Smith and Gorz, 1965). Thus, it should exhibit many of the same characteristics and patterns of gene expression as alfalfa. Kneen and LaRue (1988) recognized sweetclover as a species which was ideal for studying the symbiosis of *Rhizobium meliloti* and its host.

Genome size and genetic complexity

The genome size of *Melilotus alba* is almost 10 times larger than the model plant *Arabidopsis*: 1254 Mbp versus 193 Mbp per haploid genome when Feulgen microdensitometry is used to measure DNA levels (Bennett and Leitch, 1995). When other legumes are compared using flow cytometry methods (Arumuganathan and Earle, 1991), *M. officinalis* is intermediate in genome size (Table 1). The *M. alba* genome size has not yet been measured using this method.

Using a cDNA clone corresponding to the gene *MsENOD40* as a probe against genomic digests of both alfalfa and sweetclover, we detected only one band in sweetclover and two bands in alfalfa DNA (M. Lum and A.M. Hirsch, unpublished results; Asad *et al.*, 1994). Similarly, when a lectin cDNA was used as a probe (Brill *et al.*, 1995), only one band was detected in sweetclover whereas three were found in alfalfa genomic digests (L.M. Brill and A.M. Hirsch, unpublished results). *Medicago truncatula* (barrel medic), which has been proposed as a model legume because of its small genome size and because it is not as genomically complex as alfalfa (Barker *et al.*, 1990), also has three lectin genes (Bauchrowitz *et al.*, 1992). Thus, for genes important for the establishment of the symbiosis, the sweetclover genome may be simpler than the medic genomes.

A linkage map has been generated for both *M. alba* and *M. polonica* using

Table 1. Nuclear DNA content of various legumes

Plant	Ploidy level (x)	DNA content (Mbp/1C)
Arabidopsis thaliana	2	145
Arachis hypogaea	4	2813
Glycine max	4	1115
Medicago sativa	4	1510
Medicago truncatula	2	454–526
Melilotus officinalis	2	1086
Phaseolus vulgaris	2	637
Pisum sativum	2	3947, 4397
Trifolium pratense	2	468
Vigna radiata	2	579

All data from Arumuganathan and Earle (1991).

morphological, isozyme, and RAPD markers (N.F. Weeden and T. Cargnoni, unpublished results). Both maps are surprisingly short: 252 centimorgans with 62 markers for *M. alba* and 314 centimorgans with 81 markers for *M. polonica*.

Non-symbiotic mutants

Early work (pre-1965) on the genetics and cytogenetics of sweetclover is described in Smith and Gorz (1965). In addition to producing sweetclovers low in coumarin content (*Cu/cu* alleles), and lacking β-glucosidase activity (*B/b* alleles), as well as identifying genes important for seed colour (*Y/y*) and for seed and seedling colour (*C/c*), Gorz and Haskins and their colleagues developed 47 different *M. alba* lines. These lines were generated from either biennial lines or from crosses between U389 and a biennial form of *M. alba*.

U389 was subsequently used as the parent line for mutagenesis with ethyl methanesulphonate treatment (EMS) (Table 2). The mutants obtained by the University of Nebraska Laboratory are indicated by the prefix 'U'. A large number of chlorophyll mutants was obtained (Gengenbach *et al.*, 1970; Ronnenkamp *et al.*, 1975). Another important mutant from the Nebraska lines is U390, a short-internode dwarf (Figure 1) (Gengenbach *et al.*, 1969).

Kneen and LaRue (1988) utilized U389 for a large-scale mutagenesis programme with EMS, azide, or neutron/γ-radiation; these mutants are designated by the prefix 'BT' (Table 2). Thirty-five stable chlorophyll-deficient lines were independently isolated, but allelism tests with each other and with the Nebraska lines have not been done. In addition, 15 independently isolated stable Antho$^-$ (anthocyanin-minus) mutants were isolated using either EMS, γ-irradiation, or nitroso ethyl or methyl urea (Table 2). We crossed U390, a super-dwarf mutant, with BT25, a super-tall *Melilotus* with green stems (see Table 2). We found that stem colour showed a 3:1 ratio in the F2 generation, but height was semi-dominant (W. Yang and A.M. Hirsch, unpublished results). Crossing can be difficult because of the small size of the sweetclover flowers, but methods have been developed to make this operation easier to accomplish (Miller *et al.*, 1991b; W. Yang and A.M. Hirsch, unpublished results).

Nodulation mutants

Kneen and LaRue (1988) also isolated nine non-nodulating lines of sweetclover. Miller *et al.* (1991a) demonstrated that the nine represented five independent loci (Table 2). The mutations in genes designated as *sym* are monogenic and recessive. BT35, BT58, and BT62 (*sym1*) were found to be allelic. Mutations at the *sym3* locus (BT61, BT69, and BT70) were the result of two different types of mutagens, EMS and fast neutron bombardment. It was interesting that these different agents induced non-nodulation at the same locus (Miller *et al.*, 1991a).

The different *sym* mutants exhibit varying responses to wild-type *R. meliloti*

Table 2. Sweetclover monogenic mutants (excluding chlorophyll mutants); symbiotic mutants appear in bold typeface

Designation (author)	Type of mutant	Phenotype	Comments
U390 (Gorz)	EMS	Short-internode dwarf	Recessive (*dwdw*); wild-type for nodulation
U362 (Gorz)	EMS	Folded leaflet	Recessive (*ff*)
U363 (Gorz)	EMS	Elongated stem	Recessive (*elel*)
U367 (Gorz)	EMS	Short-petiole dwarf	Recessive ($d_{sp}d_{sp}$)
U391 (Gorz)	EMS	Multifoliate leaf	Dominant (*MfMf*)
U392 (Gorz)	EMS	Curled leaf	Recessive (*clcl*)
U393 (Gorz)	EMS	Cotyledonary branching	Recessive (*cbcb*)
BT9 (LaRue)	EMS	Antho$^-$, dark seed	
BT19 (LaRue)	EMS	Antho$^-$	
BT20 (LaRue)		Antho$^-$	
BT25 (LaRue)	γ-Irradiation	Late nodulation, small roots	
BT26 (LaRue)	γ-Irradiation	Antho$^-$, tall, small roots, few laterals	
BT52 (LaRue)	EMS	Antho$^-$	
BT58 (LaRue)	**EMS**	**Nod$^-$**	**Recessive; *sym1***, allelic with BT35, BT58, BT62
BT59 (LaRue)	**EMS**	**Nod$^-$**	**Recessive; *sym2***
BT60 (LaRue)	EMS	Few laterals	
BT61 (LaRue)	**EMS**	**Nod$^-$**	**Recessive; *sym3***, allelic with BT69 (neutron), BT70 (neutron)
BT65 (LaRue)	Nitroso ethyl urea	Nod$^+$, Lut$^-$, Antho$^-$, pale seed	
BT67 (LaRue)	Nitroso methyl urea	Late nod, small roots, leaves slow to green	
BT68 (LaRue)	**Neutron radiation**	**Nod$^-$**; pale leaflets, short stature	**Recessive; *sym4***
BT71 (LaRue)	**Neutron radiation**	**Nod$^-$**	**Recessive; *sym5***

inoculation. Originally, *sym1* was thought to be less responsive than *sym3* mutants to *R. meliloti* (Utrup *et al.*, 1993). However, one of the *sym1* mutants, BT62, was found to represent a weak allele; there was a low level (12%) of nodulation found on the roots of this mutant (Wu *et al.*, 1996). Although all the mutants exhibit some degree of root hair deformation, there was no evidence for infection thread formation in *sym3*, *sym1* and *sym5* mutants. In addition, several of the *sym* mutants are affected in their responses to mycorrhizal fungi (A.M. Hirsch, T.A. LaRue and J. Norris, unpublished observations). This finding matches with pea, in which Duc and colleagues found that pea mutants blocked at infection thread formation (Inf$^-$) are also Myc$^-$ (Duc *et al.*, 1989).

Treatment of the five different nodulation mutants with the auxin transport

Figure 1. A mature plant (6 weeks old) of U390, a dwarf mutant of sweetclover (*Melilotus alba* Desr.) generated by Gorz and Haskins.

inhibitor, *N*-(1-naphthyl)phthalamic (NPA), results in the formation of nodule-like structures that contain transcripts for the early nodulin gene *ENOD2* (Wu *et al.*, 1996). This shows that, although these mutants are unable to undergo cell divisions in response to *Rhizobium* inoculation, they are not blocked in the later stage of nodule development, i.e. expression of the *ENOD* genes.

We found that one of the Antho⁻ mutants (BT65), which was shown to be deficient in luteolin (T.A. LaRue and G. Hrazdina, unpublished observations), also formed pseudonodules in response to NPA. *In-situ* hybridization analysis of *ENOD40* gene expression showed that this gene is expressed in the pericycle of the NPA-induced nodule-like structures (Hirsch *et al.*, 1997). As both the *ENOD2* and *ENOD40* probes were derived from alfalfa cDNA clones, it is likely there is significant homology between alfalfa and sweetclover genes. Indeed, Utrup and Norris (1996) have determined that sweetclover nodules formed in response to wild-type and mutant rhizobia exhibit nodulin patterns similar to those for alfalfa, and we have determined that a number of alfalfa cDNAs hybridize with *Melilotus* DNA (M. Lum, L.M. Brill, and A.M. Hirsch, unpublished results).

Transformation

Preliminary results indicated that sweetclover is transformable by *Agrobacterium rhizogenes* (P. van Rhijn and A.M. Hirsch, unpublished observations). However, we wanted to use other methods of transformation because hairy root transformation often leads to disturbed hormone levels in transformed tissues. Our goal also is to use U390 instead of U389, as starting material for mutagenesis because U390 is very short compared to U389. It is not much larger than *Arabidopsis* and, like this model plant, utilizes significantly less space in a greenhouse or growth cabinet, making it easier to screen for mutant plants.

First, we examined the responses of sweetclover to the antibiotic kanamycin which was dissolved in a plant nutrient–agar medium. From these experiments we determined that sweetclover was kanamycin-tolerant even up to levels as high as 125 μg/ml (W. Yang and A.M. Hirsch, unpublished results). This meant that we could not easily use kanamycin as a selectable marker for transformation events. On the other hand, we found that sweetclover is not resistant to the herbicide Basta® (phosphinothricin) (Y. Li and A.M. Hirsch, unpublished results). We found that 12.5 mg Basta®/L is sufficient to kill sweetclover seedlings completely. Basta® was applied either in the agar medium or in water and used to soak the roots of seedlings (Figure 2).

We are now in the process of testing various transformation methods for the purpose of obtaining transgenic sweetclover lines. Our preliminary results indicate that the method described by Larkin *et al.* (1996) is generating transformants (M. Lum and A.M. Hirsch, unpublished results).

Acknowledgements

The authors are grateful to Dr H.J. Gorz, who generously gave sweetclover seeds and knowledge. Part of this research was performed by the first author as part of a sabbatical leave in the laboratory of E.M. Meyerowitz at Caltech. I am grateful to Dr Meyerowitz for his hospitality and encouragement, and to Drs M. Frohlich and S. Jacobsen of the Meyerowitz laboratory for their help with some of the experiments. My thanks also to M. Lum, L. Brill, Y. Li, and W. Yang for their unpublished results. This research was supported in part by a Council on Research grant from the UCLA Academic Senate.

Additional information about sweetclover can be obtained at <www.lifesci.ucla.edu/mcdbio/Faculty/Hirsch/sweetclover/sweetclover.html>.

Figure 2. Basta® at 12.5 mg/L was applied directly to the roots of germinated sweetclover seedlings every day for a week. All of the seedlings died in the flat treated with Basta® whereas the untreated control flat contains numerous healthy seedlings.

References

Arumuganathan K, Earle E D 1991. Nuclear DNA content of some important plant species. Plant Mol. Biol. Rep., 9, 208–218.

Asad S, Fang Y, Wycoff F, Hirsch A M 1994. Isolation and characterization of cDNA and genomic clones of *MsENOD40*; transcripts are detected in meristematic cells. Protoplasma, 183, 10–23.

Barker D G, Biachi S, Blondon F, Dattee Y, Duc G, Essad S, Flament P, Gallusci P, Genier G, Guy P, Muel X, Tourneur J, Denarie J, Huguet T 1990. *Medicago truncatula*, a model plant for studying the molecular genetics of the *Rhizobium*–legume symbiosis. Plant Mol. Biol. Rep., 8, 40–49.

Bauchrowitz M A, Barker D G, Nadaud I, Rougé P, Lescure B 1992. Lectin genes from the legume *Medicago truncatula*. Plant Mol. Biol., 19, 1011–1017.

Bennett M D, Leitch I J 1995. Nuclear DNA amounts in angiosperms. Ann. Bot., 76, 113–176.

Brill L M, Konyalian V R, Hirsch A M 1995. A 1.7-kilobase genomic fragment of *Medicago sativa* DNA contains the lectin gene *Mslec1*. Plant Physiol., 108, 1311–1312.

Duc G, Trouvelot A, Gianinazzi-Pearson V, Gianinazzi S 1989. First report of non-mycorrhizal plant mutants (Myc⁻) obtained in pea (*Pisum sativum* L.) and fababean (*Vicia faba* L.) Plant Sci., 60, 215–222.

Gengenbach B G, Haskins F A, Gorz H J 1969. Genetic studies of induced mutants in *Melilotus alba*. I. Short-internode dwarf, curled leaf, multifoliate leaf, and cotyledonary branching. Crop Sci., 9, 607–610.

Gengenbach B G, Haskins F A, Gorz H J 1970. Genetic studies of induced mutants in *Melilotus alba*. II. Inheritance and complementation of chlorophyll-deficient mutants. Crop Sci., 10, 154–156.

Hirsch A M, Fang Y, van Rhijn P *et al.* 1997. Studies on *MsENOD40* gene expression in alfalfa (*Medicago sativa* L.) and white sweetclover (*Melilotus alba* Desr.) In: Legocki A, Bothe H and Pühler A. Biological Fixation of Nitrogen for Ecology and Sustainable Agriculture. NATO ASI Series, Vol. G39. Springer-Verlag, Berlin, Heidelberg. pp. 33–36.

Kneen B E, LaRue T A 1988. Induced symbiosis mutants of pea (*Pisum sativum*) and sweetclover (*Melilotus alba annua*). Plant Sci., 58, 177–182.

Larkin P J, Gibson J M, Mathesius U, Weinman J, Gartner E, Hall E, Tanner G J, Rolfe B G, Djordjevic M A 1996. Transgenic white clover. Studies with the auxin-responsive promoter, GH3, in root gravitropism and lateral root development. Transgen. Res., 5, 325–335.

Miller J E, Viands D R, LaRue T A 1991a. Inheritance of nonnodulating mutants of sweetclover. Crop Sci., 31, 948–952.

Miller J E, Viands D R, LaRue T A, Gorz H J 1991b. Cross-pollination technique for spontaneously self-pollinated sweetclover. Crop Sci., 31, 475–476.

Ronnenkamp R R, Gorz H J, Haskins F A 1975. Genetic studies of induced mutants in *Melilotus alba*. IV. Inheritance and complementation of six additional chlorophyll-deficient mutants. Crop Sci., 15, 187–188.

Smith W K, Gorz H J 1965. Sweetclover improvement. Adv. Agron., 17, 163–231.

Utrup L J, Norris J H 1996. Nodulin gene expression in effective root nodules of white sweetclover (*Melilotus alba* Desr.) and in ineffective nodules elicited by mutant strains of *Rhizobium meliloti*. J. Exp. Bot., 47, 195–202.

Utrup L J, Cary A J, Norris J H 1993. Five nodulation mutants of white sweetclover (*Melilotus alba* Desr.) exhibit distinct phenotypes blocked at root hair curling, infection thread development, and nodule organogenesis. Plant Physiol., 103, 925–932.

Wu C, Dickstein R, Cary A J, Norris J H 1996. The auxin transport inhibitor *N*-(1-naphthyl)phthalamic acid elicits pseudonodules on nonnodulating mutants of white sweetclover. Plant Physiol., 110, 501–510.

SECTION V

Molecular probes in taxonomy and ecology of diazotrophs

23
Isolation and identification of diazotrophic bacteria from rice, wheat and kallar grass

UZMA HASSAN, M. SAJJAD MIRZA, SAMINA MEHNAZ, GHULAM RASUL AND KAUSER A. MALIK

Abstract

Three bacterial isolates were obtained from the roots of rice, wheat and kallar grass. These isolates fix nitrogen in pure culture as confirmed by acetylene reduction assay in semi-solid nitrogen-free growth media. Production of growth hormones by these isolates was confirmed by colorimetric method. The isolates from wheat (strain Wb-3) and rice (strain N-4) were identified as *Azospirillum brasilense* and *A. lipoferum*, respectively, on the basis of their morphology, carbon utilization pattern and biochemical tests. The isolate K-1 from kallar grass, previously identified as *Azospirillum brasilense*, was confirmed to belong to a newly named genus *Azoarcus* (Reinhold-Hurek *et al.*, 1993) by using 16S rRNA-targeted oligonucleotide probes and some morphological and physiological characteristics.

Introduction

After the discovery of a very active association between *Azospirillum* species and the root system of various graminaceous plants (Dobereiner and Day, 1976), the use of these bacteria as biofertilizers fixing atmospheric nitrogen and producing phytohormones has attracted the attention of agronomists. Bacteria of the genus *Azospirillum* and other nitrogen-fixing soil bacteria associate with the roots of many plants including important crops such as wheat, rice, maize and sugarcane (Tarrand *et al.*, 1978; Dobereiner *et al.*, 1976; Elmerich *et al.*, 1992; Malik *et al.*, 1994). The organisms now classified in the genus *Azospirillum* were originally described in 1925 as *Spirillum lipoferum* (Beijerinck, 1925). At present five species of the genus *Azospirullum*, namely *A. lipoferum, A. brasilense, A. irakense, A. halopraeferens* and *A. amazonense*, have been recognized (Tarrand *et al.*, 1978;

K.A. Malik et al. (eds.), Nitrogen Fixation with Non-Legumes, 197–205.

Khammas *et al.*, 1989; Reinhold *et al.*, 1987; Dobereiner, 1992). In the rhizosphere of grasses diazotrophic bacteria belonging to at least 11 genera have been found (Elmerich *et al.*, 1992). These include *Azoarcus*, an isolate from kallar grass which grows on saline sodic, alkaline soils having low fertility (Sandhu and Malik, 1975). These diazotrophic Gram-negative rods could not be assigned to any previously described taxon on the basis of phenotypic characteristics (Reinhold *et al.*, 1986). As a result of a polyphasic study, a new genus *Azoarcus* was proposed to accommodate these organisms (Reinhold-Hurek *et al.*, 1993). These bacteria constituted a separate rRNA branch in rRNA superfamily III, which corresponds to the beta subclass of the Proteobacteria. This affiliation of *Azoarcus* to the beta subdivision of the Proteobacteria was further confirmed by reverse transcriptase sequencing of 16S rRNA (Hurek *et al.*, 1993).

In the present study isolation and identification of three diazotrophs from wheat, rice and kallar grass on the basis of morphological and physiological characteristics and 16S rRNA-targeted oligonucleotide probe hybridizations was carried out.

Materials and methods

Isolation of diazotrophs

The roots of wheat were thoroughly washed with sterile water to remove adhering soil. One gram root pieces were homogenized in 10 ml sterile water and serial dilutions were prepared. These dilutions were used to inoculate combined carbon medium (CCM; Rennie, 1981) and nitrogen free malate medium (NFM; Okon *et al.*, 1977) and incubated at 30°C. The vials showing bacterial growth and acetylene reduction activity were used to inoculate plates of the same media to obtain pure colonies. Isolation of the strain N-4 from rice and strain K-1 from kallar grass has been previously described (Bilal *et al.*, 1990; Malik *et al.*, 1994).

Acetylene reduction assay

Acetylene reduction activity of the vials inoculated with bacterial isolates was measured by injecting 10% acetylene after 48 h of growth on a gas chromatograph (Gasukuro Kogyo model 370) using Porapak N column. To measure specific activity of the cultures protein estimation was carried out by the method described by Lowry *et al.*, (1951).

Indoleacetic acid (IAA) production

For detection and quantification of IAA production by the bacterial isolates, cultures were grown in Okon's malate medium (Okon *et al.*, 1977) enriched with sucrose 5 g, mannitol 5 g and ammonium chloride 1 g per litre of the medium. Tryptophan 100 mg/L was added as precursor of IAA. After 4 weeks of growth, qualitative estimation of indoleacetic acid was performed by Fe-$HClO_4$ and Fe-H_2SO_4 reagents (Gordon and Weber, 1951). For quantitative estimation of IAA by HPLC, the ethyl acetate extraction method was used (Tien *et al.*, 1979).

Antibiotic resistance pattern

Resistance of the isolates to antibiotics was tested on LB plates.

Physiological and biochemical tests

The isolates were first segregated into two groups, Enterobacters (facultatively anaerobic, Gram-negative, and with the ability to produce acid and gas from glucose) and the non-enterobacters.

Physiological and biochemical tests were performed using the QTS-20 miniaturized identification system (DESTO Laboratories, Karachi, Pakistan). The oxidation fermentation test was performed as described by Hugh and Leifson (1953). Catalase was identified by the MacFaddin method (1980), using H_2O_2 and pure culture colonies from nutrient agar plates. The isolates found to be non-enterobacters (cytochrome-oxidase positive, oxidation-fermentation negative or positive) and morphologically similar to *Azospirillum* or *Azoarcus* were further tested for carbon source utilization for comparison with these genera.

Dot-blot hybridization

Bacterial cells of the two isolates from rice, i.e. K-1 and *Zoogloea* Ky-1 (Bilal and Malik, 1987) were grown in LB for 24 h at 30°C and centrifuged at 13 000 rpm for 5 min. The cell pellets from 1.5 ml cultures were washed with TE buffer (10 mM Tris. Cl; 1 mM EDTA, pH 8) and then dissolved in 200 µl of TE. Cell lysis was obtained at 37°C for 30 min with lysozyme (2 mg/ml; final concentration) and by using SDS (1%). The lysate was extracted twice with phenol/chloroform followed by two extractions with chloroform/isoamyl alcohol (24:1). After adding 1/10 volume of sodium acetate (3 M, pH 5.2) and 0.5 volume of isopropanol, the supernatant was kept at –20°C for 30 min. The nucleic acids were then precipitated by centrifugation at 13 000 rpm for 20 min and the pellet was washed with 70% ethanol before drying under vacuum. The nucleic acid pellets were then dissolved in 100 µl TE. The nucleic acids (100 ng) were applied on GeneScreen filters (Dupont) using Hybri.Dot manifold (BRL), immobilized by UV light and hybridized according to Church and Gilbert (1984). The eubacterial primer (Embley *et al.*, 1988; AGGGTTGCGCTCGTTG) was used as universal probe. This probe and a 16S rRNA targeted oligonucleotide probe

(TH15: GACATCGGCCGCTCCAATCGCG) developed against *Azoarcus* sequence (Hurek *et al.*, 1993) was 5'-labelled using phage T4 polynucleotide kinase (BRL) and 10–20 μCi of [γ-^{32}P] adenosine-5'-triphosphate (3000 Ci/mmol; Amersham) according to Maniatis *et al.* (1982).

Results and discussion

Bacterial growth became visible within 48 h in the nitrogen-free semi-solid media inoculated with homogenized roots of wheat. Initially growth was visible as a veil-like pellicle just below the surface of the media. This pellicle formation is considered a characteristic of *Azospirillum*, which finds suitable oxygen concentration just below the surface due to its microaerophilic nature (Tarrand *et al.*, 1978). The vials showing pellicle formation and high ARA activity were used to inoculate NFM, CCM and LB plates to obtain single cell colonies. Pure cultures were again subjected to ARA. All cultures showed nitrogen-fixing ability and produced IAA in liquid media (Table 1). Production of phytohormones by plant-associated bacteria and their beneficial effects on plant growth have been described (Costacurta and Vanderleyden, 1995).

Isolation and identification of Azospirillum brasilense *Wb-3 from wheat*

The isolate Wb-3 obtained from wheat formed subsurface white pellicle in semi-solid NFM and CCM media which spread to the surface of the medium as the culture grew older. The cells were short Gram-negative rods (2–3 μm long) and showed helical shape and motility characteristic of genus *Azospirillum*. The ARA detected in semi-solid NFM and CCM was 1.1 and 1.6 μmol C_2H_4/mg protein per hour, respectively. Among the antibiotics tested, the isolate was found resistant only to ampicillin (Table 2). The isolate could grow on the organic acids malate and lactate, but failed to utilize glucose, sucrose and mannitol as carbon source

Table 1. Acetylene reduction activity (ARA) and indoleacetic acid (IAA) production by the isolates from rice, wheat and kallar grass

Bacterial strains	NFM (μmol C_2H_4/mg proteins per hour)	CCM (μmol C_2H_4/ mg proteins per hour)	IAA (μg/ml)
Azospirillum brasilense (WB-3)	1.1	1.64	16.1
Azospirillum lipoferum (N-4)	0.81	1.89	6.3
Azoarcus sp. K-1	1.6	1.96	10.5

NFM = Nitrogen free malate medium (Okon *et al.*, 1977); CCM = Combined carbon medium (Rennie, 1981)

Table 2. Antibiotic resistance pattern of isolates from rice, wheat and kallar grass

Antibiotics (100 µg/ml)	K-1	N-4	WB-3
Ampicilin	–	+	+
Gentamycin	–	+	–
Kanamycin	–	–	–
Rifampicin	+	+	–
Spectinomycin	–	–	–
Streptomycin	–	+	–
Tetracyclin	–	–	–

(Table 3). Based on the morphological and physiological characteristics (Tables 3 and 4) the strain Wb-3 was identified as *Azospirillum brasilense*, to which it shows maximum resemblance (Tarrand *et al.*, 1978). The isolate accumulated polyhydroxy butryate (PHB) granules which were also identified in azospirilla by other workers (Reinhold *et al.*, 1986).

Isolation and identification of Azospirillum lipoferum *N-4 from rice*

The isolate N-4 obtained from histoplane fraction of roots formed off-white colonies on solid LB and CCM media. In semi-solid NFM medium it formed fine subsurface white pellicle within 24 h which gradually spread to the surface. The colour of the semi-solid NFM medium turned blue due to the rise in pH of the medium. Examination of the wet mounts of the culture in NFM and CCM semi-solid media under a phase contrast microscope showed plump, highly motile

Table 3. Physiological characterization of the isolates from rice, wheat and kallar grass

Characteristics	*A halopraeferens*[1]	*A. irakense*[2]	*A. lipoferum*[3]	*A. brasilense*[3]	WB-3	N-4	K-1
Catalase	+	+	+	+	+	+	+
Cytochrome oxidase	+	+	+	+	+	+	+
Biotin requirement	+		+	–	–	+	–
Acidification of peptone–based glucose medium	–	–	+	–	–	+	–
Sole carbon source for growth in N-free semi-solid medium							
Glucose	–	+	+	–	–	+	–
Mannitol	+	+	+	–	–	+	–
Sucrose	–	+	–	–	–	–	–
Maltose	–	+	–	–	–	–	–
Malate	+	+	+	+	+	+	+

[1] Reinhold *et al.*, 1987; [2] Khammas *et al.*, 1989; [3] Tarrand *et al.*, 1978.

Table 4. Biochemical characterization of the isolates from rice, wheat and kallar grass

Biochemical tests	*A. irakense*[1]	*A. brasilense*[2]	*A. lipoferum*[2]	*A. halo-praeferens*[3]	WB-3	N-4	K-1
ONPG	–	+	+	+	+	–	+
Sodium citrate	–	+	+	–	+	+	–
Sodium malonate	+	+	+	+	–	+	–
Lysine decarboxylase	+	+	+	+	+	+	–
Arginine dihydrolase	–	+	+	–	+	+	–
Ornithine decarboxylase	+	+	+	–	+	+	+
H_2S production	–	–	–	–	–	–	–
Urea hydrolysis	+	+	+	–	+	+	–
Tryptophan deaminase	–	–	–	+	–	–	–
Indole	–	–	–	–	–	–	–
Acetoin (VP)	–	–	–	–	–	–	–
Gelatine hydrolysis	–	–	–	–	–	–	+
Acid from glucose	–	–	+	–	–	+	–
Nitrate reduction	+	+	+	+	+	+	+
Acid from maltose	–	–	–	–	–	–	–
Acid from sucrose	–	–	–	–	–	–	–
Acid from mannitol	–	–	+	–	–	+	–
Acid from arabinose	–	–	–	–	–	–	–
Acid from rhamnose	–	–	–	–	–	–	–
Acid from sorbitol	–	–	–	–	–	–	–
Acid from inositol	–	–	–	–	–	–	–

[1] Khammas *et al.*, 1989; [2] Tarrand *et al.*, 1978; [3] Reinhold *et al.*, 1987.

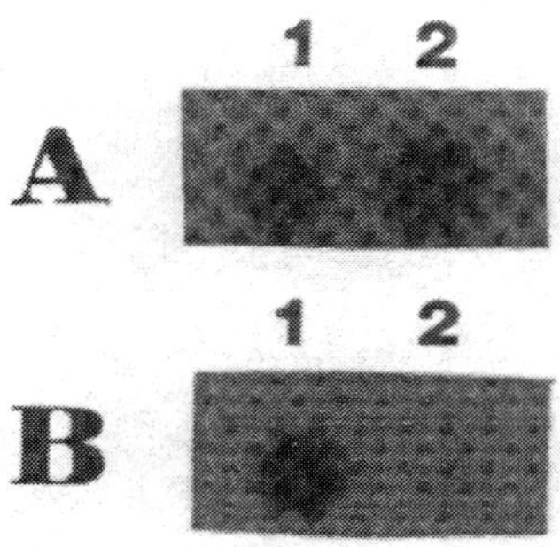

Figure 1. Dot–blot hybridization of 16S rRNA-targeted oligonucleotide probes. **A**: Hybridization with the universal eubacterial probe (p1115), indicating presence of rRNA target on the blot. **B**: Hybridization with oligonucleotide probe specific for *Azoarcus* (TH15; Hurek *et al.*, 1993). This probe strongly hybridized with rRNA of *Azoarcus* K-1 but did not hybridize with the rRNA of *Zoogloea* strain Ky-1. **1** = rRNA of *Azoarcus* K-1; **2** = rRNA of *Zoogloea* Ky-1.

helical cells resembling *Azospirillum*. The cells range in length from about 5 µm in CCM and NFM to 30 µm in LB liquid medium. This variability in cell size has also been noted in other strains of *A. lipoferum* (Krieg and Dobereiner, 1984).

Among the antibiotics tested in the present study, strain N-4 showed resistance to ampicillin, gentamycin, rifampicin and streptomycin, while it was sensitive to kanamycin, spectinomycin and tetracyclin (Table 2). The resistance of strain N-4 to antibiotics gentamycin, rifampicin and streptomycin can prove useful in differentiation of this *Azospirillum* strain from *A. brasilense* strain Wb-3 in ecological studies. The isolate could grow on glucose, mannitol, lactate and malate, while it could not utilize sucrose, maltose, etc. (Table 3) and required biotin for its growth. The morphological characteristics and physiological tests (Tables 3 and 4) showed that this strain belongs to *Azospirillum lipoferum* (Tarrand *et al.*, 1978). Isolation of *A. lipoferum* has often been reported from maize, while *A. brasilense* has been commonly obtained from wheat and rice (Baldani and Dobereiner, 1980).

Isolation and identification of Azoarcus *K-1 from kallar grass*

The isolate K-1 from kallar grass forms thin subsurface pellicle in semi-solid NFM and CCM media. On LB plates it formed off-white colonies. The organism is an aerobic, Gram-negative, highly motile straight rod (cell width 0.5–0.8 μm; length 2–4 μm). In NFM and CCM the strain was able to reduce acetylene at the rate of 1.6 and 1.96 μmol C_2H_4/mg protein per hour, respectively. The isolate could grow well on salts of organic acids such as lactate and malate, but could not utilize carbohydrates such as glucose, sucrose and mannitol, as carbon source. The morphological characteristics and carbon utilization pattern of the isolate K-1 resembled *Azoarcus* described by Reinhold-Hurek *et al.* (1993). Previously, due to the non-availability of the description of *Azoarcus* at that time, the strain K-1 was classified as *Azospirillum brasilense* mainly on the basis of its ability to fix nitrogen and carbon source utilization pattern (Bilal *et al.*, 1990). The genus *Azoarcus* was first isolated from kallar grass, and later its association with kallar grass was confirmed by fluorescent antibodies (Reinhold *et al.*, 1987) and immunogold electron microscopy (Hurek *et al.*, 1991). The isolate K-1 was further confirmed as *Azoarcus* by using a 16S rRNA-targeted oligonucleotide probe (Hurek *et al.*, 1993) which hybridized strongly with RNA of K-1 on a dot-blot but did not show signals with the RNA of *Zoogloea* strain Ky-1 (Figure 1). *Zoogloea* strain Ky-1 is a diazotrophic bacterium which was isolated from kallar grass by Bilal and Malik (1987).

Three isolates discussed here can fix nitrogen and produce phytohormones, and have great potential for use as biofertilizers for the important cereal crops like wheat and rice.

References

Baldani V L D, Dobereiner J 1980. Host plant specificity in the infection of cereals with *Azospirillum* species. Soil Biol. Biochem., 12, 433–440.

Beijerinck M W 1925. Uber ein *Spirillum*, welches freien Stickstoff binden kann? Centralbl. Bakt. Parasitkde., 63, 353–357.

Bilal R, Malik K A 1987. Isolation and identification of a N_2-fixing zoogloea-forming bacterium from kallar grass histoplane. J. Appl. Bacteriol., 62, 289–294.

Bilal R, Rasul G, Qureshi J A, Malik K A 1990. Characterization of *Azospirillum* and related diazotrophs associated with roots of plants growing in saline soils. World J. Microbiol. Biotechnol., 6, 46–52.

Church G M, Gilbert W 1984. Genomic sequencing. Proc. Natl. Acad. Sci. USA, 81, 1991–1995.

Costacurta A, Vanderleyden J 1995. Synthesis of phytohormones by plant-associated bacteria. Crit. Rev. Microbiol., 21, 1–18.

Dobereiner J 1992. The genera *Azospirillum* and *Herbaspirillum*. In: Ballows A, Truper H G, Working M, Harder W, Schleifer K H, eds. The Prokaryots, 2nd edn. Springer-Verlag, Berlin, Heidelberg, New York, 2236–2253.

Dobereiner J, Day J M 1976. Associative symbioses in tropical grasses: characterization of microorganisms and dinitrogen fixing sites. In: Proceedings of the 1st International Symposium on Nitrogen Fixation, vol. 2, 518–538.

Dobereiner J, Marriel I E, Nery M 1976. Ecological distribution of *Spirillum lipoferum* Beijerinck. Can. J. Microbiol., 22, 1464–1473.

Elmerich C, Zimmer W, Vielle C 1992. Associative nitrogen-fixing bacteria. In: Stacey G, Burris R H, Evan H J, eds. Biological Nitrogen Fixation. Chapman & Hall, New York, 212–258.

Embley T M, Smida J, Stackebrandt E 1988. Reverse transcriptase sequencing of 16S ribosomal RNA from *Faenia rectivirgula, Pseudonocardia thermophila* and *Saccharopolyspora hirsuta*, three Wall type IV actinomycetes which lack mycolic acids. J. Gen. Microbiol., 134, 961–966.

Gordon S A, Weber R P 1951. Colorimetric estimation of indoleacetic acid. Plant Physiol., 26, 192–195.

Hugh R, Leifson E 1953. The taxonomic significance of fermentative versus oxidative metabolism of carbohydrates by various Gram-negative bacteria. J. Bacteriol., 66, 24–26.

Hurek T Buurggraf S, Woese C R, Reinhold-Hurek B 1993. 16S rRNA-targeted polymerase chain reaction and oligonucleotide hybridization to screen for *Azoarcus* spp., grass-associated diazotrophs. Appl. Environ. Microbiol., 59, 3816–3824.

Hurek T, Reinhold-Hurek B, Montagu M van, Kellenberger E 1991. Infection of intact roots of kallar grass and rice seedling by '*Azoarcus*'. Dev. Plant Soil Sci., 48, 235–242.

Khammas K M, Ageron E, Grimont P A D, Kaiser P 1989. *Azospirillum irakense* sp. nov., a nitrogen-fixing bacterium associated with rice roots and rhizosphere soil. Res. Microbiol., 140, 679–693.

Krieg N R, Dobereiner J 1984. Genus *Azospirillum*. In: Krieg N R, ed. Bergey's Manual of Systematic Bacteriology, vol. 1. Williams & Wilkins, Baltimore, 94–104.

Lowry H O, Rosebrough N J, Farr A G, Randall R J 1951. Protein measurement with Folin phenol reagent. J. Biol. Chem., 193, 265–275.

MacFadin J F (ed.) 1980. Biochemical Tests for Identification of Medical Bacteria. Williams & Wilkins, Baltimore, pp 51–54.

Maniatis R, Fritsch E F, Sambrook J 1982. Molecular Cloning. Cold Spring Harbor Laboratory, Cold Spring Harbor, New York.

Malik K A, Rasul G, Hassan U, Mehnaz S, Ashraf M 1994. Role of N_2-fixing and growth hormones producing bacteria in improving growth of wheat and rice. In: Hegazi N A, Fayez M, Monib M, eds. Nitrogen Fixation with Non-Legumes. Proceedings of The Sixth International Symposium on BNF with Non-Legumes, Ismailia, Egypt, 6–10 September 1993. Cairo University, Giza, Egypt, 409–422.

Okon Y, Albercht S L, Burris R H 1977. Methods for growing *Spirillum lipoferum* and for counting it in pure culture and in association with plants. Appl. Environ. Microbiol., 33, 85–88.

Reinhold B, Hurek T, Fendrik I 1987. Cross-reaction of predominant nitrogen-fixing bacteria with enveloped, round bodies in the root interior of kallar grass. Appl. Environ. Microbiol., 53, 889–891.

Reinhold B, Hurek T, Niemann E G, Fendrik I 1986. Close association of *Azospirillum* and diazotrophic rods with different root zones of kallar grass. Appl. Environ. Microbiol., 52, 520–526.

Reinhold B, Hurek T, Fendrik I, Pot B, Gillis M, Kersters K, Thielemans D, De Ley J 1987. *Azospirillum halopraeferens* sp. nov. A nitrogen fixing organism associated with roots of kallar grass (*Leptochloa fusca* (L) Kunth). Int. J. Syst. Bacteriol., 37, 43–51.

Reinhold-Hurek B, Hurek T, Gillis M, Hoste B, Vascanneyt M, Kersters K, De Ley J 1993. *Azoarcus* gen. nov., nitrogen fixing proteobacteria associated with roots of kallar grass (*Leptochloa fusca* L. Kunth), and description of two species. *Azoarcus indigens* sp. nov. and *Azoarcus communis* sp. nov. Int. J. Syst. Bacteriol., 43, 574–584.
Rennie R J 1981. A single medium for the isolation of nitrogen fixing bacteria. Can. J. Microbiol., 27, 8–14.
Sandhu G R, Malik K A 1975. Plant succession – a key to the utilization of saline soils. Nucleus (Karachi), 12, 35–38.
Tarrand J J, Krieg N R, Dobereiner J 1978. A taxonomic study of the *Spirillum lipoferum* group, with description of a new genus, *Azospirillum* gen. nov., and two species *Azospirillum lipoferum* (Bejerinck) sp. nov. and *Azospirillum brasilense* sp. nov. Can. J. Microbiol., 24, 967–980.
Tien T M, Gaskins M H, Hubbell D H 1979. Plant growth substances produced by *Azospirillum brasilense* and their effect on the growth of pearl millet (*Pennisetum americanum* L.). Appl. Environ. Microbiol., 37, 1016–1024.

SECTION VI

Applicability and economic importance of associative diazotrophs

24
Diazotrophs associated with non-legumes grown in sandy soils

N.A. HEGAZI, M. FAYEZ, G. AMIN, M.A. HAMZA, M. ABBAS, H. YOUSSEF AND M. MONIB

Abstract

This study was carried out to monitor the effect of introducing single and multiple strains of diazotrophs to plant systems (wheat and barley) under gnotobiotic conditions. The relative scarcity of nutrient supply and maintenance requirements under the present experimental conditions resulted in slower growth and much longer doubling time compared to those found in their respective batch cultures. Single inoculations significantly enhanced the efficiency of nitrogen fixation in both plant systems compared to non-inoculated controls. Among tested diazotrophic strains, cells of *Azospirillum* spp. and *Azotobacter* spp. were superior in their ARAs, being higher with barley than with wheat seedlings. Increases in total dry weight of plants between 16.8% and 78% were achieved in the wheat system and between 54.5% and 68% in the barley system, with the highest-value being for *Azospirillum* spp. and *Pseudomonas* spp. treatments, respectively. It was observed that the type of plant host has a pronounced effect on cell growth and the type of interactions between the two diazotrophic candidates inoculated jointly into wheat and barley systems. Similar to dual inoculation, the doubling time of each candidate increased when inoculated as a mixed inoculum of three or four diazotrophs into the wheat system, and decreased in the barley system.

Introduction

Inevitably, as populations grow, dryland agriculture will have to provide an increased proportion of total agriculture output – nearly all from semi-arid and arid deserts. The importance of drylands will grow, rather than diminish, over time. So will the need to develop and promote appropriate technologies to raise the productivity and support sustainable development of these areas. Such

K.A. Malik et al. (eds.), Nitrogen Fixation with Non-Legumes, 209–222.

sustainable development conserves resources and is environmentally non-degrading. Therefore, management of such lands requires further attention to attain sustainability, secure economic return and build up soil fertility.

A unique approach in this respect is the concept of 'integrated fertilizers management' (IFM) which comprises the proper complementing use of chemical and biological fertilization. It implies, among other things, the judicious use of chemical fertilizers together with maximum manipulation of biofertilization with functional rhizospheric microorganisms (RMO), mainly biological nitrogen fixers. Therefore, efforts are made to enrich the soil–plant system with representatives of nitrogen fixers. This includes diazotrophs associated to roots of non-legumes; a system that is less specific than the symbiotic one of rhizobia–legume and more subject to the varying effects of the environment. To cope with the situation, heavy inocula, in quantitative and qualitative terms, as well as repeated introduction, are proposed to secure positive results of inoculation. The present study reports on some of the results obtained through a continuing research campaign to heavily introduce diazotrophs to the plant-soil system under conditions prevailing in the Egyptian eastern deserts and Sinai. The concept of heavy introductions of diazotrophs in competitive and heterogeneous populations is explored at laboratory, greenhouse and field levels.

Interaction of single and multi-strain inocula with barley and wheat under gnotobiotic conditions

The main objective of this study was to monitor the effect of introducing single and multiple strains of diazotrophs to plant systems under gnotobiotic conditions. This is an attempt to study the close interaction of tested diazotrophs with host plants. Seedlings of wheat and barley were surface-sterilized and germinated onto agar plates. Two seedlings were transferred to a model gnotobiotic system with 124 ml tubes containing the 25 ml of the semi-solid basal medium of Reinhold *et al.* (1985). To each tube, 4–5 g glass beads and 1 ml of pure or mixed broth cultures of tested diazotrophs (*A. brasilense, A. chroococcum, K. pneumoniae, P. putida* and *Bacillus* spp.) were added. Culture tubes were incubated in a growth chamber with a 16/8 h light/dark cycle at 23–25°C. To measure acetylene-reducing activity (ARA), tubes were sealed with suba seal and 10% acetylene was injected. ARA measurements using a gas chromatograph (DELSI DN 200/250) were done at 3, 10, 17, 24 and 33 days following bacterial inoculation. Results are expressed as nmol ethylene produced per plant. Values are the means of six replicates. At the same intervals, the bacterial load on plant roots was measured. Plant shoots were cut and roots left in the tubes, vigorously mixed with the glass beads using a vortex for 15 min. The plate count technique and a modified combined carbon source (CCM) culture medium (basal salt supplemented with glucose 2 g, sucrose 2 g, malic acid, 2 g and sodium lactate 2 ml/L) was employed. By the end of the experiment, root, shoot and total dry weight of plants were determined and expressed as mg per two plants. Results obtained can be summarized as follows.

Introduction of single inoculum

Growth kinetics of single and mixed combinations of a number of potent associative diazotrophs grown as batch cultures were as from an earlier study (Hamza *et al.*, 1994). It was indicated that a synergetic effect occurred among interactants as doubling time substantially decreased in mixed populations. To obtain further information on such kinetics in rhizospheres of wheat and barley the present study was conducted. Figure 1 shows an example for doubling time of

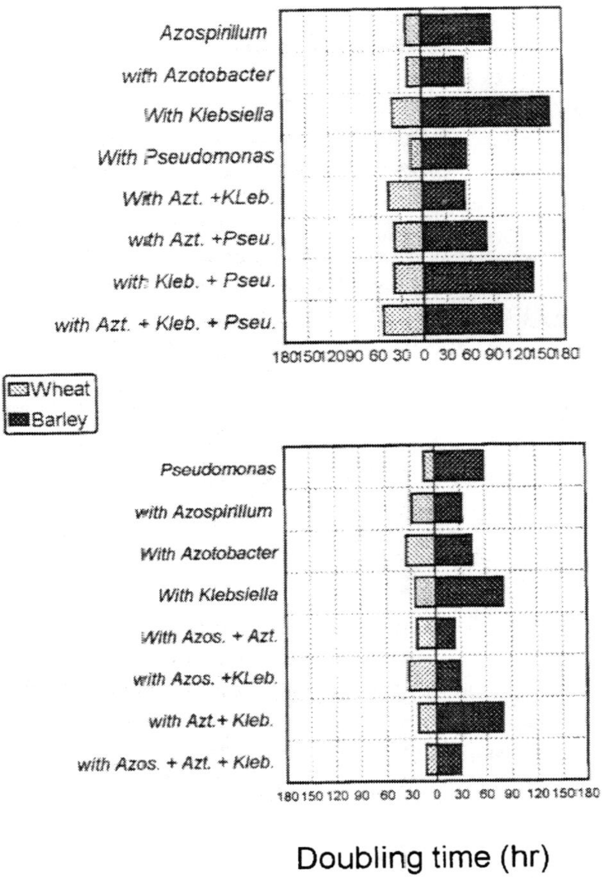

Figure 1. Doubling time for various diazotrophs inoculated singly or in mixed populations in the rhizosphere of barley and wheat in gnotobiotic systems.

tested strains when grown singly or in various combinations under gnotobiotic conditions. Under these particular conditions, tested diazotrophs did survive in appreciable populations, and also maintained their activities. However, the relative scarcity of nutrient supply and maintenance requirements resulted in slower growth and much longer doubling time compared to those found in their respective batch cultures. Also, growth performances of bacterial cells were totally different in the rhizosphere of wheat compared to barley.

Doubling times of cells grown in rhizosphere of wheat ranged between 12.7 h (*Pseudomonas* spp.), and 20 h (*Azospirillum* spp.); and in barley between 29.5 h (*Klebsiella* spp.) and 90.3 h (*Azospirillum* spp.). Such large differences in bacterial doubling time might indicate a significant variation in nature and concentrations of root exudates between the two crops (Vancura and Stotzky, 1976). Similarly, a wide range of doubling time for microorganisms grown in the rhizosphere of plant seedlings was reported by several investigators and under different experimental conditions (DeMing and Alexander, 1986; Loper *et al.*, 1985; Quatrini *et al.*, 1992). Although the slow growth rate of diazotrophs, under tested conditions, single inoculation enhanced significantly the efficiency of nitrogen fixation in both plant systems compared to non-inoculated system. Acetylene-reducing activity (ARA) increased with time and reached its maximum levels on day 10 (Figures 2 and 3). Among tested diazotrophic strains, cells of *Azospirillum* spp. and *Azotobacter* spp. were superior in their ARAs, being higher with barley than with wheat seedlings. The maximum ARA of 510 and 496.6 nmol C_2H_4 plant were recorded for the barley system, and only 253.4 and 235.4 nmol C_2H_4/plant for the wheat system. Much lower ARAs were observed for cells of *Klebsiella* spp. and *Pseudomonas* spp. in both systems. As a result of single inoculation, the growth of both wheat and barley plants was greatly improved, compared to non-inoculated plants (Figure 4). Increases in total dry weight of plants between 16.8% and 78.8% were achieved in the wheat system and between 54.5% and 68.5% in the barley system, with the highest being for *Azospirillum* spp. and *Pseudomonas* spp. treatment respectively.

Introduction of two-strain mixed inocula

As shown in Figure 1, the type of plant host has a pronounced effect on cell growth and type of interaction between the two diazotrophic candidates introduced jointly into wheat and barley systems. An antagonistic type of interaction might have occurred among diazotrophic candidates, dually grown in wheat rhizosphere, as their respective doubling time was greatly increased compared to single inoculation. On the contrary, the doubling time for each candidate, grown dually in the barley rhizosphere, was significantly reduced in most treatments, which might reflect a mutualistic type of interaction. Halsall and Goodchild (1986) described the nature of the microbial interaction, being variable and depending on interacting microorganisms and environmental conditions. Either the supply of beneficial metabolites or the presence of inhibitory products might

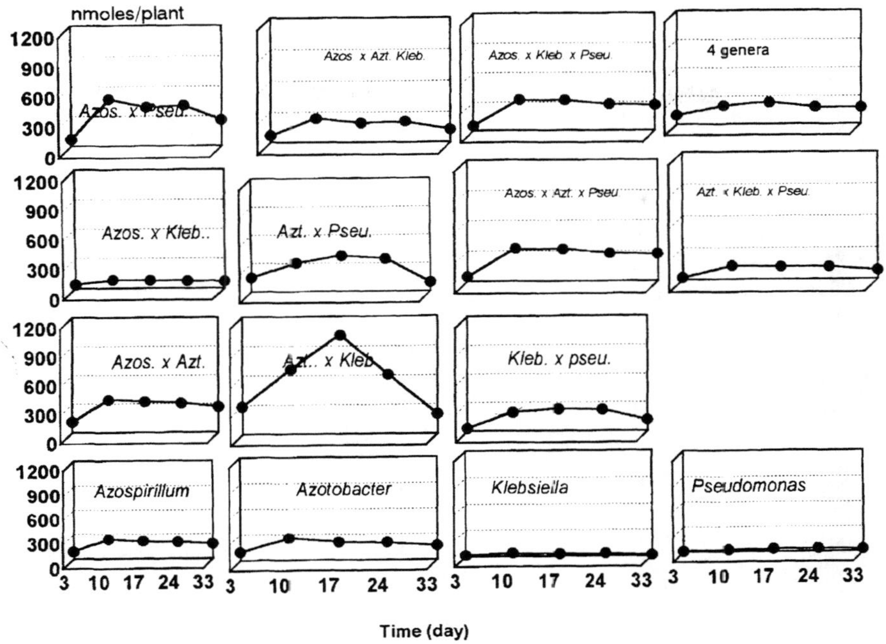

Figure 2. Nitrogenase activity on roots of wheat plants inoculated with various combinations of diazotrophs in gnotobiotic systems.

detemine the type of interaction (Lippi *et al.*, 1992). The obtained results strongly demonstrate that the type of host plant might also be considered to be an additional factor in determining the type of interaction among mixed bacterial inoculation. As reported by Khammas *et al.* (1994), introduction of mixed inoculum of two diazotrophic candidates markedly stimulated the ARA of wheat and barley systems regardless of their type of interaction (Figure 2). Much higher levels of activities were recorded compared to single inoculations. Interestingly, cells of *Klebsiella* spp. and *Pseudomonas* spp., which showed the lowest ARA, when introduced singly to wheat seedlings exhibited the highest ARA of 1012.2 and 377.9 nmol C_2H_4/plant respectively, when either of them was combined with cells of *Azotobacter* spp. It is well known that, among all living microorganisms and tissues, *Azotobacter* cells have the highest specific respiration rate of 2000 µl O_2/mg d.w./h (Schlegel, 1988). Thus, an O_2-protection effect is seriously considered to explain the relatively high levels of ARA recorded for the wheat system supplied with a dual inoculum of *Azotobacter* spp. and either of *Klebsiella* spp. or *Pseudomonas* spp. compared to single inoculum. ARAs of the barley system were greatly improved whenever *Azospirillum* spp. or *Azotobacter* spp. was a participant in dual inoculation (Figure 3). The highest values (1060 nmol C_2H_4/plant) were reported for *Azotobacter* spp. + *Azospirillum* spp. followed by *Azospirillum* spp. mixed with either *Pseudomonas* spp. or *Klebsiella* spp. Responses

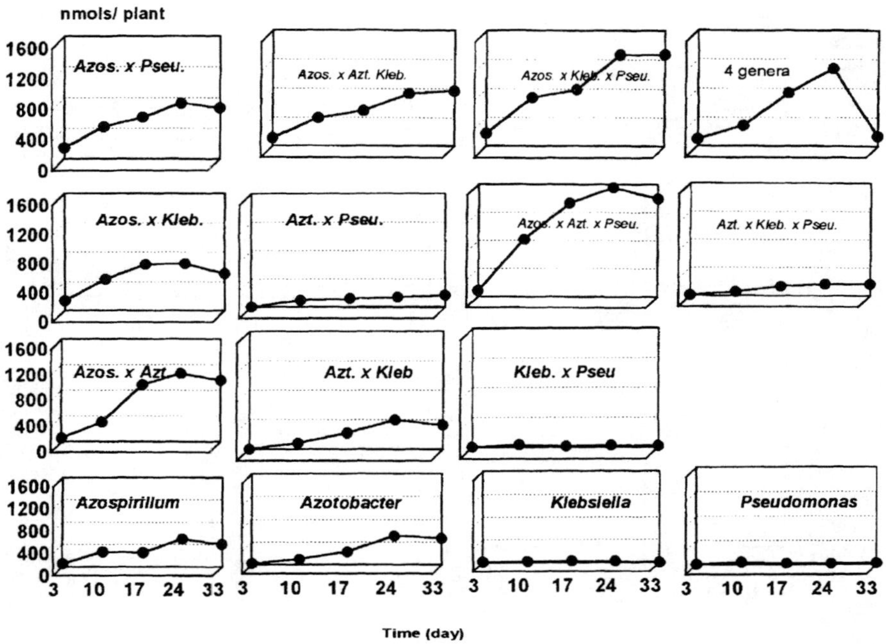

Figure 3. Nitrogenase activity on roots of barley plants inoculated with various combinations of diazotrophs in gnotobiotic systems.

of plant growth to dual diazotrophic inoculation were much better than single inoculation, particularly with the barley system (Figure 4). Generally, increases in total dry weight of plants, over non-inoculated ones, were much higher for barley than for wheat. The highest increase of 91.8% was found when a dual inoculation of *Azotobacter* spp. and *Pseudomonas* spp. was introduced into the barley system compared to only 55.6% for the wheat system supplied with *Azotobacter* spp. and *Klebsiella* spp. An explanation is the better development of barley seedlings (22–24 g/plant) compared to wheat (8–16 g/plant) under tested gnotobiotic conditions, creating a more favourable environment, in terms of nature and concentration of root exudates, for cell growth and metabolic activities of rhizospheric microorganisms (Prikryl and Vancura, 1980). Using a radioactively labelled carbon, Barber and Martin (1976) estimated the amount of organic carbon released from plant roots and found that over a cultivation period of 21 days it was much higher for barley roots (47% of root dry weight, corresponds to 13% of total dry weight of plant) compared to wheat roots (26% of root dry weight, corresponds to 9% of total dry weight of plant). Furthermore, Vancura and Stotzky (1976) demonstrated that supporting root exudates differ qualitatively and quantitatively, due not only to plant species but also individual cultivars.

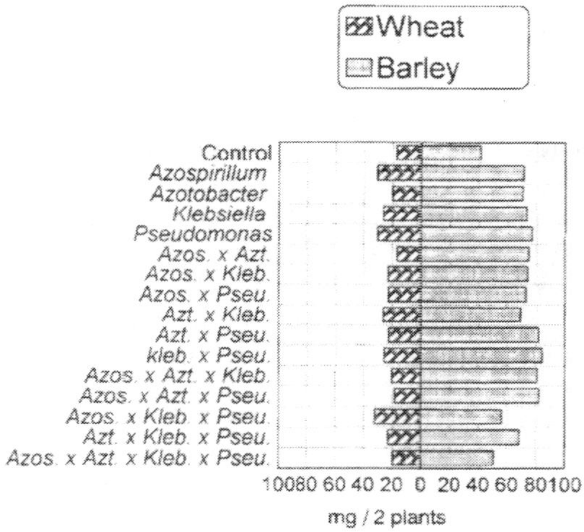

Figure 4. Total dry weight of barley and wheat plants as affected by various genera of associative diazotrophs as pure and mixed populations in gnotobiotic systems.

Introduction of more than two diazotrophic candidates

Similar to dual inoculation, the doubling time of each candidate increased when introduced as a mixed inoculum of three or four diazotrophs into the wheat system, and decreased in the barley system (Figure 1). However, a different performance was observed for ARA of various systems (Figures 2 and 3). Significantly lower levels of activities were determined in wheat systems supplied with all three- or four-strain combinations with the highest being only 274 nmol C_2H_4/plant. With barley seedlings, ARAs were markedly increased, compared to single and dual inoculation, and the highest value (1555 nmol C_2H_4/plant) was achieved when a mixed inoculum of *Azospirillum* spp., *Azotobacter* spp. and *Pseudomonas* spp. was introduced. Compared to single and dual inoculation, no significant differences were observed in plant responses (Figure 4). This might be attributed to limited C-sources and nutrients available to satisfy the cumulated populations of introduced diazotrophs. The picture might differ under a natural plant-soil system where a sufficient supply of root exudates, which might reach 50% of root dry weight (Prikryl and Vancura, 1980), supports growth of the introduced multi-strains. An exception was wheat seedlings when inoculated with a mixed culture of *Azospirillum* spp., *Klebsiella* spp. and *Pseudomonas* spp. This particular treatment has resulted in a significantly higher increase in total dry weight of plant of up to 90%, over non-inoculated plants. It should be noticed that whenever barley or wheat seedlings responded positively to inoculation treatments, cells of *Azotobacter* spp., as a slow grower, and/or *Azospirillum* spp. as a fast grower, were participants in such an inoculum. This continues the observations of Lynch and Clark (1984) and Quatrini *et al.* (1992) showing that stimulation of plant growth is not necessarily related directly to the colonization potential of bacteria used as inoculant. Our results suggest that the outcome of interactions between members of a mixed inoculum, under diazotrophic conditions, depends mainly on the relative growth rate of the whole population, and also the nitrogen fixation rate. To maximize the beneficial effect of such an inoculum, a diazotrophic candidate with high ARA should be included regardless of its growth rate. Finally, it should be stated that although multi-strain combinations in several cases might not be quantitatively supportive to plant growth, they might have an ecological advantage under natural conditions. This is true if we consider the ecological concept 'The environment selects'. In other words, inclusion of more than one neutral or mutual diazotroph members in the inocula would be an advantage, as it allows the plant-soil environment to select the most compatible microsymbiont.

Pot experiments

The individial and multi-strain inocula of associative diazotrophs that successfully contributed to the development of wheat and barley under gnotobiotic conditions were evaluated in a number of pot experiments. Non-legume plants tested were barley (cvs. Giza 123, Giza 124, California Marriot and Rihan-03), wheat (cvs. Sakha 8, Sakha 69, Giza 124, Giza 162 and Giza 163), maize (hybrid 10 and hybrid 320) and sorghum (cvs. 51007 and Dorado). Seeds were soaked in 7-day-old broth cultures of single or multi-strain tested diazotrophs (*A. brasilense, A. chroococcum, K. pneumoniae, P. putida, Bacillus* spp. or a composite of all) for 1 h. They were planted in plastic pots filled with 8 kg sandy soil plot^{-1} (TN 0.012%, WHC 20%, pH 8.9 and E.C dSm-1 0.72) collected from irrigated fields in Ismailia. Pots were irrigated with Hoagland nutrient solution having a full or half-dose of N at regular intervals, and arranged in a greenhouse in a complete randomized design with four replicates. Four-month-old plants were sampled and determined for total dry matter and nitrogen yields.

Results shown in Table 1 demonstrate that, irrespective of cultivars, all plants tested positively responded to inoculation with single-tested strains of associative diazotrophs. Total dry matter yield of plants simultaneously fertilized with a half-dose of N and inoculated with diazotrophs scored increases of total dry matter (ranging from 22% to 36%) over corresponding controls. Such yields were comparable to those receiving full N, indicating that inoculation did satisfy about 50% of plant N-requirements. Response to multi-strain preparations differed among host plants; however, positive gains were recorded for all tested host plants. In some cases such gains did overcome those obtained for a number of single-strain inoculations. Wheat was among the most responsive plants to the composite inoculum of all strains tested.

As to N-yield positive gains were generally obtained for single and multi-strain preparations. Increases were either significant or approached significant levels, and approximated yields of full N-fertilized plants. Similar N-yield

Table 1. Total biomass and nitrogen yields of various graminous plants as affected by N-fertilization, and single and multi-strain inocula of associative diazotrophs

Treatments	Barley	Wheat	Maize	Sorghum
Total biomass (g per four plants)				
1. Full N	7.72	3.57	40.8	38.8
2. *Azospirillum* spp. + 1/2 N	6.61	5.92	53.4	53.4
3. *Azotobacter* spp. + 1/2 N	8.21	4.80	62.4	68.6
4. *Klebsiella* spp. + 1/2 N	9.19	5.45	47.0	46.2
5. *Bacillus* spp. + 1/2 N	12.00	5.16	52.4	48.0
6. *Pseudomonas* spp. + 1/2 N 7-	8.31	3.86	41.2	35.4
Composite inoculum + 1/2 N	9.74	5.32	51.4	44.2
LSD 0.05	1.10	0.67	12.6	14.8
Nitrogen yield (mg per four plants)				
1. Full N	26.3	11.2	246.8	243.2
2. *Azospirillum* spp. + 1/2 N	23.1	24.9	339.0	420.2
3. *Azotobacter* spp. + 1/2 N	28.7	24.9	472.6	472.8
4. *Klebsiella* spp. + 1/2 N	29.9	24.6	261.6	338.2
5. *Bacillus* spp. + 1/2 N	31.5	12.9	338.6	435.8
6. *Pseudomonas* spp. + 1/2 N	28.5	11.9	234.8	252.4
Composite inoculum + 1/2 N	22.3	13.3	245.0	278.6
LSD 0.05	2.8	2.8	93.6	98.8

increases of diazotroph-inoculated wheat was recorded by Baldani *et al.* (1987) and De Freitas *et al.* (1987)

Gains in total dry matter and N-yields were plant-dependent. With wheat, azospirilla produced the highest total biomass. In the case of barley, and for all cultivars tested, highest magnitudes of response (increases of 47–63%) were obtained with *Bacillus* spp. This indicates a degree of host-diazotroph affinity of barley to *Klebsiella* spp., but wheat to *Azospirillum* spp. and *Bacillus* spp., a phenomenon observed by Rennie and Larson (1979). Response to inoculation with diazotrophs was not plant- but also cultivar-dependent. Data presented in Table 2 demonstrate the behaviour of a number of wheat and barley cultivars to various inoculum combinations. In the presence of a rational dose of N-fertilizer, the most responsive cultivars to composite inoculation were Giza 163 of wheat and Giza 124 and California of barley.

As a general trend, the multi-strain preparation, together with a rational dose of N-fertilizer, produced an increasing yield that approximated that of full N-fertilized plants. This signifies the use of composite inoculum as having no negative interactions among enclosed individuals. An additional advantage of the multi-strain preparation is the broad-spectrum accessibility of varying groups of diazotrophs to the pressure of selection occuring in the soil–plant environment. Therefore, the choice was to prepare multi-strain inocula for wheat and barley inoculation under field conditions of sandy soils.

Table 2. Total biomass yield (g $plant^{-1}$) of barley and wheat cultivars as affected by inoculation and nitrogen fertilization

Treatments	Barley				Wheat				
	Giza 123	Giza 124	California	Rihan	Sakha 8	Shakha 69	Giza 162	Giza 163	Giza 164
Full N	1.99	2.16	2.10	1.46	0.64	0.72	1.15	1.17	0.78
Half N	1.82	2.65	2.58	1.89	0.83	1.09	0.80	0.86	0.73
Azospirillum spp. + 1/2 N	1.92	1.49	1.92	1.28	1.37	1.32	1.53	1.87	1.32
3. *Azotobacter* spp. + 1/2 N	2.38	2.32	1.95	1.57	1.15	1.31	1.26	1.51	0.78
4. *Klebsiella* spp. + 1/2 N	2.39	2.65	2.33	1.83	1.30	1.32	1.43	1.60	1.17
5. *Bacillus* spp. + 1/2 N	3.18	3.38	3.43	2.05	1.61	1.15	0.99	1.52	1.18
6. *Pseudomonas* spp. + 1/2 N	1.88	1.63	2.95	1.85	0.86	0.99	0.75	1.25	0.97
Composite inoculum + 1/2 N	2.41	2.68	3.05	1.61	0.74	1.63	1.27	1.77	1.25
LSD 0.05		0.71					0.37		

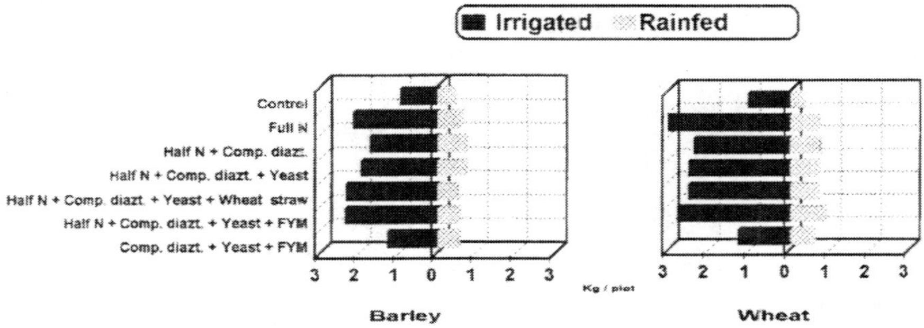

Figure 5. Total biological yield of wheat (c.v. Gemaza 1 and barley (cv. Giza 124) as affected by different inoculation and fertilization treatments under irrigated (Ismailia) and rainfed (El-Arish) conditions.

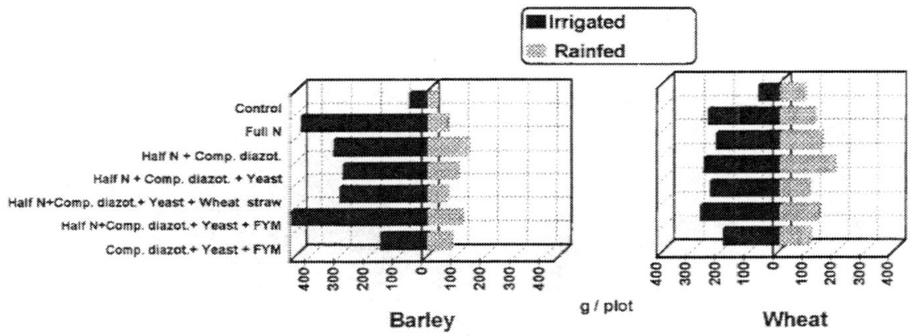

Figure 6. Grain yield of wheat (cv. Gemaza 1) and barley (cv. Giza 124) as affected by different inoculation and fertilization treatments under irrigated (Ismailia) and rainfed (El-Arish) conditions.

Field trials

Six field trials were executed, throughout the winter seasons from 1992 to 1996, in desert soils of Ismailia under sprinkler irrigation and in El-Arish under rainfed conditions. The first site lies at 33°E and 30°S at an altitude of 200 m above sea-level; ranges of rainfall and temperature were 22–45 mm and 13–23°C in winter; relative humidity was 50–70%. The second site lies at 33°E and 31°S at an altitude of 17.1 m above sea-level; ranges of rainfall and temperature were 100–120 mm, 7.6–17.9°C; relative humidity was 71–77%.

The major objectives were to evaluate the performance of the multi-strain inocula of associated diazotrophs (with or without yeast), and possible interactions with mineral and organic fertilization. Mineral N-fertilization was applied at the rates of 60 and 120 kg N per acre for wheat (cvs. Sakha 69, Gemaza 1, Benisuif 1 and Giza 155) and barley (cvs. Giza 123 and Giza 124). Organic amendments included wheat straw (C/N, 40:1) and farmyard manure (C/N, 40:1). A liquid composite inoculum

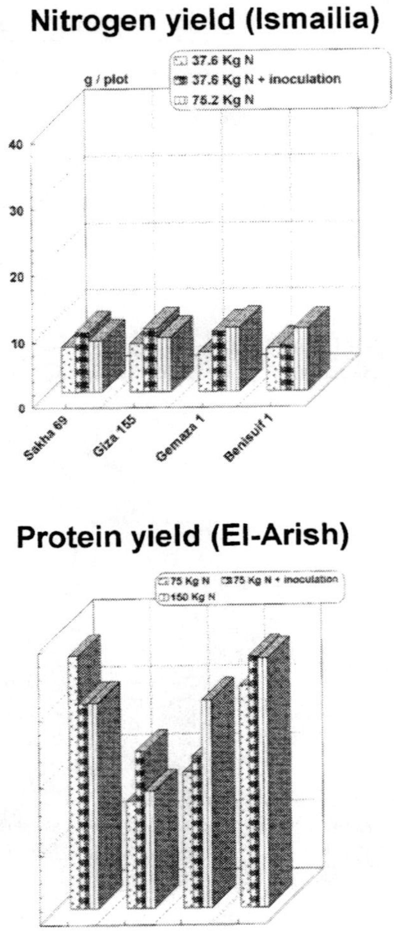

Figure 7. Nitrogen and protein yield (g/plot) of straw and seed in Ismailia and El-Arish respectively.

was prepared by mixing separately grown batch culture of *A. brasilense, A. chroococcum, K. pneumoniae, P. putida* and *Bacillus* spp. In some cases (*Saccharomyces cerevisiae*, 10^8 cells ml^{-1}) were included in the multi-strain inoculum preparation. Seeds were soaked in the prepared inoculum for 1 h at the rate of 200 g seed/500 ml, then kept in the shade. Seeds were drilled in rows 20 cm apart and 2.5 m long. A boost of 1L liquid inoculum per plot was sprayed adjacent to the roots 18 days post-sowing. The plot size was 3 m^2 (2.5 m × 1.2 m), and each treatment was replicated four times in randomized split-plot design where main plots were assigned for cultivars and sub-plots for inoculation. Plants were collected at harvest to determine total biological and grain yields. Also, nitrogen yield of straw and amino acid contents of grains were estimated in some trials.

Table 3. Average biomass and grain yields of wheat and barley cultivated under irrigated and rainfed conditions

Crop	Irrigated field		Rainfed field	
	Full N	1/2 N + inoc.	Full N	1/2 N + inoc.
Wheat				
Total biological yield (kg $plot^{-1}$)	1.99	1.74	0.54	0.58
Grain yield (g $plot^{-1}$)	202.9	201.9	120.1	140.7
Barley				
Total biological yield (kg $plot^{-1}$)	2.83	2.49	0.70	0.66
Grain yield (g $plot^{-1}$)	427.9	296.3	96.7	102.5

It is observed that under sprinkler irrigation, considerably higher biomass yield (2.2 kg $plot^{-1}$) obtained for wheat, in particular cultivar Gemaza 1, with full N-fertilization (Figure 5). Inoculation with associative diazotrophs, when combined with a rational dose of N-fertilizer (60 kg N $acre^{-1}$) resulted in a significantly similar yield (1.9 kg $plot^{-1}$). The grain yield followed a similar trend (Figure 6). This indicates that composite inoculation could save about 50% of plant N-requirements. Application of organic fertilizers did not result in a significant additional yield. This strongly emphasizes that decomposition of organic matter applied and nutrient release are impaired under the harsh and stressed environmental conditions prevailing in the experimental sites tested. On the contrary, straw amendment under fertile soil conditions of the Nile Valley significantly contributed to the productivity of inoculated cereal crops (Hegazi *et al.*, 1983). Incorporation of yeasts into the multi-strain inoculum resulted in no significant additonal yield. Variations among the different treatments followed a similar trend under rainfed conditions of El-Arish, but productivity was inferior.

Similar to wheat, simultaneous inoculation with multi-strain of associative diazotrophs and rational N-fertilization of barley resulted in statistically higher yields comparable to full N-fertilization. Organic fertilization did not significantly interact with either inoculation or N-fertilization. Again, irrigated fields of Ismailia supported better yields of barley compared to rainfed areas in El-Arish (Table 3).

Another set of field trials were carried out in the same sites to evaluate the response of several wheat cultivars to the composite inoculum of associative diazotrophs in the presence of rational dose of N-fertilizer. Beside the increases in total dry matter and grain yields attributed to inoculation in the presence of a rational dose of N-fertilizers, gains in total N-yield were also recorded. Percentage increases in straw N and seed protein yields of inoculated plants were 5.9–50% higher than those received no inocula (Figure 7). The most responsive cultivars differed among localities, being Gemaza 1 under irrigated and Giza 155 under rainfed conditions. On the other hand, cultivar Sakha 69 did not respond to inoculation. Plant cultivar–diazotroph interactions wre also reported by several investigators (Kucey, 1988; Millet *et al.*, 1984).

Acknowledgements

This study was supported by the national project on introducing agro-biotechnologies based on biological nitrogen fixation for the development of Sinai, funded by the Egyptian Ministry of Agriculture and Land Reclamation and the French Food Aid Counterpart Fund (FFACF-BLAFE).

References

Barber D A, Martin J K 1976. The release of organic substances by cereal roots into soil. New Phytol., 75, 69–80.

Baldani V L, Baldani J I, Mandel J, Rocha R, Dobereiner J 1987. Inoculation of field grown wheat with *Azospirillum brasilense* inoculants in peat and oil. 4th International Symposium on Nitrogen Fixation with Non-legumes, Abstract book, 169.

De-Freitas J R, Germida J J, Rennie R J 1987. Response of winter wheat to inoculation with diazotrophic bacteria. 4th International Symposium on Nitrogen Fixation with Non-legumes, Abstract book, 166.

De-Ming L, Alexander M 1986. Bacterial growth rates and competition effect on nodulation and root colonization by *Rhizobium meliloti*. Appl. Environ. Microbiol., 52, 807–811.

Halsall D M, Goodchild D J 1986. Nitrogen fixation associated with development and localization of mixed populations of *Cellulomonas* spp. and *Azospirillum brasilense* grown on cellulose or wheat straw. Appl. Environ. Microbiol., 51, 849–854.

Hamza A M, Youssef H H, Helmy A *et al.*, 1994. Mixed cultivation and inoculation of various genera of associative dizazotrophs. In: Hegazi, N A, Fayez M, Monib M, eds. Nitrogen Fixation with Non-Legumes. American University In Cairo Press, 319–326.

Hegazi N A, Monib M, Amer H A, Shokr E S 1983. Response of maize plants to inoculation with azospirilla and (or) straw amendment in Egypt. Can. J. Microbiol., 29, 888–894.

Khammas K M, Tran V V, Kaiser P 1994. Nitrogen fixation in rice plantlets inoculated with pure or mixed cultures of *Azospirillum* species and *Bacillus polymyxa*. In: Hegazi N A, Fayez M, Monib M, eds. Nitrogen Fixation with Non-Legumes. American University In Cairo Press, 289–293.

Kucey R M N 1988. Plant growth-altering effects of *Azospirillum brasilense* and *Bacillus* C-11–25 on two wheat cultivars. J. App. Bacteriol., 64, 187–196.

Lippi D, Cacciari I, Pietrosanti T, Pietrosanti W 1992. Interactions between *Azospirillum* and *Arthrobacter* in diazotrophic mixed culture. Symbiosis, 13, 107–114.

Loper J E, Haack C, Schorth M N 1985. Population dynamic of soil pseudomonads in the rhizosphere of potato (*Solanum tuberosum* L.). Appl. Environ. Microbiol., 49, 416–422.

Lynch J M, Clark S J 1984. Effects of microbial colonization of barley (*Hordeum vulgare* L.) roots on seedling growth. J. Appl. Bacteriol., 56, 47–52.

Millet E, Avivt Y, Feldman M 1984. Yield response of various wheat genotypes to inoculation with *Azospirillum brasilense*. Plant Soil, 80, 261–266.

Prikryl Z, Vancura V 1980. Root exudates of plants. VI. Wheat root exudation as dependent on growth concentration gradient of exudates and the presence of bacteria. Plant Soil, 57, 63–83.

Quatrini P, Vettraino A M, Carcupino M *et al.*, 1992. Colonization of sorghum roots by *Azospirillum* and three bacteria isolated from sorghum rhizosphere. Symbiosis, 13, 255–269.

Reinhold B, Hurek T, Fendrik I 1985. Strain-specific chemotaxis of *Azospirillum* spp. J. Bacteriol., 162, 190–195.

Rennie R J, Larson R I 1979. Dinitrogen fixation associated with disomic chromosome substitution lines of spring wheat in the phytotron and in the field. International Workshop on Associative N_2-Fixation, CENA, Piracicba, Brazil.

Schlegel H G 1988. In: General Microbiology, 6th edn. Cambridge University Press, p. 11.

Vancura V, Stotzky G 1976. Gaseous and volatile exudates from germinating seeds and seedlings. Can. J. Bot., 54, 518–532.

25
Effect of inoculation with nitrogen-fixing *Klebsiella* on potato yield

V.N. PISHCHIK, I.I. CHERNYAEVA, A.P. KOZHEMAYKOV, N.I. VOROBYOV, A.M. LAZAREV AND L.P. KOZLOV

Abstract

A significant positive effect on yield of potato cultivars was obtained after inoculation with *Klebsiella mobilis* strains CIAM880 and CIAM853 when low doses of nitrogen fertilizer were used. The yield of various potato cultivars increased 1.2–1.4-fold as compared with uninoculated plants. Inoculation with *Klebsiella mobilis* decreased the content of nitrates in potato tubers 1–2-fold. The differences between plant nutrition of various potato cultivars were determined by a plant genotype. Inoculation with *Klebsiella mobilis* significantly increased the content of starch, nitrogen, phosphorus and potassium in potato tubers. The studied bacteria were able to colonize actively the potato tubers and roots. After joint inoculation *Klebsiella mobilis* decreased the survival of phytopathogenic *Erwinia carotovora* CIAM884 and protected the plants against soft-rot disease. It was concluded that *Klebsiella mobilis* strains CIAM880 and CIAM853 can be considered as associative nitrogen fixers (their acetylene-reducing activity was 329–407 nmol/ml per hour) and PGPR bacteria (all strains were able to produce indoleacetic acid) and used as biopreparations for increasing the productivity of potatoes.

Introduction

Bacteria of the family Enterobacteriaceae have a great deal of practical and theoretical importance in agriculture. *Erwinia carotovora*, having a strong pectolytic activity, causes soft-rots in a wide variety of plants (Dye, 1969). On the other hand, enterobacteria form a significant component of the non-phytopathogenic epiphytic microflora on plant surfaces (Rennie *et al.*, 1982; Haahtela *et al.*, 1986). These non-phytopathogenic bacteria interact with the plant and may perhaps play some important role in modifying the courses of diseases caused by phytopathogenic *Erwinia* species (Chatterjee *et al.*, 1969).

K.A. Malik et al. (eds.), Nitrogen Fixation with Non-Legumes, 223–235.

The present report attempts to study the interaction between potato plants and bacteria belonging to *Klebsiella mobilis* and phytopathogenic *Erwinia carotovora* in pot experiments.

Material and methods

Bacterial strains

Bacterial strains (CIAM) were taken from the Culture Collection of the Research Institute for Agricultural Microbiology (St Petersburg, Russia): *Erwinia carotovora* 384, 784, *Klebsiella mobilis* 880, 853; *Enterobacter cloaceae* 264; *Serratia marcescens* 235. Strains of the genus *Klebsiella* (CIEVM): 1349, 1339; ATCC13883 were kindly provided by Dr N.R. Vasser (Pasteur Institute of Epidemiology and Microbiology, St Petersburg, Russia). Strains of the genus *Erwinia* (CIMVUA): 8807, 3528 were kindly provided by Professor R.I. Gvozdyak (Institute of Microbiology and Virology, Kiev, Ukraine). The strain of *Klebsiella planticola* CTAA91 was kindly provided by Professor V.T. Emtsev (Timiriazev Agricultural Academy, Moscow, Russia).

Microbiological analysis

Bacterial growth was measured using a colorimeter with a 540 nm filter; pH bacterial suspension was measured using a digital pH meter. Contents of $N\text{-}NO_3$ and $N\text{-}NH_4$ were measured using standard nitrate and ammonium electrodes (St Petersburg, Russia). The nitrate content was measured according to the Griess method, using sulphanilic acid and naphthalamine. The glucose concentration was measured using a colorimeter with a 540 nm filter, according to Maijorova's method (1971).

The studied enterobacteria were tested for indoleacetic acid production (IAP) using Salkovsky's reagent (Hartmann *et al.*, 1983).

Gas-chromatography analysis

The bacterial suspension (0.1 ml with density 10^9 bacterial cells per 1 ml) was added to the flask containing 20 ml selective Omelyansky medium for the determination of bacterial gas exchange by the method of gas chromatography (Shirokova and Evdokimov, 1981). Prior to determination of the ARA, each strain was grown in liquid medium for 48 h at 20°C. The bacteria were harvested by centrifugation and resuspended in sterile water. The studied strains were tested for nitrogen fixation by inoculating cultures into the flasks (five replications) containing 20 ml Vinogradsky medium with a growth-limiting quantity of combined nitrogen (25 μg nitrogen per ml as ammonium sulphate). The bacteria were incubated under aerobic conditions at 20°C. After incubation for 48 h the cotton plugs were replaced by rubber cups and 10% of air gas phase replaced by acetylene. After 24 h the ethylene content was determined on a 'Cvet' gas chromatograph.

Enumeration of bacteria in microbial cenosis

For differentiation of virulent *E. carotovora* CIAM884 and avirulent *K. planticola* CIAM880, both their growth on solid Omelyansky (1940) medium (glucose 10 g K_2HPO_4, 1 g; $MgSO_4$, 0.5 g; $(NH_4)_2PO_4$, 1 g; $CaCl_2$, 0.1 g; NaCl, $FeSO_4$, 0.1 g/l distilled water) containing 0.005% 2,3,5-triphenyl tetrazolium chloride (Friedman, 1964) and immunofluorescent procedures (Allan and Kelman, 1977) were used. There tests were also used for detection of *E. carotovora* at the time of isolation from potato tubers and roots in the pot experiments. In addition, an antibiotic-resistant mutant test was used in the pot experiments. Spontaneous resistant mutants were obtained by the gradient plate technique (Szybalski, 1952). Mutants of *E. carotovora* CIAM884, *K. mobilis* CIAM853 and CIAM880 were resistant to 40 μg/ml rifampicin and 50 μg/ml nalidixic acid.

Potato roots (approximately 1 g, three replications) were taken from the soil at various growth stages, washed by shaking in 100 ml of water five times, approximately for 1–2 min, and homogenized in 10 ml of sterile water. The bacteria in the homogenate were assumed to be rhizoplane bacteria. The homogenate was serially diluted in 10-fold steps and plated on solid Omelyansky medium (five replications) supplemented with 40 μg/ml rifampicin and 50 μg/ml nalidix acid or 0.005% 2,3,5-triphenyl tetrazolium chloride.

Pot experiments

Pot experiments were carried out in a greenhouse under plastic film (in summer, for 3 months) on potato cultivars Nevsky, Detskoselsky and Priekulsky. The potato plants were grown in pots containing 5 kg unsterile derno-podzolic soil. This soil had the following agricultural characteristics: total nitrogen 0.080%, total carbon 0.80 mg/100g; available P, 3–4.4 mg P_2O_5/100 g; available K, 7–12 mg K_2O/100 g; sum of absorbed bases 6–8 mg Eq/100 g; hydrolytic acidity 2.9 mg Eq/100g; pH 4.9–5.1. The pH of soil was adjusted to 6.5–6.7 with $CaCO_3$ to create favourable conditions for plant development. The following doses of fertilizers were applied in all experiments before planting (mg/kg soil): NH_4NO_3, 200, 400, 600, respectively; KH_2PO_4, 380; KCl, 240; MgC1, 60; H_3BO_3, 3; $MnSO_4$, 3; $ZnSO_4$, 3; Na_2MoO_4, 1.4.

Bacterial cultures used as inoculum were grown in Omelyansky (1940) liquid medium on a shaker at 25°C for 3 days. The inoculation was carried out immediately after planting, using 10 ml of bacterial suspension (in the concentration of 1 mlrd cells per 1 ml solution). Control plants were treated with autoclaved cultures. There were five pots per strain or control. The plants were supplied daily with sufficient water to maintain an average soil moisture of 60% water-holding capacity.

Total nitrogen content in plants was determined by the Kjeldahl method using a 'Kjeltek-auto' analyser. Total phosphorus content in plants was determined by the Skulachev method, and total potassium content by the sodium cobaltnitrite method, using a photoelectric concentration colorimeter, model KFK-2 (Russia). Total starch content in plants was determined by the Pochinok method. All the methods for measuring total nitrogen, phosphorus, potassium and starch content are described by Ermakov *et al.* (1987).

Statistical analysis

A calculation of the confidence intervals, the coefficient of normal correlation and a standard method of the variance analysis were used (Lakin, 1980). LSD indicates least significant difference.

Results

Analysis of bacterial metabolism

The results represented in Table 1 show that the source of nitrogen had a great effect on growth, glucose utilization and respiration rate of the strains having different pathogenicity for potato plants. The addition of $NaNO_3$ significantly decreased (2-fold) the glucose utilization by avirulent *K. mobilis* CIAM880. In this case the glucose utilization by virulent *E. carotovora* CIAM884 changed very little. As a consequence $NaNO_3$ had an influence also on bacterial growth, especially on the growth of *K. mobilis* CIAM880. At the same time *E. carotovora* were more tolerant bacteria to the presence of $NaNO_3$ in the medium.

Avirulent bacteria *K. mobilis* CIAM880 released CO_2 approximately 3-fold higher and H_2 4-fold higher than did virulent *E. carotovora* CIAM884 in the standard Omelyansky medium with reduced nitrogen source $(NH_4)_2HPO_4$. The addition of $NaNO_3$ into the medium decreased the CO_2 and inhibited H_2 production by the bacteria studied.

The nitrogen metabolism of bacteria was also changed with the availability of $NaNO_3$ in the medium. Under these conditions bacteria CIAM880 and CIAM884 utilized only 6–10% ammonium nitrogen and reduced 59–87% nitrate nitrogen, that led to the production of nitrite: 92–159 reduced mg/L (Table 1). This is to be contrasted with the growth of bacteria in the medium containing only ammonium nitrogen, when the bacteria utilized 47–60% of ammonium nitrogen.

The avirulent bacteria *K. mobilis* CIAM880 prevailed over the virulent bacteria *E. carotovora* CIAM884 in the microbial cenosis under cultivation on the Omelyansky medium. The percentage incidence of the avirulant bacteria *K. mobilis*

Table 1. Effect of the source of nitrogen in the growth medium on the metabolism of bacteria *Erwinia carotovora* CIAM884 and *Klebsiella mobilis* CIAM880

Nitrogen sources; bacteria	Optical density of bacterial suspension	Production of CO_2, (vol.%)	Production of H_2 (vol. %)	Glucose utilization (%)	Utilization of N-NH_4 (%)	Utilization of N-NO_3 (%)	Production of N-NO_2 (mg/L)
N-NH₄							
CIAM880	8.80	20.5	7.6	88	60	0	0
CIAM884	0.60*	5.4	1.9	46	47	0	0
$LSD_{0.05}$	0.10	2.6	2.4	5	7		
N-NH₄+N-NO₃							
CIAM880	0.68	6.7	0	56	8	87	154
CIAM884	0.72	4.6	0	60	6	59*	92*
$LSD_{0.05}$	0.08	2.4		5	4	15	17
N-NO₃							
CIAM880	0.30	5.8	0	32	0	94	158
CIAM884	0.40*	3.9	0	44*	0	75*	144
$LSD_{0.05}$	0.06	1.8		8	0	8	10

Notes: (1) Initial glucose concentration was 5 mg/ml^{-1}; (2) N-NH_4 was added as $(NH_4)_2HPO_4$; (3) N-NO_3 was added as $NaNO_3$.
* In this table the values in each column differed significantly from those for other strains.
LSD = Least significant difference

CIAM880 was about 58%. The addition of $NaNO_3$ changed the correlation between avirulent *K. mobilis* CIAM880 and virulent *E. carotovora* bacteria CIAM884. The percentage incidence of the avirulent bacteria *K. mobilis* among the mixed population was 34%.

Acetylene-reducing analysis

The acetylene-reducing activity of the avirulent isolates of *K. mobilis* CIAM880 and CIAM853 was approximately 2 times higher than that of virulent *E. carotovora* CIAM884 (Table 2). Other strains of the Enterobacteriaceae family, *Erwinia rhapontici* CIMVUA8807 and *Klebsiella planticola* CTAA91 and CIEVM1339, were capable of reducing nitrogen (their acetylene-reducing activity was 329–407 nmol (ml^{-1} h^{-1}). Neither of two *Klebsiella pneumoniae* strains CIEVM1349, ATCC13883 or *Serratia marcescens* CIAM235 was found to fix nitrogen as determined by the ARA method.

Table 2. Acetylene reduction activity and productions of phytohormones by the studied strains (in batch culture)

Bacteria	ARA, C_2H_4 formed (nmol ml^{-1} h^{-1}	IAA production (μg/ml)	Production of C_2H_4 (vol.%/10^{-5})
Erwinia carotovora			
Virulent strains			
CIAM884	215±11	4.4±0.3	29.7±2.3
CIAM784	202±30	6.5±0.3	30.4±2.8
Klebsiella mobilis			
CIAM880	329±22	5.4±0.3	12.8±1.2
CIAM853	407±11	5.9±0.2	10.9±0.8
Erwinia rhapontici			
CIMVUA8807	375±30	10.1±0.4	15.2±2.4
Erwinia chrysanthemi			
CIMVUA3528	0	6.0±0.3	30.0±1.8
Enterobacter cloaceae			
CIAM264	0	7.1±0.4	0
Klebsiella planticola			
CTAA91	334±22	4.9±0.2	12.6±0.8
CIEVM1339	339±11	5.3±0.2	10.2±1.4
Klebsiella pneumoniae subsp. *pneumoniae*			
ATCC13883	0	4.1±0.2	0
CIEVM1349	0	5.3±0.2	0
Serratia marcescens			
CIAM235	0	3.7±0.2	0

Analysis of production of plant growth-promotion substances

The studied bacteria were tested for indoleacetic acid production (IAP). Results are summarized in Table 2. All the studied strains were found to produce indoleacetic acid. The differences between IAA production by virulent *E. carotovora* CIAM 884 and CIAM784 and by avirulent *K. mobilis* CIAM880 and CIAM853 were not significant.

At the same time, virulent strains of *E. carotovora* produced two or three times more endogenous C_2H_4 than did avirulent bacteria.

Pot experiments

In the first pot experiment the competitiveness and survival of virulent *E. carotovora* CIAM884 and avirulent *K. mobilis* CIAM880, and their effect on potato cultivar Nevsky, was studied (Table 3). The studied enterobacteria were able to colonize actively the root surface, and their survival was dependent on the content of nitrates in potato tubers. Nitrate content in potato tubers increased when the level of nitrogen fertilizer increased. If the plants were inoculated with a single strain of *K. mobilis* CIAM880 the total number of introduced bacteria decreased with increase in the dose of nitrogen fertilizer. After mixed inoculation with avirulent *K. mobilis* CIAM880 and virulent *E. carotovora* CIAM884 strains the avirulent bacteria prevailed over the virulent strain in the microbial cenosis when plant nitrogen nutrition was balanced. The percentage incidence of virulent *E. carotovora* CIAM884 increased 2-fold, and they dominated over the avirulent bacteria *K. mobilis* CIAM880 when nitrogen fertilizer was added in a maximal dose (600 mg/kg). The quantity of rot tubers also increased with increase in level of nitrogen fertilizer.

In the second experiment the effect of inoculation with avirulent strains of *K. mobilis* on the yield of various potato cultivars and nitrate content in the tubers of potato was studied (Table 4). The maximal positive yield response for various potato cultivars was observed when a minimal dose of nitrogen fertilizer was added (200 mg/kg). The yield of various cultivars of potato increased 1.2–1.4-fold as compared with control plants. At a high level of added nitrogen (600 mg/kg) the differences between the inoculated plants and the uninoculated control were not significant (data not shown). The inoculation with avirulent *K. mobilis* CIAM880 and CIAM853 decreased on average 1.1–1.2-fold the content of nitrates in potato tubers.

Inoculations with avirulent enterobacteria CIAM880 and CIAM853 increased the total content of starch, nitrogen, phosphorus and potassium in the potato tubers. However, the differences between the inoculated and uninoculated plants in some cases were not significant (Table 5). It is possible that the differences between plant nutrition of various potato cultivars were determined by plant genotype. For example, the potato cv. Detskoselsky was more responsive to inoculation with the studied bacteria. The total content of starch increased on average 1.08 times; nitrogen, 1.2 times; phosphorus, 1.5 times and, potassium, 1.2 times.

Table 3. Effect of inoculation with *K. mobilis* CIAM 880 (avirulent) and *E. carotovora* CIAM884 (virulent) on yield and rot disease of potato cv. Nevsky in pot experiment

Variants of inoculation	Doses of NH_4NO_3 (mg/kg)		
	200	400	600
Potato yield (g/pot)			
1. Uninoculated control	132	164	180
2. *K. mobilis* CIAM880	187*	191*	175
3. *E. carotovora* CIAM884	95*	139*	107*
4. *K. mobilis* CIAM880 + *E. carotovora* CIAM884	127	182*	173
$LSD_{0.05}$	21	16	11
Content of nitrates in potato tubers (mg/kg)			
1. Uninoculated control	13	23	144
2. *K. mobilis* CIAM880	9*	17*	125*
3. *E. carotovora* CIAM884	11	21	138
4. *K. mobilis* CIAM880 + *E. carotovora* CIAM884	11	25	126*
$LSD_{0.05}$	3	4	12
Number of bacteria (10^4 cells/g tuber)			
2. *K. mobilis* CIAM880	35 ± 1.0	27 ± 2.5	12 ± 1.5
3. *E. carotovora* CIAM884	31 ± 1.5	29 ± 1.5	28 ± 2.0
4. *K. mobilis* CIAM880 + *E. carotovora* CIAM884	34 ± 1.2 + 15 ± 1.2	18 ± 1.5 + 14 ± 1.5	12 ± 1.2 + 18 ± 1.2
Total number	49 ± 1.8	32 ± 2.3	30 ± 1.8
Revealing of rot disease (percentage of rotted tubers)			
1. Uninoculated control	0	0	0
2. *K. mobilis* CIAM880	0	0	0
3. *E. carotovora* CIAM884	4	8	17
4. *K. mobilis* CIAM880 + *E. carotovora* CIAM884	3 (latent form)	3	7

* In this and in succeeding tables the values in each column differed significantly from those for uninoculated plants.
LSD = Least significant difference

The studied strains CIAM880 and CIAM853 were able to colonize actively the potato roots. The population curve of each strain had distinctive features (Figure 1A, B). The population size of avirulent strain CIAM880 reduced rapidly during the first 30 days after mixed inoculation with virulent bacteria CIAM884, then significantly increased during 10 days, and over the last 20 days became stable at a high level (6.3×10^5–7.9×10^5 cells per gram of potato roots). In comparison with the result of combined inoculation the population size of avirulent strain CIAM880 was more stable, and after 20 days did not reduce rapidly.

Discussion

It was shown that the non-phytopathogenic enterobacteria had a large effect on respiration rate and medium oxidation level as compared with the pathogenic

Table 4. Effect of inoculation with *Klebsiella mobilis* on the yield and nitrate content in tubers of various potato cultivars

Treatments	Doses of NH_4NO_3 (mg/kg)			
	200		400	
	Potato yield (g/pot)	Nitrate content (mg/kg)	Potato yield (g/pot)	Nitrate content (mg/kg)
Detskoselsky cultivar				
1. Uninoculated plants	168.4	25.1	184.0	138.0
2. *K. mobilis* CIAM853	210.0*	23.2	225.0*	133.0
3. *K. mobilis* CIAM880	217.5*	20.0*	234.7*	123.0*
$LSD_{0.05}$	20.6	2.8	10.2	8.4
Priekulsky cultivar				
1. Uninoculated plants	140.9	15.3	168.3	126.3
2. *K. mobilis* CIAM853	160.4*	14.1*	177.6	120.0*
3. *K. mobilis* CIAM880	172.9	12.3*	182.4	115.1*
$LSD_{0.05}$	16.4	0.5	10.2	4.2
Nevsky cultivar				
1. Uninoculated plants	132.4	20.0	164.0	130.0
2. *K. mobilis* CIAM853	164.0*	18.7	175.3	128.0
3. *K. mobilis* CIAM880	187.3*	16.0*	190.5*	115.1
$LSD_{0.05}$	24.8	2.8	20.6	4.2

*See Table 3.

bacteria. The non-pathogenic bacteria *K. mobilis* CIAM880 had highly developed sugar fermentation and nitrogen fixation complexes as compared with the phytopathogenic *Erwinia carotovora* CIAM884.

The studied bacteria were able to assure 'nitrate respiration'. It should be noted that avirulent bacteria *K. mobilis* CIAM880 possessed more active nitrate reductase and less active nitrite reductase as compared with virulent bacteria *E. carotovora* CIAM884. The comparatively higher content of toxic nitrite in the cultural medium under the growing avirulent bacteria *K. mobilis* CIAM880 may be major reason for the low tolerance of these bacteria to the presence of $NaNO_3$ in the medium. This leads to a change in the relationship between the number of avirulent bacteria *K. mobilis* CIAM880 and the number of phytopathogenic *E. carotovora* CIAM884 in the microbial community. The pathogenic bacteria dominated over avirulent bacteria under unfavourable nitrogenous conditions for plants. It was established that this unfavourable situation for plants may be improved by inoculation with avirulent strains of *K. mobilis*.

All introduced avirulent bacteria were able to colonize actively the potato tubers and roots. In pot experiments their number was high and stable (6–8 × 10^5 cells/g fresh roots) during the period of plant growth. Consequently after joint inoculation avirulent *K. mobilis* CIAM880 decreased the population of phytopathogenic *E. carotovora* CIAM884 and protected the plants against soft-rot

Table 5. Effect of inoculation of potato cultivars with avirulent *K. mobilis* on the yield and content of nutrient elements in tubers.

Treatments	Potato cultivars		
	Detskoselsky	Priekulsky	Nevsky
Yield (g/pot)			
Uninoculated control	184.0	168.3	164.0
CIAM880	225.0*	177.6	175.3
CIAM853	234.7*	182.4	190.5*
Total content of starch (%)			
Uninoculated control	16.0	13.0	14.0
CIAM880	17.2*	15.0*	14.5
CIAM853	17.4*	14.5	15.0
Total N (%)			
Uninoculated control	0.38	0.30	0.34
CIAM880	0.47*	0.34	0.34
CIAM853	0.45*	0.33	0.38*
Total P (%)			
Uninoculated control	0.16	0.13	0.15
CIAM880	0.23*	0.25*	0.20*
CIAM853	0.24*	0.27*	0.18
Total K (%)			
Uninoculated control	0.57	0.45	0.52
CIAM880	0.71*	0.54*	0.54
CIAM853	0.70*	0.52*	0.58

*See Table 3.

diseases. As reported previously (Pishchik *et al.*, 1995) electron microscopy studies showed that the non-phytopathogenic bacteria CIAM880 competed for energy-rich sites with pathogenic bacteria CIAM884 and protected the plant from rot disease (Pishchik *et al.*, 1995). Other authors have revealed that *K. mobilis* had maximal root colonization activity in comparison with other diazotrophic strains (Kravchenko *et al.*, 1993). Because of their high competitiveness, avirulent bacteria *K. mobilis* may play an important role in plant disease regulation.

All the studied strains were found to produce indoleacetic acid. The ability of enterobacteria strains to produce auxins was determined previously (Zimmer *et al.*, 1995; Turjanitsa *et al.*, 1995; Koga *et al.*, 1991; Haahtela *et al.*, 1990). It was supposed that the ability of bacteria to produce auxins is more important than nitrogen-fixing ability in their positive effect on plant growth (Fallik *et al.*, 1989). It is possible that the effect of the present avirulent bacteria *Klebsiella* sp. on plant growth and mineral nutrition is due to growth promotion substances secreted by the bacteria.

It is known that the excess of endogenous ethylene inhibited the growth of plant roots (Jackson, 1991; Esashi, 1991). We have found that the phytopathogenic bacteria

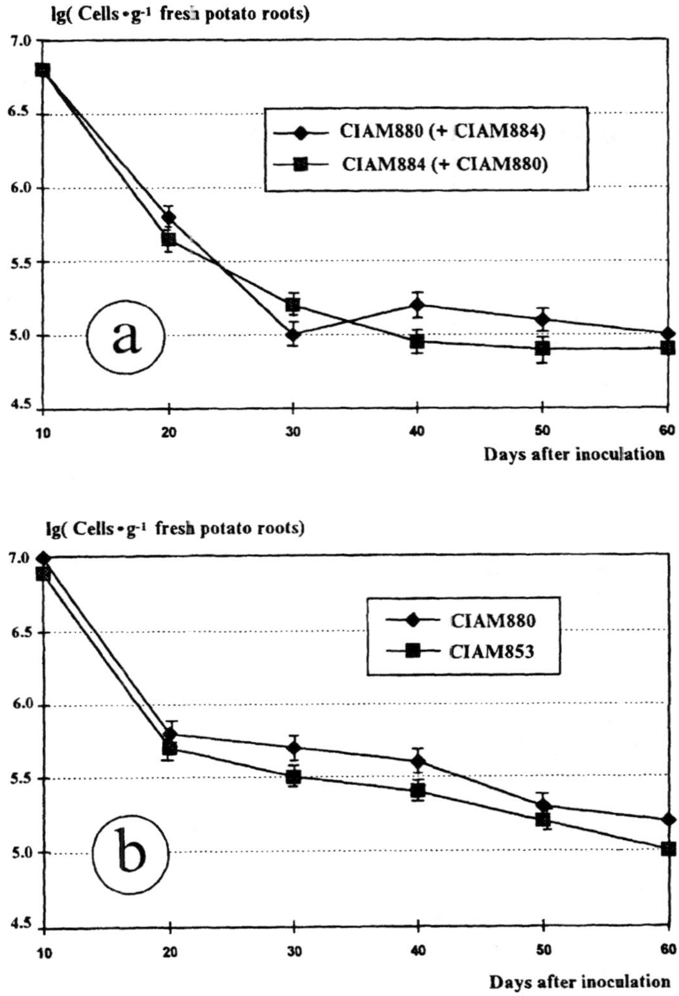

Figure 1. Survival of the introduced bacteria on potato roots (cv. Nevsky): (**a**) mixed inoculation with *Klebsiella mobilis* CIAM880 and *Erwinia carotovora* CIAM884; (**b**) single inoculation with *Klebsiella mobilis* CIAM880 and CIAM853.

E. carotovora produced the endogenous ethylene more than did non-phytopathogenic bacteria from the genus *Klebsiella*. Therefore the phytopathogenic bacteria *E. carotovora* could inhibit the growth of potato roots due to heightened production of endogenous ethylene. This should be considered as an additional effect which can determine the complex pathogenicity mechanism of the studied bacteria.

The avirulent bacteria *K. mobilis* had a beneficial effect on plant nutrition. It is possible that introduced bacteria increased fertilizer recovery by the plants, because the total content of nitrogen, potassium and phosphorus increased in the inoculated plants. The response of plants to inoculation was dependent on plant genotype.

As was shown in the pot experiment, the non-phytopathogenic *K. mobilis* prevailed over the phytopathogenic *E. carotovora* and protected the plants against soft rot. It was earlier shown that no differences between the character of rhizosphere colonization by avirulent CIAM880 or virulent bacteria CIAM884 have been revealed (Pishchik *et al.*, 1995). In addition to this, it was shown that bacteria *K. mobilis* had maximal root colonization activity (Kravchenko and Makarova, 1993) in comparison with other diazotrophic strains.

It was concluded that *K. mobilis* CIAM 880 and CIAM853 can be considered as associative nitrogen fixers and PGPR bacteria. All strains can be used as biopreparations for increasing the productivity of potato.

Acknowledgements

We are very grateful to Professor R.I. Gvozdyak (Ukraine), Professor V.T. Emtsev (Russia), and Dr N.R. Vasser (Russia) for supplying enterobacteria strains. The authors also with pleasure express thanks to Professor R.I. Gvozdyak, Professor Eu.P. Sivolodsky (Russia), Professor Yaacov Okon (Israel), and Dr A.A. Belimov (Russia) for discussion and criticism. We are grateful to engineer L.N. Sokol (Russia) for technical assistance in the chromatography work. This work was supported by research programme 'INTERBIOAZOT-2000'.

References

Allan E, Kelman A 1977. Immunofluorescent stain procedures for detection and identification of *Erwinia carotovora* var. *atroseptica*. Phytopathology, 67, 1305–1312.

Chatterjee A K, Gibbins L N, Carpenter J A 1969. Some observations on the physiology of *Erwinia herbicola* and its possible implication as a factor antagonistic to *Erwinia amylovora* in the 'fire-blight' syndrome. Can. J. Microbiol. 15, 640–642.

Dye D W 1969. A taxonomic study of the genus *Erwinia*. II. The 'carotovora group'. N.Z. J. Sci., 12, 81–97.

Ermakov A I, Arasimovitch V V, Iarosh NP 1987. Methods for Biochemical Investigation of Plants. Agropromizdat, Leningrad (in Russian).

Esashi Y 1991. Ethylene and seed germination. In: Matoo A K, Shuttle J C, eds. The Plant Hormone Ethylene. CRC Press, Boca Raton, 133–137.

Fallik E, Okon Y, Epstein E, Goldman A, Fisher M 1989. Identification and quantification of IAA and IBA in *Azospirillum brasilense*-inoculated maize roots. Soil Biol. Biochem., 21, 147–153.

Friedman B 1964. Carbon source and tetrasolium agar to distinguish virulence in colonies of *Erwinia carotovora*. Phytopathology, 54, 494–495.

Haahtela K, Laakso T, Korhonen T 1986. Associative nitrogen fixation by *Klebsiella* spp.: adhesion sites and inoculation effects on grass roots. Appl. Environ. Micromiol., 52, 1074–1079.

Haahtela K, Rönkko R, Laakso T, Williams P H, Korhonen T K 1990. Root associated *Enterobacter* and *Klebsiella* in *Poa pratensis*: characterization of an ion-scavenging system and a substance stimulating root hair production. Mol. Plant Microb. Interact., 3, 358–363.

Hartmann A, Singh M, Klingmuller W 1983. Isolation and characterization of *Azospirillum* mutants excreting high amounts of indole-acetic acid. Can. J. Microbiol., 29, 916–923.

Jackson M B 1991. Ethylene in root growth and development. In: Matoo A K, Suttle J C, eds. The Plant Hormone Ethylene. CRC Press, Boca Raton, 159–181.

Koga J, Adachi T, Hidaka H 1991. IAA biosynthetic pathway from tryptophan via indole-3 pyruvic acid in *Enterobacter cloaceae*. Agric. Biol. Chem., 55, 701–705.

Kravchenko L V, Makarova N M 1993. Kinetic of the root colonization for cereals after introduction of associated bacteria. Microbiology, 62, 524–529.

Krieg N R, Holt J A (eds). 1984. Bergey's Manual of Systematic Bacteriology, 9th edn. Williams and Wilkins, London.

Lakin G F 1980. Biometrics. Higher School, Moscow, (in Russian).

Maijorova I V 1971. The determination of glucose content in liquid cultural medium using tetrasolium. Appl. Biochem. Microbiol., 7, 494–495 (in Russian).

Omelyansky V L 1940. The Manual of Microbiological Methods. Academy of Sciences, Moscow, SU (in Russian).

Pishchik V N, Chernyaeva I I, Lukyanenko E O, Lebsky V K, Vorobyov N I, Lazdrov A M, Kozolov L P 1995. Biological diversity and competitiveness of nitrogen-fixing bacteria *Erwinia carotovora*. In: Tikhonovich I A, Prororov N A, eds. Nitrogen Fixation: Fundamental and Applications Proceedings, 10th International Congress on Nitrogen Fixation. St. Petersburg, 28 May–3 June. Kluwer, Dordrecht, 774.

Rennie R J, de Freitas J R, Ruschel A P, Vose P B 1982. Isolation and identification of N_2-fixing bacteria associated with sugar cane (*Saccharum* sp.). Can. J. Microbiol., 28, 462–467.

Shirokova L N, Evdokimov S N 1981. The gas-chromatography investigations of the metabolism of the microorganisms. Leningrad, 27–30 (in Russian).

Szybalski W 1952. Microbial selection. 1. Gradient plate technique for study of bacterial resistance. Science, 116, 46–48.

Turjanitsa A I, Petak A M, Nichik M M, Petrosova V P 1995. Study of capacity of *Klebsiella* bacteria to fix air nitrogen and to produce plant growth hormones. Mikrobiol. Zh. (Ukrainian). 57, 28–34 (in Russian).

Zimmer W, Kloos K, Hundeshagen B, Neiderau E, Bothe H 1995. Auxin biosynthesis and denitrification in plant growth promotion bacteria. In: Fendrik J, de Gallo, Vandeleyden J, de Zamoroczy M, eds. *Azospirillum* VI and related microorganisms. Series G: Ecological science, Vol. 37. Springer, Berlin, 120–141.

26
Responses of rice plants to inoculation with *Azospirillum* sp. under field conditions

NGUYEN THI PHUONG CHI, HA HONG THANH AND NGUYEN NGOC DZUNG

Abstract

Germinated rice seeds were inoculated with *Azospirillum* strains Mat 2–1b and DA 9–3a, then directly brought into the fields at several places of the Redriver Delta, North Vietnam.

After 35–50 days, the treated seedlings appeared to be taller and more vigorous than the non-treated ones. The inoculated seedlings showed a rather higher survival rate under conditions of 10°C lower air temperature for a long time (8 days) in comparison with non-inoculated seedlings. Corn yields obtained in the years 1991, 1992, 1993 and 1995 had been increased by treatment with *Azospirillum* inoculates; it could reach 105–125%. Additionally, 3–5 days earlier corn ripening of the inoculated rice plants had been observed.

Introduction

Vietnam is an agricultural country with a water rice cultivation tradition of 1000 years. In order to increase rice production in Vietnam it is necessary to import a large proportion of the chemical fertilizers. Due to uncontrolled use of chemical fertilizers for many years the area of marginal land is on the increase. In such a situation Vietnam has been obliged to search for biofertilizers in order to reduce the use of chemical fertilizers without a decrease in plant yields.

Azospirilla are a group of bacteria capable of plant growth stimulation (Michiels *et al.*, 1989). In Vietnam these bacteria have been isolated and characterized (Nguyen, 1992; Nguyen *et al.*, 1990, 1995). In this chapter the responses of rice plants to inoculation with *Azospirillum* preparations under field conditions will be reported.

K.A. Malik et al. (eds.), Nitrogen Fixation with Non-Legumes, 237–240.

Figure 1. Rice seedlings with and without *Azospirillum* inoculation DC: Control; TN: inoculated.

Materials and method

Mat 2-lb and Da 9–3a are *Azospirillum* strains which had been isolated from the rice roots unsterilized or sterilized by using a H_2O_2 10% solution for 20 s, respectively. They are capable of associative nitrogen-fixation with rice seedlings and indoleacetic acid production (Nguyen, 1992; Nguyen *et al.*, 1995). After pre-cultivation in MAZ medium (Albrecht and Okon, 1980) in which biotin had been replaced by yeast extract (1 g/L), the culture was added to peat as a carrier. After 7–10 days of growth the preparations could reach 10^9 cells/g wet weight.

Fifty grams of *Azospirillum* preparation was used for the inoculation of 6 kg freshly germinated rice seeds.

According to the fifth or tenth month crop, after 35–50 days of seed sowing the rice seedlings had been observed and transplanted in farmers' fields.

Results and discussion

Appearance of rice seedlings before transplanting

It was reported by the farmers that the inoculated plants were taller, more vigorous, with a larger root system, and the colour of their leaves appeared to be greener than the controlled ones (Figure 1). Particularly, the inoculated young rice plants of the fifth month crop had survived at a higher percentage when the air temperature was 10°C lower for a long time (8 days).

Rehabilitation after transplant

If roots of rice seedlings had been inoculated again before transplanting, their rehabilitation appeared to be better than the non-treated ones, especially for the *Azospirillum* preparation using the strain DA 9–3a.

Corn yield

Because of its market significance the rice variety CR203 has been used relatively widely in the Redriver Delta, North Vietnam. Except for the difference in the treatment with *Azospirillum* preparations, all regimes of fertilization, watering, etc. were the same for both the controls and inoculated treatments. Results are presented in Table 1.

A 3–5 day earlier ripening of rice seeds had been observed for the experiments (Figure 2).

Table 1. Effect of *Azospirillum* inoculation on corn yield of rice

Date	Place*	Crop	Rice variety	Area (m^2)	Corn yield* (%)
1991	Vinhngoc, Dongda, Hanoi	5th month	CR203	360	107
	Daimo, Tuliem, Hanoi	5th month	CR203	360	109
	Ankhanh, Hoaiduc, Hatay	5th month	CR203	30	106
	Ankhanh, Hoaiduc, Hatay	5th month	Nep	30	119
	Ankhanh, Hoaiduc, Hatay	5th month	CN2	30	117
	Ankhanh, Hoaiduc, Hatay	5th month	V14	30	110
1992	Dongngac, Tuliem, Hanoi	5th month	CR203	360	111
	Dongngac, Tuliem, Hanoi	5th month	Nep	360	106
	Hoanglien, Tuliem, Hanoi	5th month	CR203	360	116
	Hoanglien, Tuliem, Hanoi	5th month	Nep	360	115
	Trungvan, Tuliem, Hanoi	5th month	CR203	10 000	125
1993	Dongngac, Tuliem, Hanoi	5th month	CR203	360	119
	Dongngac, Tuliem, Hanoi	5th month	Nep	360	117
	Thuyphuong, Tuliem, Hanoi	5th month	CR203	360	105
	Thuyphuong, Tuliem, Hanoi	5th month	Nep	360	106
	Trungvan, Tuliem, Hanoi	5th month	CR203	10 000	121
	Dongngac, Tuliem	10th month	CR203	360	117
	Thuyphuong, Tuliem, Hanoi	10th month	CR203	360	108
	Thuyphuong, Tuliem, Hanoi	10th month	Nep	360	108
1995	Kimphu, Yenson, Tuyenquang	5th month	CR203	7200	108
	Kimphu, Yenson, Tuyenquang	10th month	baothailun	14 000	125

* As compared to 100% value given to controls.

Figure 2. Rice ears of the variety CR203 with and without *Azospirillum* inoculation. (Every bunch consisted of 30 ears; left: control; right: experiment).

Conclusion

Rice plants gave positive responses under natural conditions in the Redriver Delta, North Vietnam when inoculated by preparations using *Azospirillum* strains Mat 2–1b and DA 9–3a.

References

Albrecht S I, Okon J 1980. Culture of *Azospirillum*. In: San Pietro, ed. Methods in Enzymology, vol. 69. Academic Press, New York.

Michiels K, Vanderleyden J, Gool A 1989. *Azospirillum*–plant root association: a review. Biol. Fertil. Soils, 8, 356.

Nguyen N D 1992. The ability of the production of indoleactic acid (IAA) by *Azospirillum* sp. isolated from rhizosphere of rice. J. Biol., 14.

Nguyen N D, Nguyen T P C, Ha H T 1990. Associative N_2-fixation activity of *Azospirillum* spp. with rice seedlings. J. Biol., 13, 3.

Nguyen N D, Kraut M, Klingmueller W 1995. Isolation and characterization of *Azospirillum* strains from soils and rice plants in North Vietnam. In: Fendrik I, Del Gallo M, Vanderlayden J, de Zamaroczy M, eds. *Azospirillum* VI and Related Microorganisms: Genetics, Physiology, Ecology. Nato ASI Series G, vol. 37. Springer Verlag, Göttingen.

Posters

27 Poster

BGA technology for rice cultivation in northern India

R.P. GUPTA, N.S. CHAUDHARY AND M.S. PANDHER

Introduction

Blue-green algae (BGA) are known to contribute up to 80 kg N/ha per season in the rice ecosystem (Kannaiyan *et al.*, 1983). However, due to heavy use of N fertilizers and high temperature prevailing during rice cultivation in the Punjab, BGA technology could not be popularized. Therefore, there is a need to obtain local efficient strains and to study the effect of BGA substitution with N fertilizers at various growth stages of rice and also to test their efficacy in basmati rice which requires a lower input of N fertilizer and is sown in the month of July when the temperature is lower.

Materials and methods

BGA composite culture, obtained from IARI, New Delhi, and locally isolated efficient BGA cultures were used in the present study.

The experiment during 1994–95 and 1995–96 was conducted to examine the effect of locally isolated efficient nitrogen-fixing strains of BGA in combination with N levels at transplanting (TP), tillering (Till) and panicle initiation (PI) on grain yield of paddy succeeding wheat in a sequence. Twenty-four-day-old seedlings of cv. PR106, raised in the nursery, were transplanted in the summer season at 20 × 5 cm spacing in a plot size 3 × 10 m. The treatments consisted of 10 combinations of N doses, BGA and time of application.

The nitrogen was given as per-treatment in the form of urea, phosphorus at 30 kg P_2O_5/ha and potash at 30 kg K_2O/ha in the form of single superphosphate and muriate of potash, respectively, were applied as basal. BGA at 2 g m^{-2} was applied at different stages as required in the treatment. To study the residual effect of BGA on the succeeding crop, after the harvest of paddy, wheat cv. HD 2329 was sown in the winter season in the same plots.

K.A. Malik et al. (eds.), Nitrogen Fixation with Non-Legumes, 243–245.

Nitrogen use efficiency (NUE) is defined as units of grain produced per unit of N applied. The NUE was thus calculated by dividing grain yield by the total amount of nitrogen applied through fertilizer for each treatment (Hussain *et al.*, 1996).

$$NUE = TGY/TNA,$$

where TGY = total grain yield and TNA is total nitrogen applied.

Results and discussion

The establishment of BGA was confirmed visually and through dominance studies using microscopic examinations. BGA was found capable of working even in the presence of chemical fertilizers, as seen from the enhanced yield of rice (Tables 1 and 2). The response to BGA, however, was more pronounced at lower levels of N (80 kg N/ha). This was perhaps due to the fact that rice under these conditions responds only up to the 120 kg N/ha level, and any increase in the N status of soil through BGA application beyond this level produced little beneficial effect on the grain yield of rice. BGA application at the panicle initiation stage or tillering stage has been found to perform better than at the transplanting stage, which could be on account of the low temperature prevailing at these stages. The yield with 40 + 40 + BGA treatment was statistically similar to that obtained with treatments receiving a total of 120 kg N/ha, thereby indicating a possibility of a saving in N fertilizer of up to 40 kg N/ha. BGA application also increased the NUE over nitrogen-alone treatments. NUE showed a decreasing trend with increase in N application.

Data analysis (Tables 1 and 2) showed a residual effect of BGA application to rice on grain yield of following wheat crop grown in sequence. The grain yield of

Table 1. Effect of substitution of BGA (composite) with different levels of nitrogen at different stages of growth of rice on grain yield (kg/ha) and nitrogen use efficiency (1994–95)

Treatments (N kg/ha)	Wheat	Paddy		Wheat	
(TP + Till + PI)		Yield	NUE	Yield	NUE
BGA + 40 + 40	90	25.45	31.8	48.52	53.9
C + 40 + 40	120	23.73	29.7	48.43	40.4
40 + BGA + 40	90	27.83	34.8	44.72	49.7
40 + C + 40	120	24.70	30.9	46.07	38.4
40 + 40 + BGA	90	28.35	35.4	41.74	46.4
40 + 40 + C	120	24.10	30.1	46.82	39.0
40 + 40 + 40	120	30.50	25.4	47.94	40.0
40 BGA + 40 + 40	90	31.10	25.9	42.06	46.7
40 + 40 BGA + 40	90	31.17	26.0	43.74	48.6
40 + 40 + 40 BGA	90	31.50	26.3	44.98	50.0
LSD (0.05)		4.41		NS	
CV (%)		9.21		7.52	

TP, at transplanting; Till, at tillering; PI, at panicle initiation

Table 2. Residual effect of BGA (composite) inoculation on yield and NUE of subsequent crop (1995–96)

Treatments (N kg/ha)	Wheat	Paddy		Wheat	
(TP + Till + PI)		Yield	NUE	Yield	NUE
BGA + 40 + 40	90	42.30	52.9	43.2	48.0
C + 40 + 40	120	38.96	48.7	45.5	37.9
40 + BGA + 40	90	46.07	57.9	47.4	52.7
40 + C + 40	120	40.42	50.5	47.2	39.3
40 + 40 + BGA	90	49.89	62.4	46.4	51.6
40 + 40 + C	120	39.04	48.8	48.1	40.1
40 + 40 + 40	120	45.17	37.6	45.7	38.1
40 BGA + 40 + 40	90	43.01	35.8	47.0	52.2
40 + 40 BGA + 40	90	44.29	36.9	49.0	54.4
40 + 40 + 40 BGA	90	45.56	38.0	45.9	51.0
Control	120	32.03	0.0	45.9	38.3
BGA	90	35.21	0.0	44.9	49.9
LSD (0.05)		8.39		NS	
CV (%)		11.97		7.14	

succeeding wheat with 90 kg N/ha in BGA-applied plots was statistically at par with yields obtained with 120 kg N/ha in no-BGA plots, again indicating a saving of 30 kg N/ha in wheat.

Acknowledgements

The authors are grateful to the Department of Biotechnology, Government of India, for providing the necessary funds to carry out this research work. The authors are also grateful to the National Facility for BGA, IARI, New Delhi, for providing the BGA composite culture.

References

Hussain G, Al-Jalod A A, Karimullah S 1996. Effect of treated effluent irrigation and nitrogen on yield and nitrogen use efficiency of wheat. Agric. Water Manag., 30, 175–184.

Kannaiyan S, Thangaraju M, Oblisami G 1983. Rice crop response to biofertilizer inoculation. Madras Agric. J., 70, 65–67.

28 Poster

Modified combined carbon N-deficient medium for isolation, enumeration and biomass production of diazotrophs

N.A. HEGAZI, M.A. HAMZA, A. OSMAN, S. ALI, M.Z. SEDIK AND M. FAYEZ

Introduction

Dinitrogen fixation associated with non-legumes has been well documented (Dobereiner, 1983; Hegazi, 1983; Fayez, 1990) and is now generally considered to make a significant contribution to the nitrogen economy of several ecosystems. This intensifies the use of field inoculation with various diazotrophs. Therefore, the need arises to monitor the populations and individuals of several associative diazotrophs inhabiting the plant-soil system. The ability of several diazotrophs to multiply and reduce acetylene on numerous carbon sources encourages the search for a single medium suitable for collective growth and biomass production of a number of diazotrophs. The present study evaluates a new medium modification based on basal salts and a mixture of five carbon sources (glucose, malic acid, mannitol, sodium lactate and sucrose) to fulfil the requirements for a single medium on which the majority of associative diazotrophs would exhibit good growth and biomass production.

Materials and methods

The combined carbon-source medium (CCM) was prepared in a basal salt solution containing the following: K_2HPO_4, 0.4 g; KH_2PO_4, 0.6 g; NaCl, 0.1 g; $FeCl_3$, 0.015 g; $CuSO_4$, 0.08 mg; $ZnSO_4$, 0.25 mg; yeast extract, 0.2 g; fermentol (a local product of corn-steep liquor), 0.2 g; KOH, 1.5 g; $MgSO_4$, 0.2 g; $CaCl_2$, 0.02 g; $MnSO_4$, 0.01 g; $NaMoO_4$, 0.002 g and distilled water, 1000 ml. The following carbon sources were included (g l^{-1}): glucose, 2.0; malic acid, 2.0; mannitol, 2.0; sucrose, 1.0. Sodium lactate was included as 0.6 ml L^{-1} (50% v/v) and pH adjusted at 7.0. The culture medium was autoclaved at 121°C for 20 min, while filter-sterilized biotin (5 μg L^{-1}) and *para*-aminobenzoic acid (10 μg L^{-1}) were added

K.A. Malik et al. (eds.), Nitrogen Fixation with Non-Legumes, 247–253.

after sterilization. Several selective culture media were included in this study: Rodriguez Caceres (1982) for *Azospirillum* spp., Hegazi and Niemela (1976) for *Azotobacter* spp., Dixon *et al.*, (1980), for *Klebsiella* spp., Haahtela *et al.*, (1983) for *Pseudomonas* spp. and Hino and Wilson (1958) for *Bacillus* spp.

Experiments were carried out to compare growth of tested diazotrophs *(A. brasilense, A. chroococcum, K. pneumoniae, Ps. putida* and *Bacillus* spp.) on their selective culture media and the proposed CCM. Also, the combined carbon-source culture medium of Rennie (1981) was included for comparison. The first experiment included preparation of batch cultures of tested diazotrophs by inoculating both selective and CCM culture media with 10 ml of 10^5–10^6 cells m^{-1}. Incubation took place at 30°C with shaking 100 rpm. Growth and population of tested diazotrophs were followed by surface-plate counting and optical density (at 560 nm). Morphological characteristics of colonies of each tested diazotroph developed on agar plates of either culture media were carefully examined. In the second experiment, tested diazotrophs were collectively grown in the CCM liquid culture medium. Then, counts of colonies appeared on surface inoculated agar plates of Rennie's and CCM were obtained at 0, 1, 3, 5 and 7-day interval.

Root samples of various medicinal, forage and gramineous plants naturally grown in sandy soils of the El-Arish area, Sinai, were collected to isolate various diazotrophs on CCM culture medium. Root systems with adhering soils were thoroughly shaken for 1 h in 450 ml bottles containing 90 ml saline solution. Serial dilutions were prepared and inoculation took place in both agar (surface-plates) and semi-solid medium. Colony counts of diazotrophs were periodically determined during incubation at 30°C applying both surface-plate count and MPN procedures. Representative colonies were selected on the basis of cell morphology and Gram stain. Further single colony isolation and purification of the selected isolates were performed on CCM culture medium. N_2-ase activity of the purified strains was measured in semi-solid media by acetylene reducing assay.

Results and discussion

It was evident that tested diazotrophs grew well on the proposed CCM culture medium with population densities very comparable to selective culture media. This was confirmed by the non-significant differences among estimates of doubling time (Figure 1) obtained for each tested strain in both media. Also, a relatively positive correlation was calculated among counts obtained on CCM and selective culture media, using surface plate counting (>0.473–0.993) and optical density (>0.427–0.912).

In comparison with the combined carbon medium of Rennie (1981), all the tested diazotrophs, when present in mixed population, successfully grew on the proposed CCM (Table 1). The latter seems to be more suitable for development of tested diazotrophs as it provides sufficient quantities of all major carbon sources and high level of yeast extract.

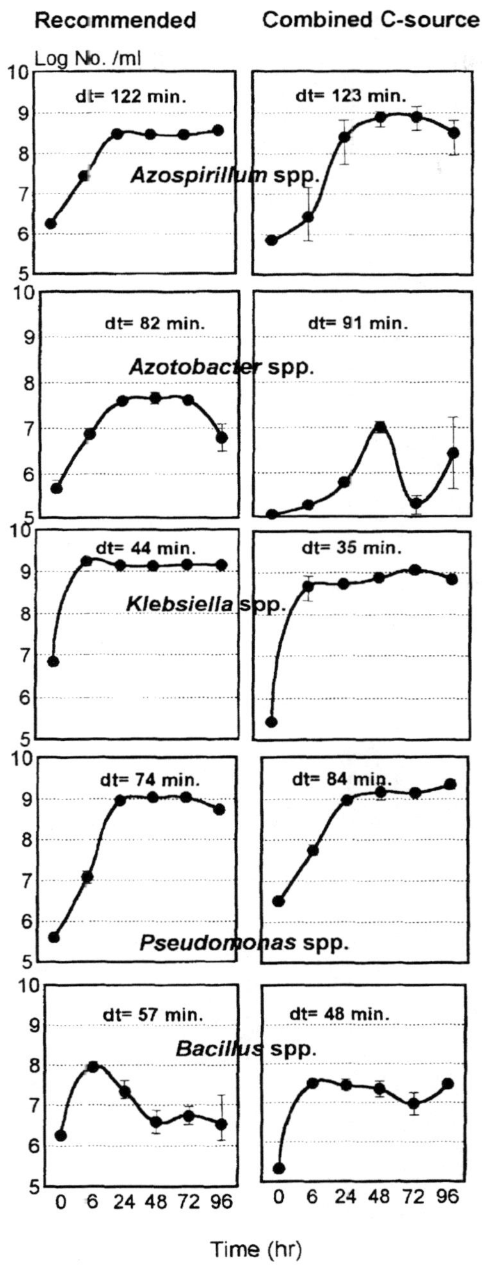

Figure 1. Growth curves and doubling time of various diazotrophs grown in recommended and combined C-source culture media.

Table 1. Average colony counts ($\times 10^4$ cfu L^{-1}) obtained on Rennie's and CCM culture media for tested diazotrophs grown in mixed population

Days of incubation	*Azospirillum* spp.		*Azotobacter* spp.		*Bacillus* spp.		*Klebsiella* spp.		*Pseudomonas* spp.	
	A	B	A	B	A	B	A	B	A	B
0	150	120	1	0.5	120	70	120	70	100	60
1	100	140	0.2	0.6	150	130	280	180	80	90
2	280	460	0.06	0.09	550	480	830	650	270	170
3	770	1660	0.3	0.1	860	960	790	600	860	480
5	1100	930	1	0.6	1200	1860	500	420	550	930
7	160	290	3	1.5	240	220	190	60	400	520

A, Rennie's; B, CCM culture media

A major advantage of CCM is that colonies of diazotrophs tested were exhibiting distinguished characteristics (Table 2 and Figures 2 and 3), enabling easy discrimination. The rigid deep red colour colonies of *Azospirillum* clearly appeared on agar plates supplemented with Congo red. *Azotobater* spp. were easily distinguished by deep brown pigmented and rather confined colonies. *Bacillus* spp. grew well and the colonies were tear-like, slimy and transparent. Colonies of *Klebsiella* appearing on CCM agar plates were slimy and confined in size (≤6 mm), not watery and overspread other colonies developed, as was the case with Rennie's agar plates. Fluorescency of *Pseudomonas* spp. was easily distinguished against bright daylight and UV.

Various diazotrophs associated with roots of desert plants developed well on the proposed CCM culture medium (Table 3). Irrespective of plant, the medium supported better growth of *Klebsiella* spp. (mean of 5.5×10^7 cfu g^{-1} root) followed by *Pseudomonas* spp. (1.4×10^7 cfu g^{-1} root). *Bacillus* spp. were the lowest in numbers compared to other diazotrophs (1.0×10^5–22×10^6 cfu g^{-1} root).

Single-colony isolation was applied to isolate all developed colonies (429 colonies) on the CCM agar plates. Based on active acetylene reduction, isolates

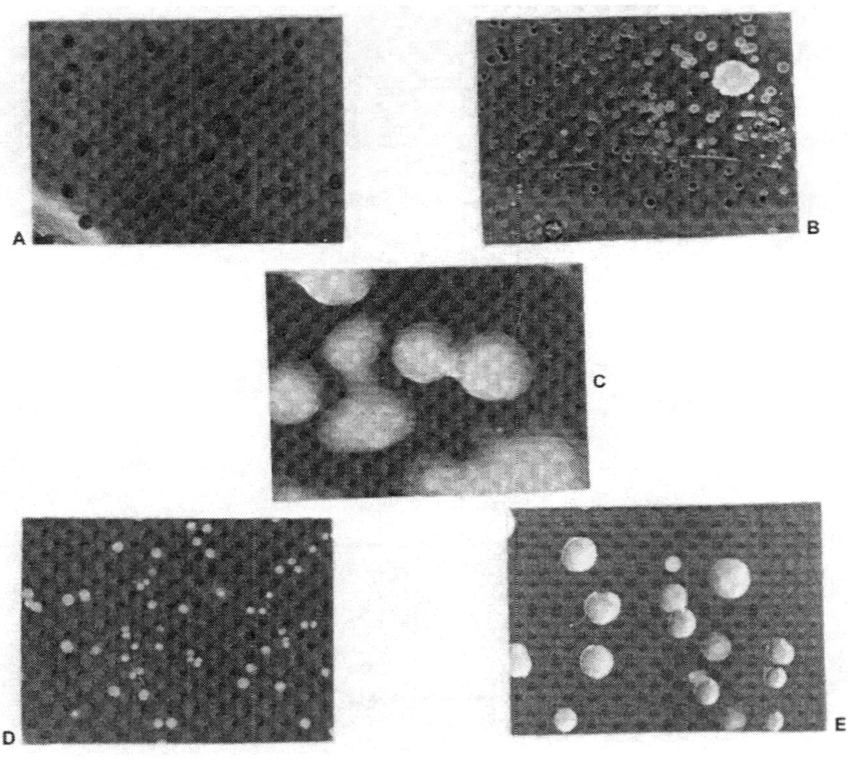

Figure 2. Colony characteristics of *Azospirillum* spp. (**A**), *Azotobacter* spp. (**B**), *Klebsiella* spp. (**C**), *Pseudomonas* spp. (**D**) and *Bacillus* spp. (**E**).

Table 2. Colony characteristics of diazotrophs developed on the proposed CCM

Character	*Azospirillum* spp.	*Azotobacter* spp.	*Bacillus* spp.	*Klebsiella* spp.	*Pseudomonas* spp.
Appearance	Circular	Circular	Circular	Irregular	Circular
Elevation	Flat	Convex	Tear-like	Convex	Flat
Margin	Undulating	Undulating	Entire	Undulating	Undulating
Transparency	Opaque	Opaque	Transparent	Translucent	Transparent
Spreading	Confined	Confined	Semi-spreading	Spreading	Confined
Consistency	Rigid	Paste-like	Mucoid	Mucoid	Paste-like
Colour	Pink with Congo Red	Dark brown	White	White	Fluorescent greenish-yellow
Diameter (mm)	≤2.0	≤2.0	≤3.0	≤6.0	≤2.0

were selected for further purification. According to their morphological, cultural and some physiological characteristics, the pure isolates belong to the genera *Azospirillum* spp., *Bacillus* spp., *Klebsiella* spp. and *Pseudomonas* spp..

Regarding acetylene-reducing activity of the identified diazotrophs, it was reported that up to 80% of the total isolates of diazotrophs did possess the ability to reduce acetylene actively. Such a high percentage indicates the suitability of the proposed medium in supporting growth and activity of true not putative diazotrophs. Among the total diazotrophs, *Pseudomonas* spp. represented 54% followed by *Bacillus* spp. (23%), *Azospirillum* spp. (12%) and *Klebsiella* spp. (11%).

In conclusion, the proposed combined carbon-source (CCM) medium proved to be appropriate and is highly recommended for ecological (isolation and enumeration) and physiological (biomass production and growth kinetics) studies on various groups of associative diazotrophs. Certainly it saves money, time and effort.

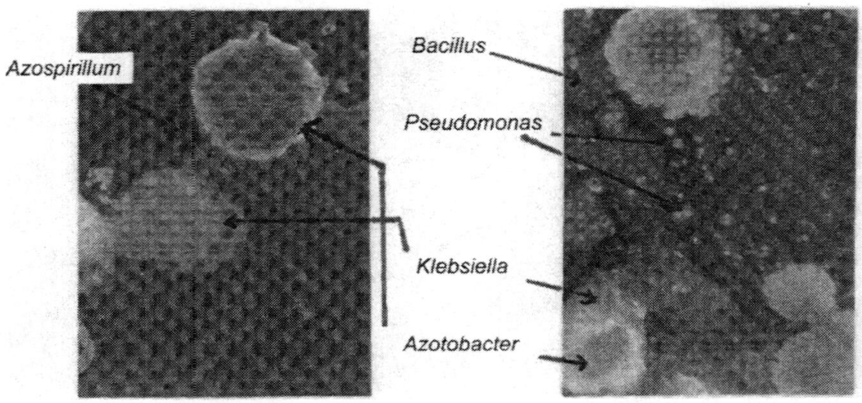

Figure 3. Seven-day-old colonies of mixed population of diazotrophs developed on CCM agar plates with or without Congo red.

Table 3. Densities of various associative diazotrophs on roots of natural plant cover of north Sinai ($\times 10^5$ cfu g^{-1} root) as developed on modified CCM

Standing crop	*Azospirillum* spp.	*Bacillus* spp.	*Klebsiella* spp	*Pseudomonas* spp.
Hordeum distichum L.	20	30	40	90
Hordeum distichum L.	30	110	110	30
Hordeum distichum L.	20	10	40	210
Hordeum distichum L.	720	220	7520	1260
Retema raetam	1	1	1	1
Hyoscyamus muticus	1	1	2	3
Agrostis vulgaris	ND	ND	20	50
Aeluropus layopoides L	2	1	3	3
Agrostis vulgaris	30	30	20	30
Pennisetum vulgaris	110	ND	110	90
Pennisetum vulgaris	4	ND	20	10
Salvia sylvestris L	60	70	250	230
Lycium showii Roem	6	6	8	8
Crotalaria aegyptiaca Benth	5	10	10	10
Artemisia herba Alba Asso	40	10	70	30

ND, not detected

Acknowledgements

This study was supported by the national project on introducing agro-biotechnologies based on biological nitrogen fixation for the development of Sinai, funded by the Egyptian Ministry of Agriculture and Land Reclamation and the French Food Aid Counterpart Fund (FFACF-BLAFE).

References

Dixon R A, Eady R R, Espin G, Hill S, Laccazino M, Khan D, Merrich M 1980. Analysis of regulation of *K. pneumoniae* nitrogen fixation (nif.) gene cluster with gene. Nature (London), 286, 128–132.

Dobereiner J 1983. Ten years *Azospirillum*. In: Klingmüller W, ed. *Azospirillum* II. Genetics, Physiology, Ecology. Brinkhaüser Verlag, Basel, 9–23.

Fayez M 1990. Untraditional N_2-fixing bacteria as biofertilizers for wheat and barley. Folia Microbiol., 35, 218–226.

Haahtela K, Helander I, Nurmiaho-Lassila E L, Sundman V 1983. Morphological and physiological characteristics and lipopolysaccharide composition of N_2-fixing (C_2H_2-reducing) root-associated *Pseudomonas* sp. Can. J. Microbiol., 29, 874–880.

Hegazi N A 1983. Contribution of *Azospirillum* spp. to asymbiotic N_2-fixation in soil and on roots of plants in Egypt. In: Klingmüller, W, ed. *Azospirillum* II. Genetics, Physiology, Ecology. Brinkhaüser Verlag, Basel, 171–189.

Hegazi N A, Niemela S 1976. A note on the estimation of *Azotobacter* densities by filter technique. J. Appl. Bacteriol., 41, 311.

Hino S, Wilson P W 1958. Nitrogen fixation by a facultative *Bacillus*. J. Bacteriol., 75, 403–408.

Rennie R J 1981. A single medium for the isolation of acetylene reducing (dinitrogen-fixing) bacteria from soil. Can. J. Microbiol., 27, 8–14.

Rodriguez Caceres E A 1982. Improved medium for isolation of *Azospirillum* spp.. Appl. Environ. Microbiol., 44, 990–991.

29 Poster

Contribution of integrated symbiotic and associative systems of biological nitrogen fixation to N-status of intercropped legumes and non-legumes in sandy soils of Egypt

M. MONIB, M.E. HASSAN, H.H. ABO-TALEB, A. METWALLY AND N.A. HEGAZI

Introduction

Intensive cropping culture is a popular practice among small-scale farmers in tropical and subtropical areas, mainly in Africa. In Egypt, several years ago, Galal and Abdul-Rasool (1962) introduced intercropping as an intensive agronomic practice of possible economic impact on Egyptian agriculture. The system soybean-corn was adopted, and research compared different intercropping patterns, selection of tolerant genotypes for each partner crop and effects of population densities of plants and their distribution in the unit area. Yields of intercropping as high as 90% for corn and 60% for soybean of respective monocrops were scored. A number of field trials in sandy and fertile Nile Valley soils were carried out to study the role of associative and symbiotic BNF in the system of corn-soybean. The present study focuses on results obtained for the field trial applied in sandy soils.

Materials and methods

The field experiment in sandy soils was executed in Ismailia to investigate the interaction of N-fertilization and bacterial inoculation. Three types of inocula were tested; rhizobia (*Bradyrhizobium japonicum*), composite inoculum of associative diazotrophs (*Azospirillum brasilense* + *Azotobacter chroococcum* + *Klebsiella pneumonia* + *Bacillus* spp., as well as Mycorrhiza) and a mixture of both; on nodulation, growth and yield of soybean as well as corn sown under an intercropping system. Soybean cv. Crawfard was drilled in rows and corn (cv. Cairo 1) planted in hills 35 cm apart using an intercropping pattern of 2:2 alternative rows. The fertilization included superphosphate (100 kg acre^{-1}) and potassium sulphate (50 kg acre^{-1}) as well as ammonium sulphate (20.5% N) added as 25

K.A. Malik et al. (eds.), Nitrogen Fixation with Non-Legumes, 255–260.

and 75 kg N acre^{-1} for soybean and 60 and 200 kg N acre^{-1} for corn. Inocula were applied by two methods: the first was seed coating of rhizobia with peat-based inoculum and the second was mixing wet soil with the peat-based preparation of associative diazotrophs then distributing it in rows and/or hills. The experimental design was randomized split-plot with 12 treatments and five replicates.

Results and discussion

For soybean, rhizobia inoculation was successful (increase of 371% over non-treated plants) and application of a rational dose of N to such rhizobia-inoculated plants supported good nodulation (Table 1), plant dry weight and yield (Table 2). Increasing the level of N-fertilization to 75 kg N acre^{-1} did not suppress performance of introduced rhizobia as they continue to exert their significant effect (46% increases over fully N-fertilized plants, Table 3). Response of soybean plants to inoculation with associative diazotrophs was rather limited compared to rhizobial effect, and the only significant increases of *ca.* 24% was reported in the presence of a limited dose of N. When associative N_2-fixers were included in the composite inocula together with rhizobia, significant increases in nodulation, as well as yield parameters of soybean, were reported only in the presence of N-fertilization. Nodule number and weight increased by 88–125% and 71–103%, respectively (Singh and Subba Rao, 1979; Ragab, 1991). Increases

Table 1. Effect of N-fertilization and inoculation with various groups of diazotrophs on nodulation and development of 60-day-old intercropped soybean plants

Treatments					
Inoculation	N-fertilization (kg N acre^{-1})	Number of nodules (no. plant^{-1})	Dry weight of nodules (mg plant^{-1})	Dry weight of plant (g plant^{-1})	N-content (mg plant^{-1})
None	0	0	0	2.66	25.89
None	25	0	0	3.97	41.85
None	75	0	0	4.39	49.14
Rhizobium	0	7.4	67.0	4.22	41.57
Rhizobium	25	7.2	79.0	4.59	66.91
Rhizobium	75	7.6	76.0	4.89	86.21
Associative	0	0	0	2.54	25.34
Associative	25	0	0	4.14	53.18
Associative	75	0	0	4.83	63.07
Rhizobium + Associative	0	6.0	50.4	3.81	42.06
Rhizobium + Associative	25	15.0	135.0	4.62	69.03
Rhizobium + Associative	75	18.0	154.0	4.98	87.27
LSD (0.05)		2.2	19.1	0.79	10.81

Table 2. Effect of N-fertilization and inoculation with various groups of N_2-fixers on yield parameters of intercropped soybean and corn plants

Treatments		Total dry weight (kg row^{-1})		Seed weight (g row^{-1})		Protein content (%)			
						Seed		Straw	
Inoculation	N-fertilization kg N acre^{-1}	Soybean	Corn	Soybean	Corn	Soybean	Corn	Soybean	Corn
None	0	0.187	1.302	12.6	143.4	31.9	7.0	8.5	2.57
None	25	0.396	1.863	90.6	252.6	32.7	7.8	12.2	3.73
None	75	1.396	3.522	275.7	583.4	35.1	9.5	14.2	4.49
Rhizobium	0	0.881	1.426	137.7	227.4	32.7	7.1	8.9	3.51
Rhizobium	25	1.151	2.656	323.7	354.9	34.8	8.0	13.7	4.30
Rhizobium	75	2.033	4.017	348.3	614.1	35.3	9.8	15.6	5.68
Associative	0	0.145	1.571	12.9	208.9	31.4	7.0	8.8	2.49
Associative	25	0.561	1.643	93.3	269.4	32.3	7.4	10.5	2.63
Associative	75	1.360	3.863	256.2	618.9	34.5	9.5	13.5	4.49
Rhizobium + Associative	0	0.488	1.737	142.2	208.9	32.4	7.0	11.4	3.45
Rhizobium + Associative	25	1.387	2.573	320.4	327.3	33.9	8.4	14.3	4.53
Rhizobium + Associative	75	2.340	4.344	387.3	775.3	35.9	10.2	18.4	5.55
LSD (0.05)		0.181	0.456	40.1	87.0	0.7	0.2	1.2	0.41

Table 3. Relative increases in total dry matter yield (g row^{-1}) of intercropped corn and soybean grown in sandy soils of Ismailia as affected by various inoculation and N-fertilization treatments

Fertilization levels (kg N $acre^{-1}$)	Rhizobia inoculation	Increases (%)	Rhizobia + Associative diazotrophs	Increases (%)	Associative diazotrophs	Increases (%)	Non-inoculated plants
Soybean							
0	881	371*	489	161*	145	–29(n.s.)	187
25	1151	191*	1387	250*	561	24(n.s.)	369
75	2033	46*	2340	68*	1360	–3(n.s.)	1396
Corn							
0	1571	21(n.s.)	1737	33*	1246	10(n.s.)	1302
60	1643	–12(n.s.)	2573	38*	2656	43*	1862
200	3863	10(n.s.)	4344	23*	4017	14*	3522

* Statistically significant (p=0.05).
n.s., Not statistically significant.

in total dry matter yield attributed to associative N_2-fixers in the presence of 25 and 75 kg N acre^{-1} were 60% and 22% over corresponding treatment receiving only rhizobia inocula.

For corn (Tables 2 and 3) increases attributed to inoculation with associative N_2-fixers ranged from 10% to 43%, and were statistically significant in the presence of 60 (43%) and 200 (14%) kg N acre^{-1}.

On the other hand, increases attributed to rhizobia inoculation ranged from 10% to 21%, and did not reach significant levels. This indicates a certain degree of host specificity among RMO and plant roots, and expresses the affinity of associative diazotrophs to cereals (Monib *et al.*, 1981; Hegazi *et al.*, 1983). A positive cumulative effect was reported when corn plants were inoculated with a mixture of either rhizospheric microorganisms (RMO) groups; increases were as high as 23–38%. The highest agronomic yield of corn was obtained by intensive inoculation in the presence of the highest levels of N-fertilization. This signifies that economic management of Ismailia soils – which represent a part of the sandy soils of the Egyptian desert – require complementation of intensive bacterial inoculation with RMO together with an adequate regime of N-fertilization. Under those particular conditions the calculated LER for intercropped corn and soybean approximated 1.35.

In quantitative terms, inocula of rhizobia and associative N_2-fixers contributed to the corn-soybean system in sandy soils with an estimated average of 17.5 and 55 kg N acre^{-1}, respectively. An actively growing legume root system may provide N for neighbouring cereal plants and for rhizosphere microflora. Under experimental conditions of the present study, differences in N-content among N-fertilization and inoculation treatment were used to derive an estimate of N-transfer. An average amount of 20 kg N acre^{-1} was calculated to be released from nodulated soybean to neighbouring corn plants. This figure was *ca.* 33 kg N acre^{-1} in fertile Giza fields (Abo-Taleb, 1991).

Acknowledgements

This work was supported by the national project on introducing agro-biotechnologies based on biological nitrogen fixation for the development of Sinai, funded by the Egyptian Ministry of Agriculture and Land Reclamation and the French Food Aid Counterpart Fund (FFACF-BLAFE).

References

Abo-Taleb H Y 1991. The effect of inoculation with specific rhizobia on N_2-fixation, nodulation and yield of beans. MSc thesis, Faculty of Agriculture, Cairo University, Egypt.

Galal S Jr, Abdel-Rasool S F 1952. Possibilities of intercropping corn and soybean in Egypt. Official report, Technical Communication for Oil Crops, Ministry of Agriculture, Egypt.

Hegazi N A, Monib M, Amer H A, Shokr S E 1983. Response of maize plant to inoculation with *Azospirilla* and/or straw amendment in Egypt. Can. J. Microbiol., 29, 888–894.

Monib M, Hegazi N A, Eid M, Shokr S E 1981. Yield increases of grain sorghum (*Sorghum vulgare*) after inoculation with nitrogen-fixing *Azospirilla*. Research Bulletin No. 1690, Faculty of Agriculture, Ain Shams University.
Ragab A A H 1991. Response of faba bean to inoculation with *Rhizobium leguminosarum* in old and newly reclaimed soils. MSc thesis, Faculty of Agriculture, Cairo University, Giza, Egypt.
Singh C S, Subba-Rao N S 1979. Associative effect of *Azospirillum brasilense* with *Rhizobium japonicum* on nodulation and yield of soybean (*Glycine max*). Plant Soil, 53, 387–392.

30 Poster

Some physiological studies of diazotrophic bacteria associated with maize in a temperate and a tropical soil

F.B. MWAURA AND D. WIDDOWSON

Introduction

Previous studies on *Azospirillum* spp. (Day and Dobereiner, 1976) and *Klebsiella oxytoca* isolated from wheat roots in Turkey (Cakmacki *et al.*, 1981) indicated little activity below 24°C and high activity at 30–40°C. In this study the effect of temperature on the growth and acetylene reduction activity, (ARA) of two temperate-soil bacteria and a tropical-soil isolate is reported.

Materials and methods

The temperate soil bacteria (A and C) and the tropical soil bacterium (L) were isolated from the rhizosphere of field-grown maize in Sweden (Mwaura and Granhall, 1986) and Kenya (Mwaura, 1986), respectively. The cultures were grown on glucose agar plates (Day and Dobereiner, 1976) for 4 days at 28°C and the colonies were then suspended in sterile 0.05 M phosphate buffer pH 7 to make three inoculants of the same optical density. For each isolate, 30 replicate vials containing 4 ml Nfb-glucose medium (0.05% agar w/v) were inoculated, and incubated at 24°C, 28°C and 37°C. Triplicate cultures were randomly picked and assayed for ARA over a 72 h period using a gas chromatograph (Mwaura and Granhall, 1986). The optical densities of the cultures were then measured using a spectrophotometer.

Results

In cultures of the temperate soil isolates the highest rates of ARA were recorded at 28°C (Figures 1A and 2A). The highest mean activities observed were 144.7 and 321.3 nmol C_2H_4/culture per hour for isolates A and C respectively. Cultures incubated at 37°C had very low ARA.

K.A. Malik et al. (eds.), Nitrogen Fixation with Non-Legumes, 261–264.

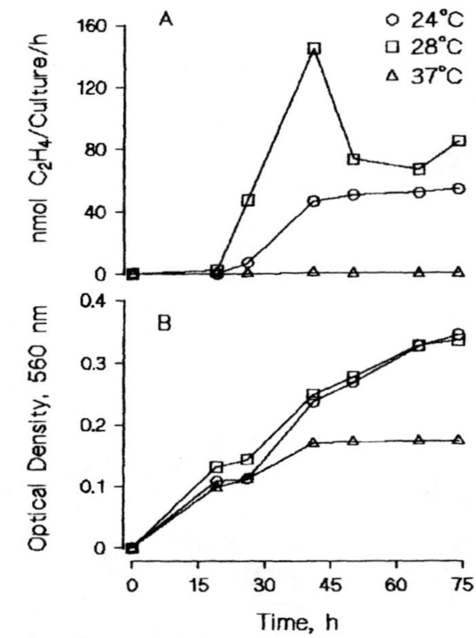

Figure 1. Nitrogenase activity (**A**) and growth (**B**) of the temperate soil isolate A.

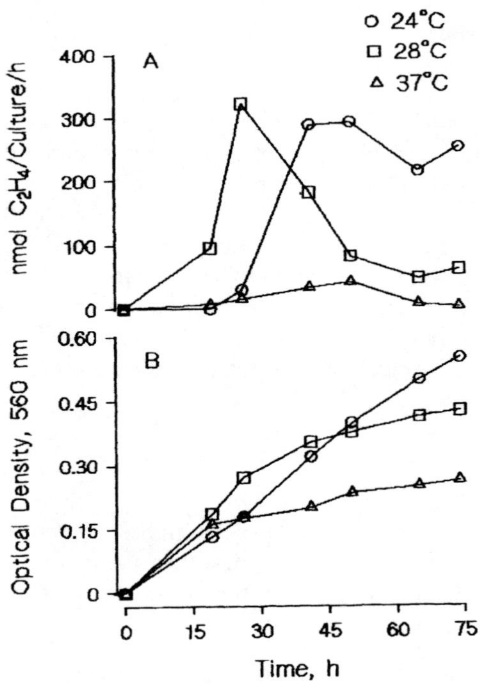

Figure 2. Nitrogenase activity (**A**) and growth (**B**) of the temperate soil isolate C.

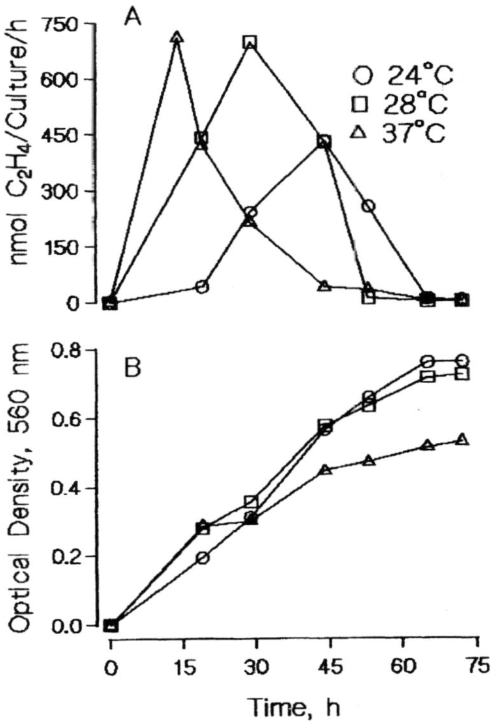

Figure 3. Nitrogenase activity (**A**) and growth (**B**) of the tropical soil isolate L.

The highest ARA (710.2 nmol C_2H_4/culture per hour) and growth rate in cultures of the tropical soil isolate L was recorded at 37°C during the first 20 h of incubation (Figure 3A). The growth rate then declined to a level below that recorded at 24°C and 28°C (Figure 3B).

Discussion

The optimum temperature for N_2-dependent growth of the temperate soil isolates was around 28°C. This compares well with 30°C reported for N_2-fixing bacteria isolated from the roots of *Digitaria sanguinalis* (Barber and Evans, 1976) and wheat (Larson and Neal, 1978) in temperate environments. The optimum temperature (37°C) for growth and ARA of the tropical soil isolate was in the same range (32–40°C) with that of N_2-fixing *Azospirillum* spp. isolated from the roots of a tropical grass *D. decumbens* (Day and Dobereiner, 1976). These temperature optima differences between temperate and tropical soil isolates may be of ecological significance.

Acknowledgement

We thank the International Foundation for Science for financial support.

References

Barber L E, Evans H J 1976. Characterization of a nitrogen-fixing bacterial strain from the roots of *Digitaria sanguinalis*. Can. J. Microbiol., 22, 254–260.

Cakmacki M L, Evans H J, Seidler R J 1981. Characteristics of nitrogen-fixing *Klebsiella oxytoca* isolated from wheat roots. Plant Soil, 61, 53–63.

Day J M, Dobereiner J 1976. Physiological aspects of N_2 fixation by *Spirillum* from *Digitaria* roots. Soil Biol. Biochem., 8, 45–50.

Larson R I, Neal J L 1978. Selective colonisation of the rhizosphere of wheat by nitrogen-fixing bacteria. In: Granhall U, ed. Environmental Role of Nitrogen-fixing Blue-green Algae and Asymbiotic Bacteria Ecol. Bull.(Stockholm), 26, 331–342.

Mwaura F B 1986. Response of maize (*Zea mays* L.) to straw amendment and inoculation with N_2-fixing bacteria in a tropical soil. In: Proceedings, IFS Workshop on Biological Improvement of Soil Fertility, 19–25 March, Dakar, Senegal, 176–184.

Mwaura F, Granhall U 1986. Nitrogen fixation (C_2H_2 reduction) associated with maize (*Zea mays* L.) in a Swedish soil. Swedish J. Agric. Res., 16, 49–56.

31 Poster

In-vitro propagation using phytohormones produced by associative diazotrophs

H.M. EL-KHAWAS, I.A. IBRAHIM, H.M. ANWAR AND N.A. HEGAZI

Introduction

Plant tissue culture techniques and micropropagation depend mainly upon synthetic plant growth regulators, e.g. auxins, cytokinins and gibberellins. Such compounds are known to be produced and released by various rhizospheric microorganisms (Tien *et al.*, 1979, Inze *et al.*, 1984; Fett *et al.*, 1987).

Therefore, the present study aims to survey rhizosphere bacteria for their ability to produce plant growth regulators, determining the biological effect of microbial growth regulators using bioassay and chemical analysis. Also, the effect of microbial plant growth regulators on micropropagation of tested plants using tissue culture techniques was examined.

Materials and methods

Isolations of major groups of associative diazotrophs from the rhizosphere of some crops

One hundred and twenty isolates of various groups of N_2-fixers were obtained from the roots of maize, sugar cane, sorghum and rye grass. Twenty representative strains were selected for their fast growth and high N_2-ase activity. API microtube system, API 20B and API 20E, were used as standard micromethods (Logan and Berkley, 1984) for strain identification.

Determination of plant growth regulators

To determine the bacterial plant hormones, the isolated strains were grown on minimal medium with 0.1 mg/ml tryptophan. After 3 days, bacterial cultures

K.A. Malik et al. (eds.), Nitrogen Fixation with Non-Legumes, 265–269.

(100 ml) were centrifuged at 4000 rpm for 15 min. The extraction was carried out according to the method of Shindy and Smith (1975). The aqueous residue was partially purified by partition with ethyl acetate. The acidic ethyl acetate fraction was then collected and dried under vacuum at 37°C to dryness, to determine auxins and gibberellins by using gas-liquid chromatography (Shindy and Smith, 1975), whereas the alkaline fraction was used to determine cytokinins by using a bioassay technique (Conrad, 1974) and high-performance liquid chromatography (Cacciari *et al.*, 1989).

Plant tissue culture bioassay

Ms basal medium of Muramshige and Skroog (1962) was found to be satisfactory for plant propagation through tissue culture techniques. Consequently all the following experiments were conducted using MS basal medium supplemented by either *Azospirillum brasilense* A_2 supernatant as a source of cytokinin at 25, 50, 100, and 200 ml/L for shoot formation or with *Klebsiella pneumoniae* K_9 supernatant as a source of auxins at 50, 75, 150, and 300 ml/L for root formation.

After 6 weeks the following parameters were determined: number of shoots, weight of plant, number of leaves, height of shoots, length of roots and number of roots. Shoots for the first experiment (banana and cordyline), and root formation percentage was recorded for the second experiment (kiwi and date palm).

Results and discussion

One hundred and twenty isolates of sugar cane, maize, sorghum and rye grass were placed under the species of *Klebsiella pneumoniae, Azospirillum brasilense* and *Enterobacter cloacae*. Twenty isolates were selected according to their high rates of N_2 fixation and studied for their ability to produce microbial growth regulators. Results in Table 1 revealed that all the tested strains had the ability to produce IAA, GA and cytokinin-like substances in variable quantities.

Data presented in Tables 2 and 3 illustrate the effect of different concentration of *Azospirillum brasilense* supernatant added to ½ MS basal medium, compared to multiplication medium for banana and cordyline. The rate of multiplication (number of shoots) of cordyline was higher in medium No. 1 and medium No. 7.

The new shoots produced from medium No. 7 were more suitable for further multiplication than the small one (medium No. 1). Results indicate that the highest average shoot length was from treatments No. 3, 5, 6 and 7 compared to other treatments. This may be due to the hormonal effect of *Azospirillum brasilense* A2 supernatant where cytokinins were 0.86 μg/ml and auxins (IAA) were 18μg/ml.

Data given in Table 3 indicate that *A. brasilense* supernatant at 100 ml/L added to ½ MS basal medium increased banana multiplication rate compared to other

Table 1. Indoleacetic acid, gibberellic acid and cytokinin-like substances produced by identified isolates

Isolates	Indoleacetic acid (μg/ml)	Cytokinin (μg/ml)	Gibberellic acid (μg/ml)
Azospirillum brasilense			
A1	32	0.43	0.25
A2	18	0.86	0.28
A3	54	0.45	0.21
A4	49	0.40	0.23
A5	38	0.52	0.22
Klebsiella pneumoniae			
K1	36	0.23	0.16
K2	33	0.41	0.13
K3	28	0.31	0.18
K4	36	0.35	0.14
K5	41	0.28	0.15
K6	29	0.36	0.22
K7	38	0.42	0.15
K8	66	0.18	0.12
K9	79	0.14	0.18
K10	32	0.33	0.16
Enterobacter cloacae			
E1	18	0.24	0.13
E2	15	0.33	0.18
E3	22	0.26	0.12
E4	24	0.39	0.15
E5	19	0.31	0.16

Average value of three replicates

Table 2. Effect of several concentrations of *A. brasilense* A2 bacterial supernatant (BS) and MS salt strength on the development cordyline plants grown *in vitro*

1	2	3	4	5	6	7
3.2	1.4	1.3	1.3	1.3	1.9	2.2
8.9	6.6	6.6	6.9	6.8	9.7	13.5
8.5	8.8	7.8	7.2	8.5	7.8	8.3
5.5	6.6	7.2	6.4	7.3	7.2	7.4
3.6	3.7	3.3	6.5	6.5	5.3	5.9
6.5	7.1	7.6	11.8	12.7	16.8	14.8

(1) ½ MS + 3 mg/L kinitin; (2) ¾ MS + 3 mg/L kinitin; (3) 1 MS + 3 mg/L kinitin; (4) ½ MS + 25 ml BS; (5) ½ MS + 50 ml BS; (6) ½ MS + 100 ml BS; (7) ½ MS + 200 ml BS.

treatments. Murashige and Skoog basal medium supplemented with 200 ml bacterial supernatant produced the highest shoots compared to other treatments. Fresh weight of banana microshoots increased by adding 200 ml of *A. brasilense* to ½ MS basal medium compared to MS basal medium without bacterial supernatant.

Data presented in Table 4 clearly show the effect of *K. pneumoniae* K9 supernatant at different concentrations compared to MS medium on root and callus

Table 3. Effect of several concentrations of *A. brasilense* A2 bacterial supernatant (BS) and MS salt strength on the development cordyline plants grown *in vitro*

1	2	3	4	5	6	7
1.3	1.4	1.8	1.4	1.4	2.4	2.2
1.3	1.6	1.6	0.9	1.6	1.4	2.1
5.4	5.3	4.3	4.9	5.8	4.8	4.3
6.1	5.4	6.3	4.3	4.7	4.7	6.8
4.7	4.4	6.3	5.5	6.5	6.5	7.7
6.6	6.3	6.7	6.5	7.4	7.1	8.2

(1) ½ MS + 5 mg/L BA; (2) ¾ MS + 5 mg/L BA; (3) 1 MS + 5 mg/L BA; (4) ½ MS + 25 ml BS; (5) ½ MS + 50 ml BS; (6) ½ MS + 100 ml BS; (7) ½ MS + 200 ml BS.

Table 4. Effect of different concentrations of *Klebsiella pneumoniae* and MS salt strength on percentage of root formation of kiwi and date palm grown *in vitro*

K. pneumoniae	Strength of medium (MS)			
supernatant (ml)	MS	¾ MS	½ MS	Mean
Kiwi				
0	15	09	00	08.0
50	56	78	81	71.7
75	80	88	84	84.0
150	78	100	100	92.7
300	56	79	87	74.0
Control + IBA	66	70	76	70.7
Mean	58.5	70.7	71.3	
Date palm				
0	00	07	10	05.7
50	11	16	22	16.3
75	24	48	44	38.7
150	45	62	60	55.7
300	36	48	56	46.7
Control + NAA	38	42	45	41.7
Mean	25.7	37.2	39.5	

formation of kiwi (*Actinidia chinensis*) grown *in vitro*. Also, results indicated that *K. pneumoniae* supernatant (150 ml) gave the best results for root formation (92.7%) compared to control (70.7%) and other treatments regardless of MS salt strength.

In addition, it is well demonstrated that MS half-strength produced the highest value for root formation where it was 71.3% regardless of *K. pneumoniae* supernatant concentration. The roots obtained on ½-strength medium are normal whereas the roots induced on full-strength and ¾-strength are weak and fibrous. The same trend was found with date palm plants.

In the present study kiwi and date palm cultured on ¾ , ½-strength MS medium supplemented with 150 ml *Klebsiella pneumoniae* supernatant gave a good rooting response and reduced callus formation. These observations indicate the possibility of large-scale use of such bacterial supernatant for propagation of kiwi and date palm.

References

Cacciari I, Lippi D, Pietrosanti T, Pietrosanti W 1989. Phytohormone-like substances produced by single and mixed diazotrophic cultures of *Azospirillum* and *Arthrobacter*. Plant Soil, 115, 151–153.

Conrad K 1974. A sensitized amaranthus cytokinin bioassay (AT74). Biochem. Physiol. Pflanzen (BPP), Bd., 165, S.531–535.

Fett W F, Osman S F, Dunn M F 1987. Auxin production by plant-pathogenic Pseudomonads and Xanthomonads. Appl. Environ. Microbiol., 53, 1839–1845.

Inze D, Follin M, Van Iijsebettens M, Simoens C, Genetello G, Van Montagu M, Schell J 1984. Genetic analysis of the individual T-DAN genes *Agrobacterium tumefaciens*, further evidence that two genes are involved in IAA synthesis. Mol. Gen. Genet., 194, 265–274.

Logan N A, Berkeley R G W 1984. Identification of *Bacillus* strains using the API system. J. Can. Microbiol., 130, 1871–1882.

Murashige I, Skoog F 1962. A revised medium for rapid growth and bioassays with Tabacco tissue culture. Physiol. Plants, 15, 473–497.

Shindy W W, Smith O E 1975. Identification of plant hormones from cotton ovules. Plant Physiol., 55, 550–554.

Tien T N, Gaskins M H, Hubbell D H 1979. Plant growth substances produced by *Azospirillum brasilense* and their effect on the growth of pearl millet (*Pennisetum americanum* L.) Appl. Environ. Microbiol., 37, 1016–1019.

32 Poster

Effect of inoculation of *Azotobacter* and PSM on fertilizer economy, plant growth and yield of winter maize

PUNEET K. SOHAL, R.P. GUPTA AND M.S. PANDHER

Introduction

Most of the fertilizers applied are rendered unavailable for crop uptake, often as much as 90% in the case of P-fertilizer and 50% in the case of N-fertilizer (Stevenson, 1986). However, crop productivity can be improved by manipulating the rhizosphere microorganisms as microbial availability of plant nutrients is non-controversial. The microbial inoculants are inexpensive, simple to use and have no side-effects. Evidence is growing of the favourable effects of *Azotobacter* and PSM in providing plant nutrients to cereals and vegetables, and supplementing the expensive inorganic fertilizers (Belimov *et al.*, 1995). The present study was conducted to see the effect of *A. chroococcum* and PSMs on fertilizer economy, plant growth and grain yield of winter maize.

Materials and methods

Twenty-four treatment combinations, consisting of three levels of P_2O_5 of SSP (0, 40 and 80 kg/ha); two levels of *Azotobacter* [with (A^+) and without (A^-) inoculation] and four PSM [control (C_0); *A. niger* 1 (C_1); *A. flavus* (C_2) and *A. niger* 2 (C_3)] were tested in a factorial split-plot design with three replications. The seed treatment method was used for inoculation with *Azotobacter* and PSM. Nitrogen at 150 kg/acre through urea was added by the split method (⅓ N at the time of sowing; ⅓ N in mid-January and ⅓ N at the tasselling stage). The data for plant growth (g/ha); percentage nitrogen and phosphorus in shoot and seed; soil available phosphorus (kg P_2O_5/ha) was recorded at maturity and grain yield (g/ha) at the time of harvest. Shoot and seed N and P were determined using the McKenzie and Wallace (1954) and Jackson method (1973), respectively, and soil P as by Olsen *et al.*, (1954).

K.A. Malik et al. (eds.), Nitrogen Fixation with Non-Legumes, 271–273.

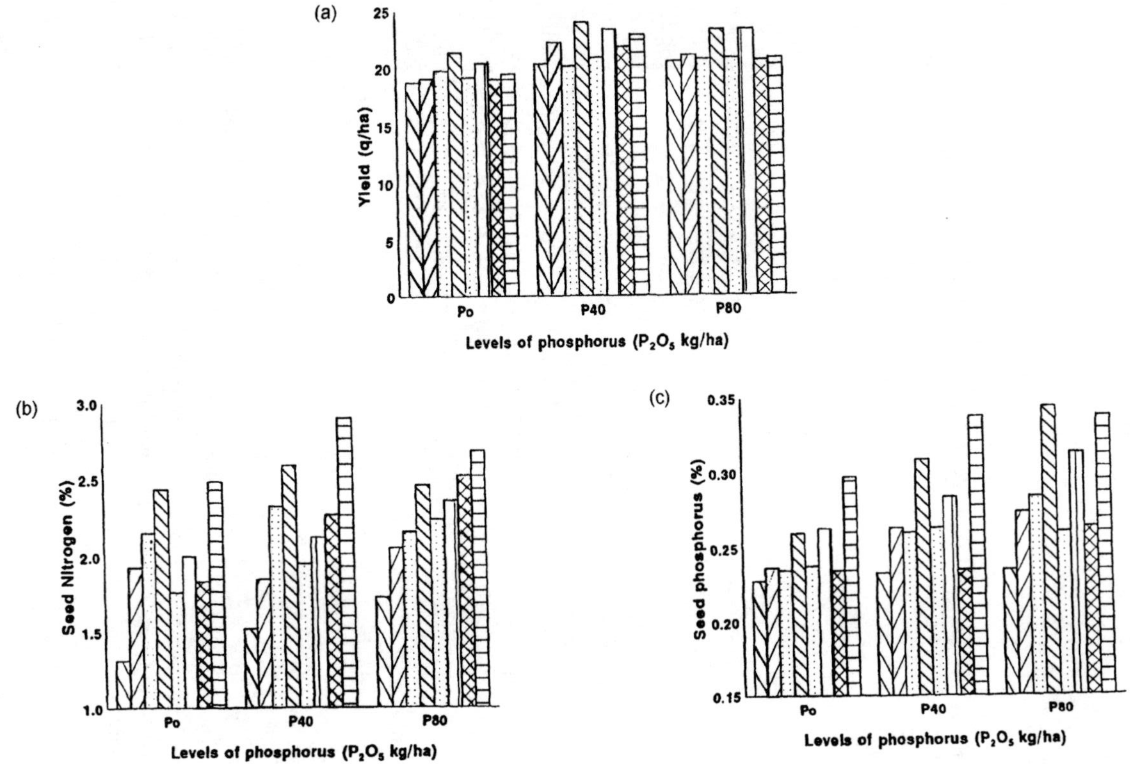

Figure 1. N and P (%) of seed and seed yield of maize.

[0]▧, C_0; ▨, C_0 + A; ⊡, C_1; ▧ C_1; + A; ⊡, C_2; ◫, C_2 + A; ⊠ C_3; □, C_3 + A

Results and discussion

Effect on plant growth and nutrient content

Single inoculation, either with *Azotobacter* or PSM, significantly increased shoot dry weight, N and P content of winter maize. The increase in these parameters was significantly high up to 40 kg P_2O_5/ha, and it became almost constant at higher doses. Dual inoculation (*Azotobacter* plus C_1/C_3) receiving one half of the usual dose of SSP, was higher or similar in influencing these parameters than in plants receiving the usual rate of SSP in the absence of inoculation. Similar results were observed in seed nitrogen and phosphorus (Figure 1b and 1c).

Effect on grain yield and P-availability

Dually inoculated plants were found superior over single inoculation at all P levels. The grain yield at 40 kg P_2O_5/ha along with inoculation was found to be at par with the yield at 80 kg P_2O_5/ha alone. The percentage increase in grain yield with *Azotobacter* was 9% and 17.7% with dual inoculation (C_1 + A^+) over control (Figure 1a).

The available soil P under unfertilized treatment with no inoculation tended to decrease when compared with initial soil P; however, single or dual combinations significantly improved the available soil P status over initial P status of soil with increasing frequency of fertilization.

The results indicated a positive role of *Azotobacter* and PSM cultures in nutrient availability and uptake, which in turn improved crop growth and hence winter maize yield and quality. The use of dual cultures shows the possibility of reducing input of 40 kg P_2O_5/ha, besides improving crop and soil health.

References

Belimov A A, Kojetniakov A P, Chuvarlyeva C V 1995. Interaction between barley and mixed cultures of nitrogen-fixing and phosphate-solubilizing bacteria. Plant Soil 173, 29–37.

Jackson M L 1973. Soil Chemical Analysis. Prentice Hall of India, New Delhi.

McKenzie H A, Wallace H A 1954. The Kjeldahl determination of nitrogen. Aust J Chem., 17, 55–79.

Olsen S R, Cole C V, Watanabe F S, Dean L A 1954. Estimation of available phosphorus in soil by extraction with $NaHCO_3$. USDA Circular No. 939, pp. 1–19.

Stevenson F J 1986. Cycles of Soil Carbon, Nitrogen, Phosphorus, Sulfur, Micronutrients. Wiley, New York.

33 Poster

Accumulation of radionuclides by associative bacteria and the uptake of ^{134}Cs by the inoculated barley plants

A.A. BELIMOV, A.M. KUNAKOVA, N.D. VASILYEVA, T.S. KOVATCHEVA, V.F. DRITCHKO, S.N. KUZOVATOV, I.R. TRUSHKINA AND YU.V. ALEKSEYEV

Introduction

Radionuclides such as radiocaesium and radiostrontium, as well as heavy metals, are among the most widespread pollutants in the surface soil layer. Radionuclides and heavy metals are easily accumulated by plants, inhibit the growth and nutrition of various agricultural crops, and can cause diseases in animals and humans. The basic modes of decreasing the uptake of radionuclides and heavy metals by plants are the processes by which the transformation of toxic elements into inassimilable forms and the improvement of soil fertility is accomplished (Prister *et al.*, 1991; Sokolov, 1995). Moreover, a wide range of soil microorganisms are resistant to radiation (Eriksen and Emborg, 1978) and heavy metals (Duxbury and Bicknell, 1983), and play a significant role in the transformation (immobilization or dissolution) of metals in soil and wastes through a number of metabolism-dependent and -independent processes (Francis, 1990; Gadd, 1990).

The aim of this study was to assess the ability of associative bacteria to immobilize radionuclides and toxic metals and to estimate the uptake of ^{134}Cs by the inoculated barley plants.

Materials and methods

Bacterial strains

The studied bacterial strains *Azospirillum lipoferum* 137, *Arthrobacter mysorens* 7, *Flavobacterium* sp. L30 and *Agrobacterium radiobacter* 10 were grown on RHM medium with the following composition (g L^{-1}): peptone, 1; sucrose, 1; glucose, 1; yeast extract, 1; NH_4NO_3, 0.5; $MgSO_4$, 0.2; $CaCl_2$, 0.02; $FeCl_3$, 0.01; pH = 6.8. After 48 h of incubation on a solid RHM medium the bacteria were

K.A. Malik et al. (eds.), Nitrogen Fixation with Non-Legumes, 275–280.

suspended in water (10^6 cells ml^{-1}) and 0.1 ml of the suspension was added to the flasks containing 10 ml of a liquid RHM medium supplemented with 6.7 Bq ml^{-1} of ^{134}Cs, 200 Bq ml^{-1} of ^{90}Sr or 10 µg ml^{-1} of $CdCl_2$. The flasks (three replications) were incubated under aerobic conditions on a rotary shaker at 28°C for 96 h. The biomass concentration in the growth medium was measured using calibration against optical density, samples were centrifuged at 10 000 ***g*** for 15 min, and the supernatants were collected. The precipitates were washed twice with 7 ml of distilled water. The content of metals in the supernatants and in the biomass was determined. Prior to determining the Cd content, the samples were evaporated at 95°C to dryness, resuspended in concentrated HNO_3, digested at 95°C until the acid evaporated, and 5 ml of 20% HNO_3 was added to dissolve the residue (Kurek *et al.*, 1982).

Pot experiments

Spring barley (*Hordeum vulgare* L.) cv. Celinney-5 plants were grown in pots containing unsterile derno-podzolic soil (humus, 1.1%, hydrolysable N, 77 mg kg^{-1}; available P, 75 mg P_2O_5 100 g^{-1}; available K, 15 mg K_2O 100 g^{-1}; hydrolytic acidity, 12.2 mg Eq 100 g^{-1}; pH = 6.0). The soil was additionally fertilized three times (before sowing, at shooting and heading stages) with (mg kg^{-1}): NH_4NO_3, 25; K_2HPO_4 80; $MgSO_4$, 15; $ZnSO_4$, 1; $CuSO_4$, 1; H_3BO_3, 1; Na_2MoO_4, 0.5. The soil was supplemented with 89 000 Bq kg^{-1} (experiment 1) or 16 600 Bq kg^{-1} (experiment 2) of ^{134}Cs. The seeds were surface sterilized with a mixture of ethanol and hydrogen peroxide (1:1) for 3 min and inoculated immediately after sowing using the water diluted bacterial suspensions in the amount of 10^7 cells per seed.

Determination of the metal content

The ^{134}Cs concentration in the medium, soil, bacterial and plant biomass was determined using a scintillation gamma-spectrometer equipped with a BDЭ-23 detector. A beta-radiometer equipped with SBT-13 detector was used for measuring the ^{90}Sr content. Cadmium concentrations were determined using an atomic absorption spectrophotometer Solar S 929 AA. The concentration ratio (CR) was computed with the following formula:

$$CR = \frac{C_b}{C_m}$$

In this formula C_b stands for metal concentration in the biomass (bacteria, plants). C_m stands for metal concentration in the medium (supernatant, soil) and comes in units of Bq g^{-1} or mg Cd g^{-1}.

Statistical analysis

Calculation of the confidence intervals and standard methods of variance analysis were used (Lakin, 1980).

Results and discussion

All of the studied bacteria were able to accumulate ^{134}Cs and ^{90}Sr by batch cultivation in a liquid nutrient medium, and caesium was concentrated to a considerably lesser extent than strontium (Figure 1). Cadmium was also taken up actively by the bacteria, with the exception of *Flavobacterium* sp. L30 which was very sensitive to the presence of Cd in the growth medium. Among four studied strains *A. lipoferum* 137 was the best for immobilization of ^{134}Cs, ^{90}Sr and Cd, having a biomass–medium concentration ratio many times greater than for the other three strains. It is well known that microorganisms and their constituent polymers are among the most efficient scavengers of metallic ions (Doyle *et al.*, 1975; Kurek *et al.*, 1982; Francis, 1990; Gadd, 1990).

There was a significant decrease in the plant–soil CR of ^{134}Cs in the roots after inoculation with *A. lipoferum* 137 on the 14th and 24th days after planting (Figure 2). *Flavobacterium* sp. L30 tended to increase ^{134}Cs uptake by the roots. However, inoculation had no pronounced effect on the uptake of caesium by shoots and the dynamics of plant–soil CRs for all treatments was similar. The

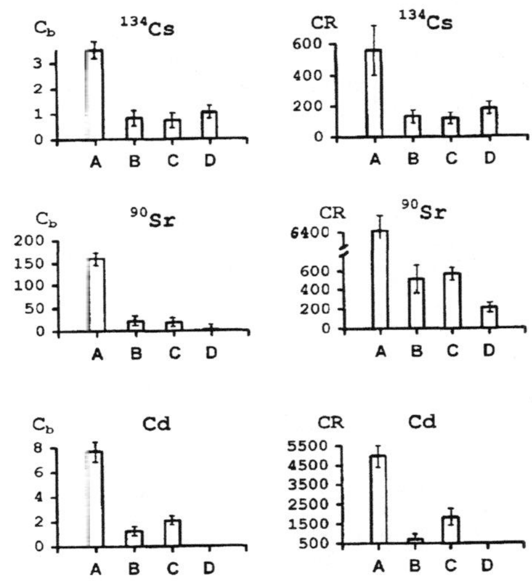

Figure 1. Accumulation of metals in the bacterial biomass by batch cultivation. C_b, concentration of ^{134}Cs, ^{90}Sr (Bq mg^{-1}) and Cd (μg mg^{-1}) in the biomass; CR, biomass–medium concentration ratio. Strains: A, *A. lipoferum* 137; B, *A. mysorens* 7; C, *A. radiobacter* 10; D, *Flavobacterium* sp. L30. Vertical bars represent confidence intervals at $p < 0.05$.

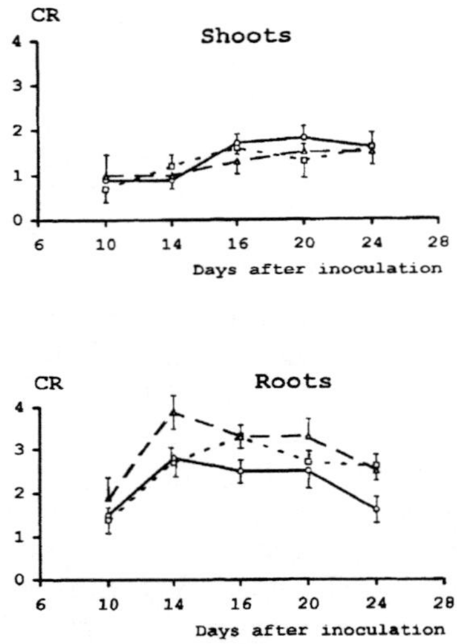

Figure 2. Uptake of ^{134}Cs by barley seedlings inoculated with associative bacteria (pot experiment 1). CR, plant–soil concentration ratio of ^{134}Cs. Treatments: □- - -□, uninoculated control; ○—○, *A. lipoferum* 137; △— —△, *Flavobacterium* sp. L30. Vertical bars represent confidence intervals at $p < 0.05$.

Table 1. Effect of inoculation with associative bacteria on barley yield (pot experiment 2)

Treatments	Number of ears (n pot^{-1})	Number of seeds (n pot^{-1})	Grain yield (g pot^{-1})	Straw weight (g pot^{-1})	Root weight (g pot^{-1})	Total plant weight (g pot^{-1})
Uninoculated control	19	161	7.9	9.1	2.4	19.4
A. lipoferum 137	23*	177	8.6	10.0*	1.8	20.4
A. mysorens 7	22	185	8.9	10.0*	2.1	21.0
A. radiobacter 10	25*	211*	9.8*	10.3*	2.4	22.5*
Flavobacterium sp. L30	24*	188	8.7	11.3*	2.3	23.2*
LSD ($p < 0.05$)	3.9	32	1.3	0.8	1.0	2.3

* In this and Table 2 the values in each column differed significantly from those for the uninoculated plants at $p < 0.05$.

roots and shoots of the inoculated and uninoculated plants were equal in dry weight (data not shown). The response of barley to inoculation with associative bacteria in the second pot experiment is shown in Table 1. A significant increase in grain yield, number of ears and seeds was achieved after inoculation with *A. radiobacter* 10. The other strains increased grain yield by 8–13% as compared with the control, however the difference was not significant. All the studied bacteria had a beneficial effect on the straw of barley. The effect of inoculation on the uptake of ^{134}Cs by the plants was in general agreement with the results of the first pot experiment (Table 2). The concentration and plant–soil CR of ^{134}Cs in the roots and straw of the plants inoculated with *A. lipoferum* 137 were 17%

Table 2. Effect of inoculation with associative bacteria on the uptake of ^{134}Cs by barley (pot experiment 2)

Treatments	Roots	Straw	Grain	Entire plant
Concentration of ^{134}Cs in plants (Bq g^{-1})				
Uninoculated control	5.2	1.39	0.43	—
A. lipoferum 137	4.3*	0.92*	0.38	—
A. mysorens 7	4.9	1.09	0.41	—
A. radiobacter 10	5.1	1.06*	0.38	—
Flavobacterium sp. L30	4.8	1.49	0.50	—
LSD ($p < 0.05$)	0.7	0.32	0.13	—
Plant–soil concentration ratio (CR)				
Uninoculated control	0.32	0.085	0.026	—
A. lipoferum 137	0.26*	0.056*	0.023	—
A. mysorens 7	0.30	0.066	0.025	—
A. radiobacter 10	0.31	0.064*	0.023	—
Flavobacterium sp. L30	0.30	0.091	0.030	—
LSD ($p < 0.05$)	0.05	0.020	0.009	—
Accumulation of ^{134}Cs in plants (Bq pot^{-1})				
Uninoculated control	12.3	12.7	3.4	28.4
A. lipoferum 137	7.3	9.1	3.2	19.6*
A. mysorens 7	10.0	11.0	3.6	24.6
A. radiobacter 10	12.3	10.9	3.7	26.9
Flavobacterium sp. L30	15.9	16.8*	4.3	37.0*
LSD ($p < 0.05$)	5.4	3.4	1.1	7.7

* See Table 1.

and 34% below that of the uninoculated control, respectively. A drop in ^{134}Cs uptake by straw was also observed after inoculation with *A. radiobacter* 10. Nevertheless, there was no evident effect of the studied bacteria on the uptake of radionuclide in barley grain. Inoculation with *A. lipoferum* 137 significantly decreased the total accumulation of caesium by the plants by 30% as compared with the uninoculated plants. In contrast to this, the strain *Flavobacterium* sp. L30 contributed to increased accumulation of ^{134}Cs by the plants, mainly through the enhancement of plant biomass (Tables 1 and 2). The results indicate that the introduced bacteria inhabiting the rhizosphere are capable of changing availability of various elements for the plants.

In conclusion, the results showed that associative bacteria are able to accumulate considerable amounts of toxic elements such as ^{134}Cs, ^{90}Sr and Cd, and can be important in influencing the availability of radionuclides for barley plants grown in contaminated soil. However, more information is needed for proper evaluation of the impact of the beneficial soil microorganisms, e.g. *Azospirillum*, for minimizing the accumulation of toxic metals by the plants.

Acknowledgements

We are very grateful to Dr V.I. Safronova and Dr K.A. Kunakov for valuable assistance in preparing this chapter.

References

Doyle J J, Marshal R T, Pfander W H 1975. Effect of cadmium on the growth and uptake of cadmium by microorganisms. Appl. Microbiol., 29, 562–564.

Duxbury T, Bicknell B 1983. Metal tolerant bacterial populations from natural and metal-polluted soil. Soil Biol. Biochem., 15, 243–250.

Eriksen W H, Emborg C 1978. Increase of radiation resistance of soil microflora exposed to long-term gamma irradiation. Appl. Environ. Microbiol., 36, 618–619.

Francis A J 1990. Microbial dissolution and stabilization of toxic metals and radionuclides in mixed wastes. Experientia, 46, 840–851.

Gadd G M 1990. Heavy metal accumulation by bacteria and other microorganisms. Experientia, 46, 834–840.

Kurek E, Czaban J, Bollag J M 1982. Sorption of cadmium by microorganisms in competition with other soil constituents. Appl. Environ. Microbiol., 43, 1011–1015.

Lakin G G 1980. Biometrics. Higher School, Moscow (in Russian).

Prister B C, Lostshilov N A, Nemetz O F, Poyarkov V A 1991. The Foundations of Agricultural Radiology. Urozaj Press, Kiev, Russia (in Russian).

Sokolov M S 1995. A possibility for obtaining ecologically safe agricultural products under environmental pollution. Agrochemistry, 6, 107–125 (in Russian).

34 Poster

The influence of chosen fungicides on *Azospirillum* dinitrogen fixation in the soil

DOROTA SWĘDRZYŃSKA AND ALEKSANDRA SAWICKA

Introduction

One of the most interesting groups of soil microorganisms fixing atmospheric nitrogen are *Azospirillum* bacteria. They live in association with roots of many plants affecting their yields (Boddey *et al.*, 1986; Boddey and Döbereiner, 1988; Chalk, 1991). Although the above-mentioned bacteria occur in soils of various climatic zones, they best develop in a tropical climate (Krieg and Döbereiner, 1986). Possibilities of practical application of these bacteria in conditions of temperate climate require isolation of strains characterized by a definite, high N_2-fixing activity as well as recognition of their response to different factors which may exert an influence during the process of plant production; among others fertilization or plant protection compounds.

The objective of our research project was to ascertain the effect of fungicide dressings commonly applied in the cultivation of cereals (Baytan Universal, Oxafun T, Vitavax) on the survival of *Azospirillum* in the soil, as well as on its nitrogen-fixing activity.

Material and methods

Experiments were carried out using an actively nitrogen-fixing strain of bacteria *Azospirillum brasilense* 65B isolated from cultivated soil in the Department of Microbiology of the Institute of Soil Cultivation, Fertilization and Soil Science in Puławy. Investigations were conducted under laboratory conditions. The soil (slightly loamy sand) was incubated at a temperature of 25°C in jars in the following combinations:

1. soil (control),
2. soil with 2% addition of saccharose,

K.A. Malik et al. (eds.), Nitrogen Fixation with Non-Legumes, 281–284.

Table 1. Influence of fungicides on *Azospirillum brasilense* number in the soil with different experimental combinations

Incubation time	Number of *Azospirillum brasilense* in soil incubated with fungicides (n 10^3/g d.m. of soil)			
	Control (without fungicides)	Baytan	Oxafun	Vitavax
Soil				
48 hours	6.60	12.10	6.60	22.00
72 hours	17.60	9.90	6.60	6.60
14 days	3.60	0.30	0.39	1.40
30 days	4.90	4.90	10.40	0.44
Soil + saccharose				
48 hours	2.80	28.00	5.04	1.00
72 hours	1792.00	784.00	336.00	224.00
14 days	1.00	0.67	0.67	3.36
30 days	280.00	3.36	6.72	6.72
Soil + Azospirillum				
48 hours	395.50	395.50	136.60	158.00
72 hours	12.43	226.00	6.78	10.17
14 days	3.95	22.60	1.01	1.08
30 days	10.73	10.73	0.45	0.45
Soil + Azospirillum + saccharose				
48 hours	10.93	12.10	28.75	10.93
72 hours	51.75	126.50	126.50	126.50
14 days	10.92	126.50	51.75	51.75
21 days	161.00	161.00	161.00	161.00
30 days	172.00	287.50	172.00	287.50

3. soil inoculated with *Azospirillum*,
4. soil with 2% addition of saccharose inoculated with *Azospirillum*.

Additionally, each of the above experimental combinations was supplemented with fungicide seed dressings:

1. Baytan Universal (a.i. 2% fuberidazole + 2.5% imazalil + 15% triadimenol).
2. Oxafun T (a.i. 37.4% carboxin + 37.5% thiuram).
3. Vitavax (a.i. 20% carboxin + 20% thiuram).

After 2, 3, 14, 21 and 30 days of incubation of soil samples, the following parameters were determined:

1. MPN of *Azospirillum* in individual combinations (Sawicka, 1983).
2. Nitrogenase activity using the acetylene to ethylene reduction method (ARA) (Sawicka, 1983).

Results and discussion

The number of *Azospirillum* in the soil in the presence of the applied fungicidal seed dressings declined, and this decline was particularly evident 2 weeks after their introduction into the soil (Table 1).

Saccharose addition into the soil resulted in proliferation of these bacteria after 3 days of incubation and a rapid decrease of their number after the next 2 weeks of incubation with saccharose. The proliferation of bacteria may have been caused by the additional carbon source in the form of indirect products of saccharose breakdown by other soil microorganisms. The increase in the number of these bacteria did not enhance fixing of molecular nitrogen in the soil. Nitrogenase activity was observed only in the treatment with the addition of saccharose to soil and inoculation with *Azospirillum*. During the first 3 days of incubation, nitrogenase activity in the soil increased in the presence of Oxafun and Baytan in comparison with the control, but later it decreased (Figure 1).

The observed enhanced nitrogenase activity need not have been caused by *Azospirillum* because in the same experimental combination a rapid growth in the number of *Azotobacter* and other soil bacteria was also observed (Swędrzyęska and Sawicka, unpublished data). Furthermore, as confirmed by our earlier investigations (Sawicka, 1991), bacteria fixing molecular nitrogen exhibit higher nitrogenase activity in the presence of other accompanying soil microorganisms.

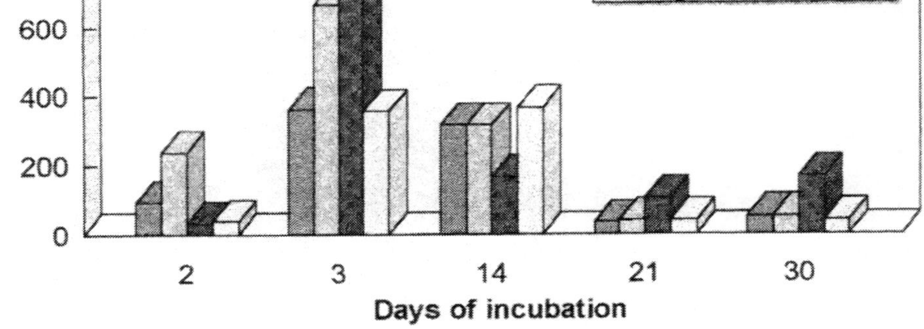

Figure 1. Effect of fungicides on nitrogenase activity in the soil inoculated with *Azospirillum brasilense* and enriched with saccharose (nM C_2H_4 h^{-1} kg^{-1} of soil)

It is evident from our investigations that such active substances as: fuberidazole, imazalil, triadimenol, carboxin and thiuram may restrict the number of *Azospirillum* and, in special cases, may enhance nitrogenase activity in the soil.

References

Boddey R M, Döbereiner J 1988. Nitrogen fixation associated with grasses and cereals: recent results and perspectives for future research. Plant Soil, 108, 53–65.

Boddey R M, Baldani V L D, Baldani J I, Döbereiner J 1986. Effect of inoculation of *Azospirillum* spp. on nitrogen accumulation by field-grown wheat. Plant Soil, 95, 109–121.

Chalk P M 1991. The contribution of associative and symbiotic nitrogen fixation to the nitrogen nutrition of non-legumes. Plant Soil, 132, 29–39.

Krieg N R, Döbereiner J 1986. The genus *Azospirillum*. In: Krieg N R, Holt J G, eds. Bergey's manual of systematic bacteriology. vol. 1. Williams and Wilkins, Baltimore, MD, 96–104.

Sawicka A 1983. Ekologiczne aspekty wiazania azotu atmosferycznego. Roczniki Akademii Rolniczej w Poznaniu. Rozprawy naukowe, 134.

Sawicka A 1991. Nitrogenase activity in soils with added root exudates. Polish J Soil Sci. 24, 179–184.

Section VII

Methodologies for assessing nitrogen fixation

35

Evaluation of biological nitrogen fixation associated with non-legumes

ROBERT M. BODDEY, BRUNO J.R. ALVES AND SEGUNDO URQUIAGA

Abstract

The method to quantify biological nitrogen fixation (BNF) that was most frequently utilized in early studies of associations between N_2-fixing bacteria and non-legume plants was the acetylene reduction (AR) assay, which subsequently was examined critically and found to be difficult to apply to produce reliable BNF estimates. However, the AR technique can produce meaningful results if *in-situ* assays are performed on undisturbed intact plant/soil systems and precautions are taken to inhibit BNF activity of cyanobacteria and tests are made to check for endogenous ethylene evolution. With the techniques based on the quantification of total N, namely total N balance, the simultaneous use of several non-N_2-fixing control plants greatly improves the chances of successful quantification of plant-associated BNF. For those with access to isotopically labelled nitrogen (^{15}N), and the facilities to analyse it, either isotopically labelled N_2 gas may be utilized or the plant growth medium may be labelled and the so-called isotope dilution technique can be applied. The use of ^{15}N-labelled atmospheres is useful to confirm BNF inputs to plants but only very short-term assays are feasible unless very sophisticated sealed chambers are available. The isotope dilution technique does not require enclosure of the plants, but application of ^{15}N-enriched fertilizers to the soil results in non-uniform labelling of the soil mineral N with depth and time. In this case as different plants (e.g. inoculated and non-inoculated) can vary considerably in their patterns of soil N uptake, they will not accumulate soil N with the same specific ^{15}N label, which is the principal source of error in this technique. The most successful applications of this technique have been achieved using soils in containers (pots or tanks) with finite depth and uniformly filled with ^{15}N labelled soil. In this review we discuss the problems associated with the use of these techniques for quantifying plant-associated

K.A. Malik et al. (eds.), Nitrogen Fixation with Non-Legumes, 287–305.

BNF, and the strategies that can be applied to ameliorate them, using examples from studies performed at CNPAB and elsewhere.

Introduction

A very large body of knowledge has been accumulated on the genetics and physiology of *Azospirillum* spp. and their association with grasses and cereals and some other non-nodulating plants, and rather less research has been conducted on other, more recently discovered diazotrophs. However, comparatively very few studies have been performed to carefully evaluate possible contributions of biological nitrogen fixation (BNF) to these plant/bacteria associations and to determine whether such contributions are of agronomic significance. There are many reports of responses to inoculation of *Azospirillum* by different grasses and cereals in terms of yield, dry matter and/or N accumulation, but in the few cases where these studies have been accompanied by investigations to quantify possible BNF contributions of the plants almost all have shown that the observed responses were due to increased uptake of soil N and not to N derived from BNF (e.g. Kapulnik *et al.*, 1985; Boddey *et al.*, 1986). It is lamentable that so few well-designed studies have been performed to evaluate actual BNF contributions to plant/diazotroph associations, and this review is intended as a contribution to assist those who wish to apply the various available techniques to make reliable estimates of BNF inputs under both controlled and field conditions.

The acetylene reduction assay

The development of the acetylene reduction technique in the 1960s (Dilworth, 1966; Schöllhorn and Burris, 1966) and its application to detect BNF with many plant/bacteria associations was largely responsible for the explosion of interest in non-legume BNF.

Some of the early work on wetland rice was performed using glass chambers driven into the soil around the plants (Balandreau *et al.*, 1974; Balandreau, 1975). Each chamber covered an area of 700 cm^2, sufficient to enclose a mature single rice plant, and some very high values of acetylene reduction activity (ARA) were recorded, ranging from 3 000 nmol C_2H_2 $chamber^{-1}h^{-1}$ at night up to 30 000 nmol C_2H_2 $chamber^{-1}h^{-1}$ in the afternoon (Figure 1). The authors suggested that this activity might be related to diurnal fluctuations in supply of root exudates to N_2-fixing bacteria, but as such high values have never been reported since, it seems probable that the peaks of activity in daytime were due to photosynthetic N_2-fixing microorganisms (presumably cyanobacteria) in the flood water. In subsequent studies using *in-situ* assays in pots (Lee and Yoshida, 1977; Boddey *et al.*, 1978) and in the field (Lee *et al.*, 1977; Watanabe *et al.*, 1978a,b; Boddey and Ahmad, 1981), either the existing surface water was removed from around the plants and/or antibiotics or Propanil (a herbicide which inhibits

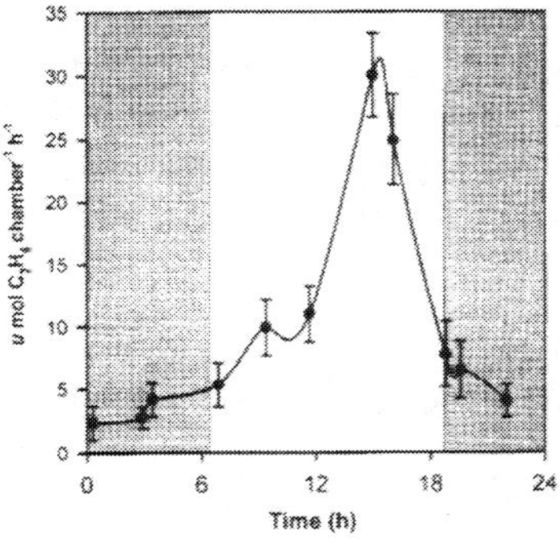

Figure 1. Diurnal variations in acetylene reduction activity of wetland rice plants estimated using an *in situ* assay in the field. Area enclosed by assay chamber 700 cm^2. Error bars indicate standard errors of the means (5 replicates). From Balandreau *et al.*, (1974).

photosynthesis but does not affect the rice plants) was added to inhibit the activity of cyanobacteria. In these studies activities reached 1 000–3 000 nmol C_2H_2 chamber $^{-1}$h^{-1}, which Watanabe *et al.*, (1978b) and Boddey and Ahmad (1981), who measured the AR activity throughout the cropping season, estimated to represent a total contribution of between 5 and 8 kg N ha^{-1} to the crop (Figure 2). One difficulty which was encountered in such assays on waterlogged systems was the diffusion of acetylene to the sites in the roots or rhizosphere or, more critically, of ethylene to the chamber atmosphere where it is sampled. Watanabe *et al.*, (1978b) vigorously stirred the surface water to release ethylene, but even then there may have been some underestimation of acetylene reduction activity.

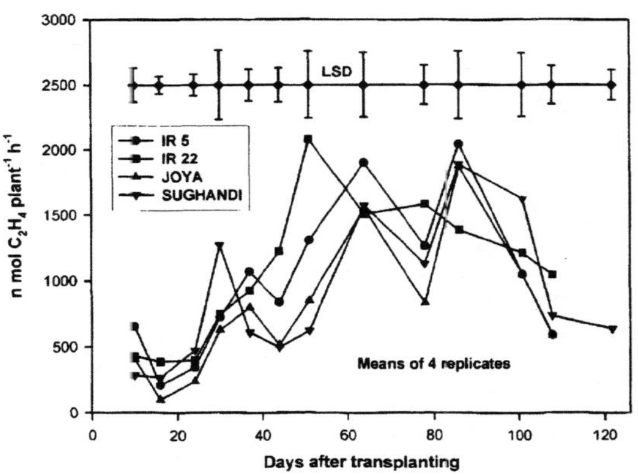

Figure 2. Acetylene reduction activity throughout plant ontogeny for four wetland rice varieties grown on a Vertisol in Trinidad, West Indies, estimated using an *in situ* assay. Area enclosed by assay chamber 250 cm^2. Error bars indicate standard errors of the means (four replicates). From Boddey and Ahmad (1981)

In other crops grown under non-flooded conditions this source of error is likely to be less serious, although the results of van Berkum and Day (1980) suggested that this would be a problem in soils of high clay content. The N_2-fixing systems of all diazotrophs are very sensitive to changes in oxygen partial pressure (pO_2). Owing to the soft texture of the soil, and its uniform anaerobic environment, the sinking of chambers or metal cylinders into waterlogged soils around rice plants is not likely to cause significant changes in pO_2 in the roots or at the soil–root interface. However, this is likely to be a major problem in aerobic soils. Investigations conducted at ICRISAT in India on the ARA of intact, 15-cm diameter, cores of millet (variety NHB-3) removed from the field at 55, 66 and 69 days after planting (DAP), and almost immediately exposed to acetylene (gas samples taken at 1 and 6 h after enclosure in sealed containers) showed that the activity ('regular core assay') ranged from 0 to 70 nmol C_2H_2 core^{-1}h^{-1} (Wani *et al.*, 1983 – Figure 3). When cores had been inserted around the plants at 15 DAP and were carefully removed from the soil at 54 DAP ('planted core assay'), the activities were almost 10 times greater, showing that physical disturbance of the cores greatly reduced ARA.

The very first reports of nitrogenase activity associated with wetland rice were based on incubations of roots which had been excised from the plants and incubated at low pO_2 (Rinaudo and Dommergues, 1971; Yoshida and Ancajas, 1971, 1973). Later this technique was used by other authors for dryland crops also, and activity was found to be initially very low for the first 6–12 h, followed by approximately linear rates of ethylene evolution (Döbereiner *et al.*, 1972a,b; Patriquin, 1978). These rates recorded after the 'preincubation' at low pO_2 were regarded as representing the original nitrogenase activity before the plants had been uprooted and exposed to air. Some comparisons with core or pot samples suggested that these post 'preincubation' rates were either similar to or less than core samples (Döbereiner *et al.*, 1972a,b; Boddey *et al.*, 1978), while other reports

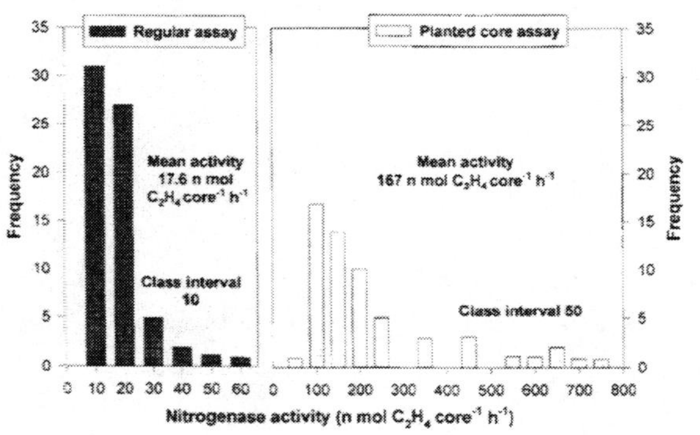

Figure 3. Acetylene reduction activity of intact cores of different millet varieties. Regular assay – intact plants removed from the soil and assayed immediately after insertion and removal of cores. Planted cores assay – cores inserted into soil, seeds planted within cores and intact plants removed from the soil with great care 54 days after planting. Data from Wani *et al.*, (1983).

suggested that the excised root technique normally gave estimates of nitrogenase activity far higher than intact core samples (Koch, 1977; Tjepkema and van Berkum, 1977). There was considerable controversy concerning the validity of the results obtained from the excised root assay (for more information see van Berkum and Bohlool, 1980; van Berkum and Day, 1980; Boddey, 1987), and the technique cannot be recommended for quantifying, or even reliably detecting, significant contributions of plant-associated BNF.

It follows therefore that ARA assays must be conducted on undisturbed plants. For field assays a procedure similar to that used by Wani *et al.*, (1983) should be adopted, and for plants grown in pots the whole pot should be enclosed in a container. Plastic bags may be used to enclose the plants, although leakage may be a problem, and thick (e.g. 125 μm) polypropylene, or materials such as Saran (Burris, 1974) of low gas permeability, are to be recommended. To calculate the ARA per pot or per core, it is necessary to know the volume of gas enclosed within the bag. For a rigid container this normally presents no problem, but for flexible containers it can be achieved using an internal standard such as propane as originally suggested by Balandreau and Dommergues (1971, 1973). In this case a known volume (e.g. 10 ml) of propane is injected into the chamber, and the concentration of this gas in the enclosure can be estimated on the same gas chromatographic column (e.g. Poropak N) as used to separate and quantify acetylene and ethylene (Boddey *et al.*, 1978). From the concentration of propane the effective volume of the enclosure can be calculated. If samples are taken shortly after the start of the assay (after mixing of gases is complete) and at the end, then any leakage of gases from the enclosure can be detected. To correctly compensate for these leakages the equations developed by Patriquin and Keddy (1978) should be utilized, which assume that leakage of ethylene is proportional to its concentration and ethylene evolution is constant with time (see Boddey, 1987, for a ready-to-use version of these equations). However, if leakage is severe, (e.g. > 40%) even then considerable errors are likely to be incurred.

Another major source of error may be that the plants or microorganisms assayed may produce ethylene in the absence of acetylene. Witty (1979) showed that such endogenous ethylene can be oxidized in soil, and is hence not apparent in control treatments without acetylene. However, acetylene inhibits this oxidation and this 'endogenously produced' ethylene will be detected during a normal ARA assay, and may be mistaken for ethylene resulting from the reduction of acetylene (nitrogenase activity). This author used radioactively labelled acetylene ($^{14}C_2H_2$) as a control, but subsequently Nohrstedt (1983) developed other control treatments which are considerably easier to apply. The two most suitable for assays of intact plant soil systems are as follows:

1. A small volume of acetylene is injected into the enclosures to produce a concentration of approximately 500 μl C_2H_2 L^{-1}. This is sufficient to inhibit ethylene oxidation, but insufficient to compete appreciably with N_2 for the reduction sites of nitrogenase. The quantity of ethylene evolved is very significant in comparison with the observed ARA, then a large proportion of

the apparent ARA is due to endogenous ethylene production and nitrogenase activity has been over-estimated by the standard assay.

2. Along with the 100–150 ml L^{-1} acetylene, normally utilized for the incubations, for this control treatment carbon monoxide is introduced to a concentration of 20 ml L^{-1}. Carbon monoxide is a potent inhibitor of nitrogenase, and Nohrstedt (1983) reported that this concentration reduced acetylene reduction by 95%.

In a study conducted at CNPAB to evaluate the ARA associated with 12 maize genotypes using an *in-situ* assay, these two control treatments were utilized (Boddey, 1987). Only minute quantitites of ethylene were evolved in the presence of 500 µl C_2H_2 L^{-1} (Table 1), and 20 ml L^{-1} of CO in the presence of 150 ml C_2H_2 L^{-1} reduced the ethylene evolution rate by between 38% and 75% (Table 2). Increasing the CO concentration to 40 ml L^{-1} increased the inhibition of C_2H_4 evolution rate to between 63% and 83%, but higher values of C_2H_4 evolution under the standard conditions were inhibited proportionately more than lower values. These results suggest that there may have been a background endogenous ethylene production of up to 20 or 40 nmol plant $^{-1}h^{-1}$, but that most of the observed ethylene evolution resulted from reduction of acetylene at least that associated with plants evolving more than 100 nmol C_2H_4 $plant^{-1}h^{-1}$.

To summarize: the most important sources of error associated with the use of the acetylene reduction technique are the interference of photosynthetic N_2-fixing bacteria, physical disturbance of the plants and endogenous ethylene production. Other sources of error are probably much less significant and were addressed in an earlier review (Boddey, 1987). Long incubations in atmospheres of acetylene (more than 5–6 h) should be avoided, as it is possible that nitrogenase could be depressed due to lack of ammonium production, although this has only been demonstrated for pure cultures of N_2-fixing bacteria (David and Fay, 1977).

After these precautions have been taken, and an *in situ* assay system has been used with suitable control treatments, what evidence exists that the AR activity detected is comparable to true inputs of BNF to plant nutrition? In the case of lowland rice, there appears to be consensus from N balance (App *et al.,* 1980, 1986) and ^{15}N dilution studies (Boddey *et al.,* 1995a; Wu *et al.,* 1995) that some long-seasoned rice varieties (e.g. IR 42) may be able to obtain as much as 20–25% of their N requirement from associated BNF. If it is assumed that a well-managed wetland rice crop without N fertilization will yield approximately 3 000 kg grain, total above-ground N accumulation will not be above 60 kg N ha^{-1}.

Table 1. Ethylene production (nmol $C_2H_4h^{-1}$) by intact soil cores of maize in an atmosphere of 150 ml L^{-1} of acetylene. Mean of four replicates. After Boddey (1987)

Maize cultivar	500 µl C_2H_2 L^{-1}	150 ml C_2H_2 L^{-1}
UnB Hs	0.8	212.7
CMS 30	8.1	210.2

Table 2 Effect of addition of carbon monoxide on the ethylene production (nmol C_2H_4 h^{-1}) of intact cores of maize in the presence of acetylene (150 ml C_2H_2 L^{-1}). Means of four replicates. After Boddey (1987)

Maize cultivar	20 ml L^{-1} CO	Without CO	40 ml L^{-1} CO	Without CO
UnB L8	45.7	74.4		
UnB L10			30.8	86.3
UnB HS			47.1	269.5
CMS 30	42.7	165.6		
CMS 36			23.7	63.3

This would suggest that at maximum rice can obtain 12–15 kg N ha^{-1} from associated BNF. The values estimated from frequent AR assays throughout the season for this crop (Watanabe *et al.*, 1978a; Boddey and Ahmad, 1981) of 5–8 kg N ha^{-1} seem to be very realistic. It should be noted that the average AR activities equivalent to these modest inputs of BNF were between 1 000 and 3 000 nmol C_2H_4 $plant^{-1}$, so that values much below 1 000 nmol $plant^{-1}h^{-1}$ for rice or other small grains are unlikely to represent any agronomically significant inputs of BNF.

Other comparisons for dryland crops are few and far between. The results of Boddey *et al.*, (1983a) and Boddey and Victoria (1986), working with the ^{15}N dilution technique, suggested inputs of BNF to the tropical grasses *Paspalum notatum* cv. batatais and *Brachiaria ruziziensis* of approximately 20 and 0 kg N ha^{-1} $year^{-1}$, respectively and acetylene reduction activity of soil cores of these grasses averaged 70 and 60 nmol C_2H_4 $core^{-1}h^{-1}$, respectively, but these activities were probably large underestimates as the cores (10 cm diameter) were assayed immediately after removal from a sandy soil where disturbance was likely to have decreased activity very considerably as described above (Wani *et al.*, 1983). In another study in the same ^{15}N-labelled soil, wheat plants, either inoculated or not with *Azospirillum brasilense*, were assayed in the same manner and presented activities between 96 and 280 nmol C_2H_4 $core^{-1}h^{-1}$ (again 'disturbed' 10 cm diameter soil cores) but no significant input of BNF was apparent from the ^{15}N data even though inoculation increased total N in the grain by over 40%. It is unfortunate that in the study of Urquiaga *et al.* (1992), where large and significant inputs of BNF were registered using ^{15}N dilution and N balance techniques, no AR assays were performed because of the difficulties of sealing such large plants in enclosures for exposure to acetylene.

Total nitrogen techniques

BNF inputs to plant/bacteria associations can be quantified by estimating gains of N to the soil/plant system by performing a total N balance. This is most easily performed in pots, or similar containers, with a known finite volume of soil that can be explored by the roots. The principle is simple: total N in the soil, seeds and any other inputs is estimated at the start of plant growth, and at final harvest soil and plant material are again analyzed for total N. If there is a net gain in N, this is regarded as being derived from the atmosphere. The problem is that the N accumulated in the plants is usually only a small proportion of the total soil N. Great accuracy is required in the sampling and analysis of the soil, so that any N gain from BNF is considerably larger than the error associated with the estimates of total soil N. This is not normally achieved with a single harvest of an annual crop, and studies which have successfully shown positive N balances have usually replanted the same crop several times in the same pots of soil. In the case of wetland rice App *et al.,* (1980) planted for successive rice crops in the same pots and demonstrated a gain in N equivalent to 18% of total plant N with the rice variety IR 28 (Table 3). In this case the authors sampled the soil by suspending the soil in water and taking subsamples using a pipette to obtain representative samples. Furthermore, in the analyses of total N, these authors included aliquots of a standard subsample in each batch of analyses (one batch refers to samples digested together in the same aluminium digestion block) to eliminate variation between batches. Another precaution utilized by Lima *et al.,* (1987) was always to include initial and final soil samples in the same batch, to eliminate the possibility of differences between batches being confused with gains and losses from the soil during plant growth. It is evident that, while no very sophisticated techniques are required for this kind of study, a great deal of very meticulous sampling and careful analysis is required, which is probably why so few studies have been performed.

One study which failed to show any significant total N gain during plant growth was the total N balance on four species reported by Loureiro and Boddey (1988 – Table 4). The pots were filled with soil which had been mixed uniformly, repetition by repetition, in a concrete mixer, and the coefficient of variation of the total soil N estimates was approximately 5%. The plants accumulated in five harvests over a 14-month period, N equivalent to approximately 12% of the N content of the soil, and the N gains (not significant at $p < 0.05$) were for *Brachiaria humidicola* and *B. decumbens*, respectively, equivalent to 48% and 38% of the total N accumulated by the plants. Even though the study was conducted with all the precautions deemed necessary to diminish sampling and analytical errors, the N gain was not significantly larger than the errors associated with the estimates of initial and final soil N.

Table 3. Total N balance on wetland rice variety IR 28 grown in pots of heavy clay soil 5 kg soil (two plants per pot) over four consecutive plantings. After App *et al.,* (1980)

Treatment	A Total N in plant	B N in seeds, insecticide, etc.	C Total N (initial)	D in Soil (final)	E N removed lost from soil (B–C)	F N Balance (A–B–E)
Planted	997	21	5 971	5 155	795	+181*
Unplanted	0	0	5 950	5 707	243	–243**

*Gain or loss of N in soil/plant systems significant at $p < 0.05$.

Table 4. Total N balance on five species of *Brachiaria* grown in pots containing 10 kg of a sandy N-deficient soil. Values are means of five replicates. Data from Loureiro and Boddey (1988) (mg N pot^{-1})

Species	Total soil N at final harvest	Total plant N (five harvests)	Total soil N at planting	Total N in soil/plant system at final harvest	N balance
B. decumbens	2 832	324	3 032	3 156	+124a
B. humidicola	2 869	379	3 066	3 247	+181a
B. arrecta	2 806	329	3 089	3 135	+46a
B. ruziziensis	2 817	331	3 133	3 149	+16a
Unplanted	2 788	–	3 107	2 788	–319b
CV(%)	4.78 n.s.	11.2 n.s.	5.33 n.s.	4.36*	

n.s. Differences between means not significant at $p < 0.05$;
* Difference between means significant at $p < 0.05$.

Obviously, where BNF inputs are larger, detection of the N gain is considerably easier. In the study on four varieties of sugar cane growth in pots of 64 kg soil, reported by Lima *et al.*, (1987), the plants were harvested after 12 (plant crop) and 22 months (ratoon crop). The error in the estimation of total soil N was marginally higher in this study compared with that of Loureiro and Boddey (1988), but over the whole experiment the N accumulated by the plants was much greater than in the study on *Brachiaria* spp., being the equivalent of between 24% and 65% of the total N in the soil at the start of the experiment (Table 5).

Table 5. Total N balance on four cultivars of sugar cane grown for 21 months in pots containing 64 kg of soil (Typic hapludulf). Means of five replicates. After Lima *et al.*, (1987)

	Nitrogen in				
		Soil			
Cultivar	Plant material[1]	Initial	Final	Loss[2]	Balance[3]
CB 47–89	34.8	53.6	46.8	6.8	+25.3
CB 47–355	16.5	49.7	44.5	4.9	+9.0
IAC 52–150	13.5	52.7	45.1	7.6	+3.2
NA 56–79	13.0	54.2	45.8	8.4	+2.0
Unplanted	–	51.1	44.1	7.0	–9.7
HSD[4] ($p = 0.05$)	12.9	8.5	6.4	7.8	13.7
Coefficient of variation (%)	36.8	7.7	6.6	49.7	*(6.4)*[5]

[1] Total N accumulated in plant crop and first ratoon, including roots.
[2] Loss of N from soil.
[3] N balance = (final N in soil + N in plant material) – (initial N in soil + 2.65 g N in urea fertilizer).
[4] Honest significant difference (Tukey; $p < 0.05$).
[5] Standard deviation.

For this reason the total N balance for the variety CB 47–89 was very significantly positive even though plant-to-plant variability was very high (coefficient of variation 37%) due to the practical restriction of having to use one plant per pot because of their large size. These data constitute unequivocal evidence of significant inputs of BNF to at least this variety of sugar cane.

Studies in larger containers than pots, such as the concrete tank used by Urquiaga *et al.,* (1992), lead to larger volumes of soil to be explored by roots of each plant and hence the ratio of N accumulated by the plant to that in the soil is lower. In this case careful sampling of the soil resulted in coefficients of variation associated with the estimates of total soil N of approximately 5%, and the total N accumulated by the different sugar cane varieties over a period of 3 years of 1.4–12.0% of total soil N (11.6–102.8 g N m^{-2}, equivalent to 116–1 028 kg N ha^{-1}). In this study significant ($p < 0.05$) gains in the soil-plant system were recorded for six of the sugar cane varieties (CB 45–79, CB 45–3, IAC 52–150, SP 70–1143, SP 79–2312 and Krakatau).

The application of this technique to plants grown in artificial media such as soil, gravel or vermiculite, of very low N content, is considerably easier, although plant growth is usually very different from soil-grown plants. Although N content of such growth media is very low, it cannot necessarily be assumed to be zero, and should be quantified before and after cropping as in studies using soil. Vermiculite has been used by several authors (Döbereiner and Day, 1975; Purchase, 1978; Dart and Wani, 1982) as this is a medium with very good properties for plant growth, but the results of Giller *et al.,* (1986) suggest that plants can remove N from interstitial sites in the 2:1 lattice of this mineral which is not removed by vigorous washing. Furthermore, N additions in seed must be taken into account, and any added bacterial inocula must be applied equally to all treatments, using dead cells in case of the control treatment.

Finally, in all these studies, whether performed in soil or low-N media, much more confidence can be placed in the conclusion that BNF inputs were significant if a control treatment with a plant known, or thought, to be unable to benefit from associated BNF is utilized. This can be an uninoculated control in the case of experiments performed in monoxenic conditions (Döbereiner and Day, 1975; Purchase, 1978) or a range of varieties of the same species in the case of experiments performed in soil (Lima *et al.,* 1987; Urquiaga *et al.,* 1992).

^{15}N isotope techniques

$^{15}N_2$ gas exposure

The incorporation of labelled N_2 gas into plants growing in an atmosphere containing ^{15}N-enriched N_2 gas can be regarded as definitive proof that the plants obtained a significant contribution from associated BNF. The only prerequisites for this conclusion are that the labelled N_2 was not contaminated with other forms of ^{15}N-labelled compounds such as $^{15}NH_3$, and that the ^{15}N enrichment of the plant after exposure was significantly higher that possible variations in the natural abundance of ^{15}N in plants not exposed to $^{15}N_2$. Purification of the $^{15}N_2$ gas can be achieved by passing it through a trap of liquid N_2, or by other means (De Polli *et al.*, 1977; Ohyama and Kumazawa, 1981).

The greatest problems associated with the use of this technique are those associated with enclosing the plants in the $^{15}N_2$-labelled atmosphere. If only roots are exposed to the $^{15}N_2$ then plant growth can continue normally, but there is usually considerable loss of labelled gas through the root/shoot interface or, more importantly, unlabelled N_2 from the surrounding air dilutes the enrichment of the $^{15}N_2$ within the enclosure. This is especially severe for wetland rice which has a well-developed aerenchyma (Yoshida and Yoneyama, 1980; Lee *et al.*, 1981).

Enclosing the entire plants in a labelled atmosphere means that if CO_2 is not replenished it disappears and photosynthesis ceases within minutes (Witty and Day, 1978). To maintain active growth when plants are enclosed in chambers for more than a few hours requires control of CO_2, O_2, transpiration, light and temperature, and such sophisticated control systems were used by Ito *et al.*, (1980) and Eskew *et al.*, (1981) for studies on wetland rice. Even then there was appreciable dilution of the $^{15}N_2$ unenriched N_2 leaking in from surrounding air.

To date there are no reports of plants exposed to ^{15}N-enriched atmospheres for more than 2 weeks, and the technique is probably best used for short-term exposures to highly enriched atmospheres using simple apparatus such as that used by Ruschel *et al.*, (1975) or De Polli *et al.*, (1977), who demonstrated significant incorporation of biologically fixed N_2 to sugar cane and *Digitaria decumbens* plants, respectively. However, such studies using this technique give little idea of the agronomic significance of such BNF inputs under field conditions.

The ^{15}N isotope dilution technique

The isotope dilution (ID) technique functions in a converse manner to the ^{15}N-enriched atmosphere technique. With the ID technique for the plant-available N in the substrate (soil or sand etc.) is labelled with ^{15}N and the quantification of BNF is achieved by estimating the degree of dilution of the ^{15}N enrichment of the N derived from the substrate. If a substrate such as washed sand or gravel is used where there is no source of N other than that applied as ^{15}N labelled nitrate, then theoretically a plant growing in this medium which obtains no BNF contribution will have a ^{15}N enrichment close to that of the labelled N added to the substrate, only being diluted by N derived from the seeds of any bacterial inoculant. If seeds and inoculant are also labelled with ^{15}N at the same enrichment as the added mineral N, then the plant at harvest will also show the same ^{15}N enrichment if there is no BNF contribution. This strategy was utilized by Lethbridge and Davidson (1983), who recorded no significant input (less than 5% of plant N) of BNF to several wheat cultivars plants grown in acid-washed sand inoculated with *Azospirillum brasilense* and other diazotrophs.

Most studies using sand or gravel as a culture medium have been performed in sterile (monoxenic) conditions. The problem here is that gases diffuse very easily in such substrates and, in contrast to non-sterile soil which has a continuous consumption of O_2 by the soil microflora feeding on soil organic matter, there is almost no O_2 consumption in the substrate, and pO_2 in the rhizosphere may be much higher than that experienced under field conditions. However, Malik and Zafar (1985) reported very significant contributions of BNF to monoxenic kallar grass (*Leptochloa fusca*) seedlings inoculated with *A. brasilense* (DSM 1691) and other diazotrophs, grown in gravel in glass tubes (20 × 3 cm) and amended with 500 µg N $tube^{-1}$ as $NH_4(SO_4)_2$ with a ^{15}N enrichment of 15.0 atom% excess. The sterile control plants (inoculated with autoclaved cells) accumulated a total of 630 µg of N, of which 216 µg were unlabelled (N in seeds, heat-treated inoculum and growth medium). The plants inoculated with *A. brasilense* accumulated a total of 1 310 µg N, of which 877 µg were unlabelled; there is little doubt that the seedlings obtained a significant contribution from BNF.

In such studies it is essential that a non-inoculated, or non-N_2-fixing, control crop is included in the design, to serve as an indicator of N contamination in the growth substrate and nutrient solution and, if inocula are used, then all plants receive the same quantity of N in the inoculum (sterilized where appropriate). Vermiculite and other 2:1 lattice substrates should be used with great caution, as evidence suggests that they are able to liberate ammonium possibly by exchange with potassium in the nutrient solution (Giller *et al.,* 1986). If inoculated plants do not accumulate significantly more N than non-inoculated controls, it is unlikely that there were significant BNF contributions even if there was a significantly lower ^{15}N enrichment in the former (Rennie, 1980, Haahtela and Kari, 1986), and it is then necessary to explain why inoculated plants were less able to absorb labelled mineral N from the media than were the controls.

When the isotope dilution technique is applied to plants grown in soil, either in pots or the field, the most serious problem is estimating the ^{15}N enrichment of the N derived from the soil (or soil + labelled fertilizer). If ^{15}N-labelled fertilizer is added to the surface of the soil, even if it is distributed evenly over the area, there will be large differences in ^{15}N enrichment in the soil mineral N with depth and with time (Fried *et al.,* 1983; Boddey and Victoria, 1986; Smith *et al.,* 1992). The decrease with time is due to continuous liberation of mineral N due to mineralization of soil organic matter. Thus it is not possible to make a single sampling of the mineral N and estimate the ^{15}N enrichment of the N available to the putative 'N^2-fixing' plant treatment. The strategy most often adopted in this case is to use a non-N_2-fixing crop as closely related as possible to the 'N_2-fixing' plant. If this control crop has an identical root distribution and temporal uptake pattern of soil N, then the ^{15}N enrichment in the N it accumulates will be the same as that removed from the soil by the 'N_2-fixing' plant. However, there is no practical method available to accurately determine the equality of patterns of soil N uptake if one of the plants obtains contributions from BNF (Boddey, 1987).

It is well known that the inoculation of cereals and other crops with *Azospirillum* spp., and all other diazotrophs so far studied, alters their rooting pattern (Brown *et al.*, 1962; Gaskins *et al.*, 1985; Okon and Hadar, 1987; Yahalom *et al.*, 1990) so that non-inoculated controls are almost certain to accumulate N with a different ^{15}N enrichment to that of the inoculated plants, and this has led to many dubious claims of significant BNF inputs to wheat and other crops under field conditions (de Freitas *et al.*, 1984; Rennie and Thomas, 1987; Kucey, 1988).

Some studies on legumes, in which several different non-N_2-fixing control crops were utilized, indicate the magnitude of possible errors that can be experienced in such open field studies. Several such studies were reviewed by Boddey (1987), and differences in ^{15}N enrichment between different control crops were frequently up to 30% or even more. In a recent study at CNPAB using several different non-N_2-fixing control crops (dryland rice, okra, non-nod soybean and sorghum) grown in soil labelled at planting with a single dose of ^{15}N-labelled ammonium sulphate, it was found that non-nod soybean accumulated N with a ^{15}N enrichment of 0.1491 atom% ^{15}N excess, while sorghum showed an enrichment of only 0.1012 atom% ^{15}N excess (Figure 4 – Boddey *et al.*, 1995b). If the lower ^{15}N enrichment of the sorghum were interpreted as due to a BNF contribution, then we would conclude that approximately 32% of the N in the sorghum was derived from BNF. Most authors consider these differences are mainly due to differences between crops in their N uptake patterns with time (Rennie, 1986; Danso, 1988; Witty *et al.*, 1988; Witty and Giller, 1991) although differences in rooting depth may be just as important (Boller and Nosberger, 1988; Boddey *et al.*, 1996). It can be concluded from the data available in the literature that the ^{15}N enrichment of different non-N_2-fixing crops grown in open field plots labelled with ^{15}N can vary considerably even when slow-release forms of labelled N are utilized (Witty, 1983; Viera-Vargas *et al.*, 1995) and only if such ID studies indicate inputs

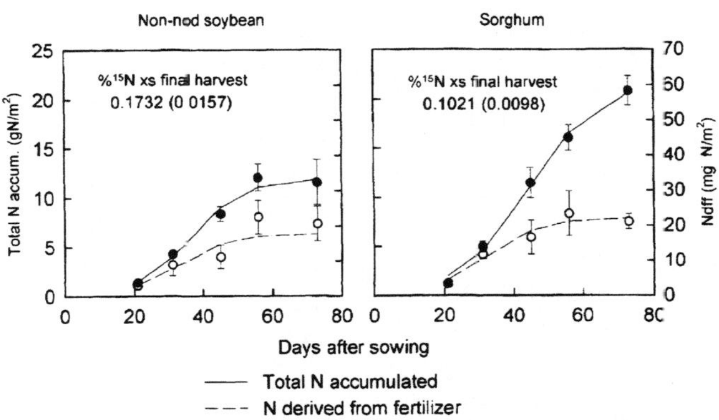

Figure 4. Accumulation of labelled and total N by two non-N_2-fixing control crops (non-nod soybean and sorghum) in soil amended with 2.0 kg N ha^{-1} of ^{15}N-labelled ammonium sulphate (21.3 atom% ^{15}N) at planting. Error bars indicate standard errors of the means (four replicates). From Boddey *et al.*, (1995b).

of above 30–40% of plant N derived from BNF, can much confidence be given to the estimates.

The solution to this problem of the interaction of different plant N uptake patterns with variations in soil ^{15}N enrichment with time and depth, is to utilize containers of finite depth which are labelled uniformly with ^{15}N in time and space. This approach has been used in several studies performed at CNPAB, and led to the detection of significant inputs of BNF to several pasture grasses, even where contributions were considerably less than 40% of plant N (Boddey *et al.,* 1983b; Boddey and Victoria, 1986; Marinda and Boddey, 1987; Miranda *et al.,* 1990). In these studies plants were grown in concrete cylinders (60 cm diameter, 50 cm depth) with a synthetic cloth at the bottom to act as a barrier to roots, or in pots. The uniform labelling of the plant-available N in the soil with ^{15}N is not straightforward. In the study to quantify BNF contributions to three *Brachiaria* species in which ^{15}N-labelled organic matter was initially incorporated uniformly in the whole depth of the cylinders (Boddey and Victoria, 1986), almost 18 months passed before the soil ^{15}N label approached stability (Figure 5) and the ^{15}N enrichment in the soil was found to increase with depth (Table 6). This was attributed to faster mineralization of soil organic matter near the soil surface, probably as this layer was subjected to more severe wetting and drying cycles than would occur at depth. However, as the control crop (*B. arrecta*, syn. *B. radicans*) had a more superficial root distribution than either the *B. decumbens* or the *B. humidicola*, the lower ^{15}N enrichment of the latter two species during the period, when the ^{15}N enrichment of the soil N was almost stabilized, was regarded as firm evidence of significant BNF contributions.

More recently studies have been conducted to quantify BNF contributions to different wetland rice varieties. In a series of experiments performed at CNPAB in a large (20 × 6 × 0.6 m) concrete tank filled with approximately 80 tons of soil mixed with ^{15}N-labelled organic matter, for the first two crops the variety IR 42, amongst others showed the lowest ^{15}N enrichment, suggesting BNF inputs of up to 30% of plant N (Boddey *et al.,* 1995a). However, the soil did not have a stable ^{15}N enrichment at this time, so these estimates could not be described as

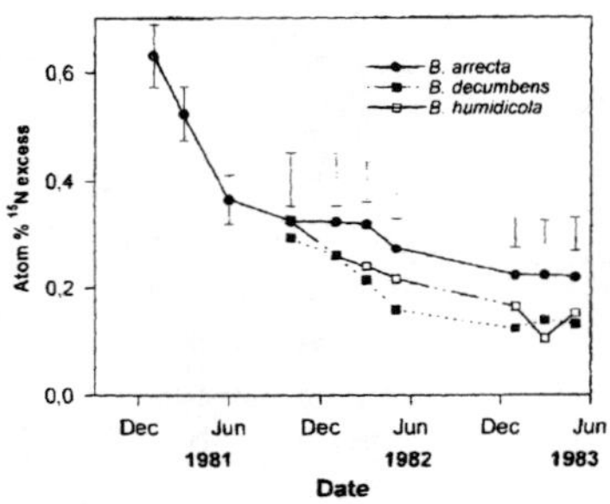

Figure 5. N-enrichment in aerial tissue of *Brachiaria arrecta, B. humidicola* and *B. decumbens* grown in soil amended with ^{15}N-labelled organic matter. Data are means of four replicates. Error bars indicate honest significant differences at $p < 0.05$ (Tukey).

Table 6. ^{15}N enrichment of N[1] mineralized from three layers of soil 30 months after addition of ^{15}N-labelled organic matter mixed uniformly to a depth of 50 cm and planted to two cultivars of *Paspalum notatum* or four species of *Brachiaria*. Means of four replicate samples. After Boddey and Victoria (1986)

	^{15}N enrichment of mineralized N (atom% ^{15}N excess)			
	Depth (cm)			
Cultivar/*species*	0–15	15–30	30–45	Mean
Batatais	0.208	0.227	0.225	0.220[AB]
Pensacola	0.253	0.234	0.257	0.248[A]
B. ruzizensis	0.241	0.233	0.259	0.244[AB]
B. arrecta	0.230	0.234	0.271	0.245[A]
B. humidicola	0.219	0.231	0.247	0.232[AB]
B. decumbens	0.210	0.200	0.214	0.205[B]
Mean	0.225b	0.226b	0.245[a]	
LSD (depth)		0.014		

[1] Determined using the methods of Cassman and Munns (1980)
[2] Least significant difference at $p < 0.05$ (Tukey).
[3] Means followed by the same letter are not significantly different at $p < 0.05$.

reliable. Wu *et al.*, (1995). working in the Philippines, used a similar strategy to select rice varieties for BNF using soil thoroughly mixed with ^{15}N-labelled ammonium sulphate 6 months prior to planting. The results of this study also showed that the variety IR-42 obtained 23% more N derived from BNF than the Palawan variety used as control. The fact that the ^{15}N enrichment of the soil mineral N was 0.294 atom% ^{15}N excess at the start of the pot experiment, and the low N_2-fixing control variety Palawan obtained a mean of 0.246 atom% ^{15}N excess during growth, shows that the fall in ^{15}N enrichment during crop growth must have been minimal. and the quantifiation of BNF can be considered as accurate.

In conclusion, therefore, it is virtually essential to apply the isotope dilution technique in soil contained in pots or other larger containers with a finite rooting depth. Time should be given for ^{15}N-enriched materials to equlibrate with the native soil organic matter, and even then it may be necessary to remix the soil in the containers before the soil can be considered of uniform ^{15}N enrichment with time and depth. The most reliable estimates of potential BNF contributions have been obtained using this strategy under conditions which are close to those experienced in the field both in the aforementioned studies on tropical pasture grasses and with wetland rice (Wu *et al.*, 1995) and with sugar cane (Urquiaga *et al.*, 1992).

References

App A A, Watanabe I, Alexander M, Ventura W, Daez C, Santiago T, De Datta S K 1980. Non-symbiotic nitrogen fixation associated with the rice plant in flooded soils. Soil Sci., 130, 283–289.

App A A, Watanabe I, Ventura T S, Bravo M, Jurey C D 1986. The effect of cultivated and wild rice varieties on the nitrogen balance of flooded soil. Soil Sci., 141, 448–452.

Balandreau J 1975. Mesure de l'activite nitrogenasique des microorganismes fixateurs libres d'azote de la rhizosphere de quelques graminees. Rev. Ecol Biol. Sol., 12, 273–290.
Balandreau J, Dommergues Y 1971. Mesure in situ l'activite nitrogenasique. C.R. Acad. Sci., 273, 2020–2023.
Balandreau J, Dommergues Y R 1973. Assaying nitrogenase (C_2H_2) activity in the field. In: Rosswall T, ed. Modern methods in the study of Microbial Ecology. Stockholm: Swedish Natural Science Research Council (NIR), 1973. (Bulletins from the Ecological Research Committee, 17).
Balandreau J P, Millier C R, Dommergues Y R 1974. Diurnal variations of nitrogenase activity in the field. Appl. Microbiol., 27, 662–665.
Boddey R M 1987. Methods for the quantification of nitrogen fixation associated with gramineae. CRC Crit. Rev. Plant Sci., 6, 209–266.
Boddey R M, Ahmad N 1981. Seasonal variations in nitrogenase activity of various rice varieties measured with an *in situ* acetylene reduction technique in the field. In: Vose P B, Ruschel A P, eds. Associative N_2 fixation, vol. 2. CRC Press, Boca Raton, FL, 219–229.
Boddey R M, Victoria R L 1986. Estimation of biological nitrogen fixation associated with *Brachiaria* and *Paspalum* grasses using ^{15}N-labelled organic matter and fertilizer. Plant Soil, 90, 265–292.
Boddey R M, Quilt P, Ahmad N 1978. Acetylene reduction in the rhizosphere of rice: methods of assay. Plant Soil, 50, 567–574.
Boddey R M, Chalk P M, Victoria R L, Matsui E 1983a. The ^{15}N-isotope dilution technique applied to the estimation of biological nitrogen fixation associated with *Paspalum notatum* cv. batatais in the field. Soil Biol. Biochem, 15, 25–32.
Boddey R M, Chalk P M, Victoria R L, Matsui E, Döbereiner J 1983b. The use of the ^{15}N isotope dilution technique to estimate the contribution of associated biological nitrogen fixation to the nitrogen nutrition of *Paspalum notatum* cv. batatais. Can. J. Microbiol., 29, 1036–1045.
Boddey R M, Baldani V L D, Baldani J I, Döbereiner J 1986. Effect of inoculation of *Azospirillum* spp. on nitrogen accumulation by field-grown wheat. Plant Soil, 95, 109–121.
Boddey R M, Oliveira O C de, Urquiaga S, Reis V M, Olivares F L, Baldani V L D, Dobereiner J 1995a. Biological nitrogen fixation associated with sugar cane and rice: contributions and prospects for improvement. Plant Soil, 174, 195–209.
Boddey R M, Oliveira O C de, Alves B J R, Urquiaga S 1995b. Field application of the ^{15}N isotope dilution technique for the reliable quantification of plant-associated biological nitrogen fixation. Fert. Res., 42, 77–87.
Boddey R M, Müller S, Alves B J R 1996. Estimation of biological N_2 fixation to two *Phaseolus vulgaris* genotypes using simulation of plant nitrogen uptake from ^{15}N-labelled soil. Fert. Res., 45, 169–185.
Boller B C, Nosberger J 1988. Influence of dissimilarities in temporal and spatial N-uptake patterns on ^{15}N-based estimates of fixation and transfer of N in ryegrass-clover mixtures. Plant Soil, 112, 167–175.
Brown M E, Burlingham S K, Jackson R M 1962. Studies on *Azotobacter* species in soil. II. Populations of *Azotobacter* in the rhizosphere and effects of artificial inoculation. Plant Soil, 17, 320–332.
Burris R H 1974. Methodology. In: Quispel A, ed. The Biology of Nitrogen Fixation. North-Holland, Amsterdam, 9–33.
Cassman K G, Munns D N 1980. Nitrogen mineralization as affected by soil moisture, temperature, and depth. Soil Sci. Soc. Am. J., 44, 1233–1237.
Danso S K A 1988. The use of ^{15}N enriched fertilizers for estimating nitrogen fixation in grain and pasture legumes. In: Beck D P, Materon L A, eds. Nitrogen Fixation by Legumes in Mediterranean Agriculture. Martinus Nijhoff, Dordrecht, 345–348.
Dart P J, Wani S P 1982. Non-symbiotic nitrogen fixation and soil fertility. In: Non-Symbiotic Nitrogen Fixation and Organic Matter in the Tropics. Symposium Papers I, Transactions, 12th International Congress on Soil Science, New Delhi, 3–27.
David K A V, Fay P 1977. Effects of long-term treatment with acetylene on nitrogen-fixing microrganisms. Appl. Environ. Microbiol., 34, 640–646.
de Freitas J R, Victoria R L, Ruschel A P, Vose P B 1984. Estimation of N_2 fixation by sugarcane, *Saccarum* sp. and soybean, *Glycine max*, grown in soil with ^{15}N labelled organic matter. Plant Soil, 82, 257–261.

de Polli H, Matsui E, Dobereiner J, Salati E 1977. Confirmation of nitrogen fixation in two tropical grasses by $^{15}N_2$ incorporation. Soil Biol. Biochem., 9, 119–123.

Dilworth M J 1966. Acetylene reduction by nitrogen fixing preparations of *Clostridium pasteurianum*. Biochem. Biophys. Acta, 127, 285–294.

Dobereiner J, Day J M 1975. Associative symbioses in tropical grasses: characterization of microorganisms and nitrogen-fixing sites. In: Newton W E, Nyman C J, Proceedings of the 1st International Symposium on Nitrogen Fixation. Washington State University Press, Pullman, 518–538.

Dobereiner J, Day J M, Dart P J 1972a. Nitrogenase activity and oxygen sensitivity of the *Paspalum notatum-Azotobacter paspali* association. J. Gen. Microbiol., 71, 103–116.

Dobereiner J, Day J M, Dart P J 1972b. Nitrogenase activity in the rhizosphere of sugar cane and some other tropical grasses. Plant Soil, 37, 191–196.

Eskew D L, Eaglesham A R J, App A A 1981. Heterotrophic $^{15}N_2$ fixation and distribution of newly fixed nitrogen in a rice-flooded soil system. Plant Physiol., 68, 48–52.

Fried M. Danso S K, Zapata F 1983. The methodology of measurement of N_2 fixation by non-legumes as inferred from field experiments with legumes. Can. J. Microbiol., 29, 1053–1062.

Gaskins M H, Albrecht S L, Hubbell D H 1985. Rhizosphere bacteria and their use to increase plant productivity. Agric. Ecosyst. Environ., 12, 99–116.

Giller K E, Wani S P, Day J M 1986. Use of isotope dilution to measure nitrogen fixation associated with the roots of sorghum and millet genotypes. Plant Soil, 90, 255–263.

Haahtela K, Kari K 1986. The role of root-associated *Klebsiella pneumoniae* in the nitrogen nutrition of *Poa pratensis* and *Triticum aestivum* as estimated by the method of ^{15}N isotope dilution. Plant Soil, 90, 245–254.

Ito O, Cabrera D, Watanabe I 1980. Fixation of dinitrogen-15 associated with rice plants. Appl. Environ. Microbiol., 39, 554–558.

Kapulnik Y, Feldman M, Okon Y, Henis Y 1985. Contribution of nitrogen fixed by *Azospirillum* to the N nutrition of spring wheat in Israel. Soil Biol. Biochem., 17, 509–515.

Koch B L 1977. Associative nitrogenase activity by some Hawaiian grass roots. Plant Soil, 47, 703–706.

Kucey R M N 1988. Alteration of size of wheat root systems and nitrogen fixation by associative nitrogen-fixing bacteria measured under field conditions. Can. J. Microbiol., 34, 735–739.

Lee K K, Yoshida T 1977. An assay technique of measurement of nitrogenase activity, in root zone of rice for varietal screening by the acetylene reduction method. Plant Soil, 46, 127–134.

Lee K K, Castro T, Yoshida T 1977. Nitrogen fixation throughout growth, and varietal differences in nitrogen fixation by the rhizosphere of rice planted in pots. Plant Soil, 48, 613–619.

Lee K K, Holst I, Watanabe I, App A A 1981. Gas transport through rice. Soil Sci. Plant Nutr., 27, 151–158.

Lethbridge G, Davidson M W 1983. Root-associated nitrogen-fixing bacteria and their role in the nitrogen nutrition of wheat estimated by ^{15}N isotope dilution. Soil Biol. Biochem., 15, 365–374.

Lima E, Boddey R M, Döbereiner J 1987. Quantification of biological nitrogen fixation associated with sugar cane using a ^{15}N-aided nitrogen balance. Soil Biol. Biochem., 19, 165–170.

Loureiro M de F, Boddey R M 1988. Balanço de nitrogênio em quatro gramíneas do gênero *Brachiaria*. Pesq. Agropec. Bras., 23, 1343–1353.

Malik K A, Zafar Y 1985. Quantification of root associated nitrogen fixation in kallar grass as estimated by ^{15}N isotope dilution. In: Malik K A, Naqui S H M, Aleem M I H, eds. Nitrogen and the Environment. Nuclear Institute of Agriculture and Biology, Faisalabad, Pakistan, 161–171.

Miranda C H B, Boddey R M 1987. Estimation of biological nitrogen fixation associated with 11 ecotypes of *Planicum maximum* grown in nitrogen-15 labeled soil. Agron. J., 79, 558–563.

Miranda C H B, Urquiaga S, Boddey R M 1990. Selection of ecotypes of *Panicum maximum* for associated biological nitrogen fixation using the ^{15}N isotope dilution technique. Soil Biol. Biochem., 22, 657–663.

Nohrstedt H O 1983. Natural formation of ethylene in forest soils and methods to correct results given by the acetylene reduction assay. Soil Biol. Biochem., 15, 281–286.

Ohyama T, Kumazawa K 1981. A simple method for the preparation, purification and storage of $^{15}N_2$ gas for biological nitrogen fixation studies. Soil Sci. Plant Nutr., 27, 263–265.

Okon Y, Hadar Y 1987. Microbial inoculants as crop-yield enhancers. CRC Crit. Rev. Biotechnol. 6, 61–85.

Patriquin D G 1978. Factors affecting nitrogenase activity (acetylene-reducing activity) associated with excised roots of the emergent halophyte *Spartina alterniflora* Loisel. Aquatic Bot., 4, 193–210.

Patriquin D G, Keddy C 1978. Nitrogenase activity (acetylene reduction) in a Nova Scotian salt marsh: its association with angiosperms and the influence of some edaphic factors. Aquatic Bot., 4, 227–244.

Purchase B S 1978. Nitrogen fixation associated with grasses. A potential source of nitrogen for Rhodesian agriculture. Rhod. Agric. J., 75, 99–104.

Rennie R J 1980. ^{15}N-isotope dilution as a measure of dinitrogen fixation by *Azospirillum brasilense* associated with maize. Can. J. Bot., 58, 21–24.

Rennie R J 1986. Advantages and disadvantages of nitrogen-15 isotope dilution to quantify dinitrogen fixation in field-grown legumes – a critique. In: Field Measurement of Dinitrogen Fixation and Denitrification, Soil Science Society of America, Special Publication, American Society of Agronomy, Madison, WI, 43–58.

Rennie R J, Thomas J B 1987. ^{15}N determined effect of inoculation with N_2 fixing bacteria on nitrogen assimilation in Western Canadian wheats. Plant Soil., 100, 213–223.

Rinaudo G, Dommergues Y 1971. Validité de l'estimation de la fixation biologique de l'azote dans la rhizosphere par la méthode de réduction de l'acétylène. Ann. Inst. Pasteur, 121, 93–99.

Ruschel A P, Henis Y, Salati E 1975. Nitrogen-15 tracing of N-fixation with soil-grown sugar cane seedlings. Soil Biol. Biochem, 7, 181–182.

Schöllhorn R, Burris R H 1966. Study of intermediates in nitrogen fixation. Fed. Proc., 24, 710.

Smith C J, Chalk P M, Hamilton S D, Hopmans P 1992. Estimating N_2 fixation by field-grown lupins (*Lupinus angustifolius* L.) using soil and plant ^{15}N enrichment. Biol. Fertil. Soils, 13, 235–241.

Tjepkema J, van Berkum P 1977. Acetylene reduction by soil cores of maize and sorghum in Brazil. Appl. Environ. Microbiol., 33, 626–629.

Urquiaga S, Cruz K H S, Boddey R M 1992. Contribution of nitrogen fixation to sugar cane: nitrogen-15 and nitrogen balance estimates. Soil Sci. Soc. Am. J., 56, 105–114.

van Berkum P, Bohlool B B 1980. Evaluation of nitrogen fixation by bacteria in association with roots of tropical grasses. Microbiol. Rev., 44, 491–517.

van Berkum P and Day J M 1980. Nitrogenase activity associated with soil cores of grasses in Brazil. Soil. Biol. Biochem., 12, 137–140.

Viera-Vargas M S, Souto S M, Urquiaga S, Boddey R M 1995. Quantification of the contribution of nitrogen fixation to tropical forage legumes and transfer to associated grass. Soil Biol. Biochem., 27, 1193–1200.

Wani S P, Dart P J, Upadhyaya M N 1983. Factors affecting nitrogenase activity (C_2H_2reduction) associated with sorghum and millet estimated using the soil core assay. Can. J. Microbiol., 29, 1063–1069.

Watanabe I, Lee K K, Alimagno B V 1978a. Seasonal change of N_2-fixing rate in rice field assayed by *in situ* acetylene reduction technique I. Experiments in long-term fertility plots. Soil Sci. Plant Nutr., 24, 1–13.

Watanabe I, Lee K, Guzman M de 1978b. Seasonal change of N_2 fixing rate in rice field assayed by *in situ* acetylene reduction technique II. Estimate of nitrogen fixation associated with rice plants. Soil Sci. Plant Nutr., 24, 465–471.

Witty J F 1979. Acetylene reduction assay can overestimate nitrogen fixation in soil. Soil. Biol. Biochem., 11, 209–210.

Witty J F 1983. Estimating N_2 fixation in the field using ^{15}N-labelled fertilizer: some problems and solutions. Soil Biol. Biochem., 15, 631–639.

Witty J F, Day J M 1978. Use of $^{15}N_2$ in evaluating asymbiotic N_2 fixation. In: Isotopes in Biological Nitrogen Fixation. International Atomic Energy Agency, Vienna, 135–150.

Witty J F, Giller K E 1991. Evaluation of errors in the measurement of biological nitrogen fixation using ^{15}N fertilizer. In: Stable Isotopes in Plant Nutrition, Soil Fertility and Environmental Studies. International Atomic Energy Agency, Vienna, 59–72.

Witty J F, Rennie R J, Atkins C A 1988. ^{15}N addition methods for assessing N_2 fixation under field conditions. In: Summerfield R J, ed. World Crops: Cool Season Food Legumes. Kluwer, Dordrecht, 715–730.

Wu P, Zhang G, Ladha J K, McCouch S R, Huang N 1995. Molecular-marker facilitated investigation on the ability to stimulate N_2 fixation in the rhizosphere by irrigated rice plants. Theor. Appl. Genet., 91, 1177–1183.

Yahalom E, Okon Y, Dovrat A 1990. Possible mode of action of *Azospirillum brasilense* strain Cd on the root morphology and nodule formation in burr medic (*Medicago polymorpha*). Can. J. Microbiol., 36, 10–14.

Yoshida T, Ancajas R R 1971. Nitrogen fixation by bacteria in the root zone of rice. Soil Sci. Soc. Am. Proc., 35, 156–157.

Yoshida T, Ancajas R R 1973. Nitrogen-fixing activity in upland and flooded rice fields. Soil Sci. Soc. Am. Proc., 37, 42–46.

Yoshida T, Yoneyama T 1980. Atmospheric dinitrogen fixation in the flooded rice rhizosphere as determined by the N-15 isotope technique. Soil Sci. Plant Nutr., 26, 551–560.

36
Use of ^{15}N isotope dilution method to quantify nitrogen fixation in legumes and its potential use for non-legumes

GUDNI HARDARSON AND CHRISTIAN HERA

Abstract

The ^{15}N isotope dilution method is potentially accurate for measuring N_2 fixation in plants. In using the method one has to ensure that the isotope ratio in the plant reflects the integrated ratio in the soil, which can be variable in time and with soil depth. The errors due to the use of an inappropriate reference plant are much higher at low levels of fixation, and these decrease with increasing rates of fixation. The ^{15}N isotope dilution method may be suitable for quantifying associative N_2 fixation, especially if the experimental error is low, or where N_2 fixation is high. Various aspects affecting the accuracy of the method will be discussed, especially potential errors of measurements of nitrogen fixation in non-legumes by associative or endophytic diazotrophs.

Introduction

The Food and Agriculture Organization of the United Nations and the International Atomic Energy Agency, through their Joint FAO/IAEA Division in Vienna and the FAO/IAEA Agriculture and Biotechnology Laboratory in Seibersdorf, Austria, have co-ordinated international programmes on biological nitrogen fixation (BNF) in developing countries for more than two decades. The main objectives of these programmes have been to enhance N_2 fixation in various cropping systems and to develop and optimize the use of the ^{15}N isotope dilution method to measure fixation in leguminous crops. A review of these international FAO/IAEA programmes on BNF has been published recently (Hardarson, 1994).

In the above-mentioned FAO/IAEA programmes considerable experience has been collected on the ^{15}N isotope dilution method and on the practical problems in its use for estimating N_2 fixation (Danso *et al.*, 1993).

K.A. Malik et al. (eds.), Nitrogen Fixation with Non-Legumes, 307–312.

This chapter discusses some of the principles of the ^{15}N isotope dilution method, some of the practical problems, and the use of the method to quantify nitrogen fixation in legumes and non-legumes, by associative and endophytic diazotrophs.

^{15}N isotope dilution method

The ^{15}N isotope dilution method involves the growth of N_2-fixing and non-N_2-fixing reference plants in soil fertilized with ^{15}N-enriched inorganic or organic fertilizer. It is based on the differential dilution in the plant of ^{15}N-labelled fertilizer by soil and fixed nitrogen (Fried and Middelboe, 1977; McAuliffe *et al.*, 1958). The use of the isotopically labelled fertilizer or tracer provides an integrated measurement of the amount of fixed N accumulated by the crop over the growing season. There are three main variations of this method, i.e. the isotope dilution method, the A-value method and the natural abundance method. Only the isotope dilution method will be discussed here.

In the isotope dilution method both fixing and non-fixing reference plants are grown in soil to which the same amount of fertilizer having the same ^{15}N enrichment has been applied (Hardarson and Danso, 1993). In the absence of any supply of N other than soil N and tracer, both plants are assumed to contain the same ratio of $^{15}N/^{14}N$, since they are taking up N of similar isotopic composition, but not necessarily the same total amount of N. In the presence of N_2 the fixing plant further lowers the ratio of $^{15}N/^{14}N$ due to incorporation of N from unlabelled air, while this does not occur in the reference plant. Calculation of percentage nitrogen derived from atmosphere (%Ndfa) can then be made by the following equation:

$$\%\text{Ndfa} = \left(1 - \frac{\%\text{Ndff}_F}{\%\text{Ndff}_{NF}}\right) \times 100 \qquad (1)$$

where $\%\text{Ndff}_F$ and $\%\text{Ndff}_{NF}$ are percentage N derived from fertilizer by fixing and non-fixing plants, respectively.

The above equation is derived from the following (Fried and Middelboe, 1977; Hardarson 1990):

$$\%\text{Ndff}_{NF} + \%\text{Ndfs}_{NF} = 100 \qquad (2)$$

$$\%\text{Ndff}_F + \%\text{Ndfs}_F + \%\text{Ndfa} = 100 \qquad (3)$$

and

$$\frac{\%\text{Ndff}_{NF}}{\%\text{Ndfs}_{NF}} = \frac{\%\text{Ndff}_F}{\%\text{Ndfs}_F} \qquad (4)$$

where $\%\text{Ndfs}_F$ and $\%\text{Ndfs}_{NF}$ are the percentage N derived from soil in the fixing and non-fixing crops, respectively. Equation (4) represents the only assumption made in the ^{15}N isotope dilution method, i.e. that both non-fixing and fixing crops take up N from soil and tracer in the same ratio. There are several factors which can affect this assumption, i.e.: (a) the tracer distribution has to be even with soil depth or the fixing and non-fixing plants have to have similar rooting patterns, and (b) the enrichment of the soil has to be constant with time or the two plants have to have similar N uptake patterns. The selection of an appropriate non-fixing reference crop for a particular fixing crop appears to be the greatest problem with the isotope dilution method.

The need for a reference crop

A reference crop is needed only for the determination of the actual amount of N_2 fixed or %Ndfa. Such measurements are needed only in some instances, such as in N-balance studies in soil or in soil–plant systems or for comparing N_2 fixed in different seasons, years or environments (Danso *et al.*, 1993). Reference crop is not needed to compare or rank treatment in terms of nitrogen fixation effectiveness. As suggested by Phillips *et al.* (1986), the greatest utility of the ^{15}N isotope dilution technique lies in its usefulness to discern relative differences in N_2 fixation capabilities. However, when reference crop is needed the accuracy and precision of the methodology depends to a great extent on selecting a suitable non-fixing reference crop. It is essential to observe the following:

1. That the reference crop does not itself fix nitrogen.
2. The rooting depths of both reference and fixing crops should be similar, or both crops should derive all of their N from the same zone.
3. Both crops should go through similar growth or physiological stages and mature at the same time.
4. Both crops should be affected in similar fashion by changes in environmental conditions during the growth period.

The following reference crops have been successfully used for estimating N_2 fixation in grain legumes:

1. Non-fixing cereal crops.
2. Non-nodulating isolines of legumes.
3. Uninoculated legumes in soil devoid of the appropriate strains of *Rhizobium*.

The sensitivity of the ^{15}N methodology

Methods based on the dilution in the plants of ^{15}N-labelled fertilizer by N derived from atmosphere and soil seem to offer potentially accurate methods to quantify symbiotic nitrogen fixation. Variations are, however, often found, depending on the non-fixing reference crop (Wagner and Zapata, 1982). It has been observed at the FAO/IAEA Agriculture and Biotechnology Laboratory in Seibersdorf that the ^{15}N methodology seems to be particularly accurate when large proportions of the N in the fixing plant are derived from atmosphere (Reichardt *et al.*, 1987; Hardarson *et al.*, 1988). This prompted us to model the percentage of N derived from atmosphere in relation to ^{15}N enrichments in the fixing and non-fixing standard crops and to investigate where major errors in estimates of N_2 fixation can be expected. Using equation (1) above, modelled curves for various $\%Ndff_F$ and $\%Ndff_{NF}$ are shown in Figure 1. In the bottom left corner of the graph the lines are very steep and it can be seen from this figure that variation in $\%Ndff_{NF}$ produces large variation in %Ndfa in that area of the graph, i.e. small variation in ^{15}N atom excess of the reference crop will produce large variation in %Ndfa. A specific example of the influence of an error as caused by the reference crop on the range of fixation is illustrated in Table 1.

It is clear from the above modelling that methods based on the dilution in the plant of ^{15}N tracer by N derived from atmosphere are potentially accurate methods to quantify N_2 fixation when a large proportion (> 70%) of the N in the fixing crop is derived from atmosphere. However, at lower fixation levels (< 30%) the methodology is less accurate, and under those conditions the selection of a standard crop and the stability of the $^{15}N/^{14}N$ ratio in the substrate is of particular importance.

It is clear from Figure 1 and Table 1 that the isotope dilution method is particularly accurate at high levels of nitrogen fixation. The methodology can therefore be easily used to quantify nitrogen fixation in legumes when fixation is more than 10–20 kg N/ha. At lower levels of fixation, which may occur in all systems, in particular among associative diazotrophs, one has to be particularly careful to eliminate all errors.

The ^{15}N isotope dilution method remains the best methodology when N_2 fixation is to be quantified in various symbiotic systems. Some of the advantages and disadvantages of conventional (N difference, acetylene reduction, xylem solute methods) and isotope dilution methods (isotope dilution, A-value, natural abundance) are compared in Table 2.

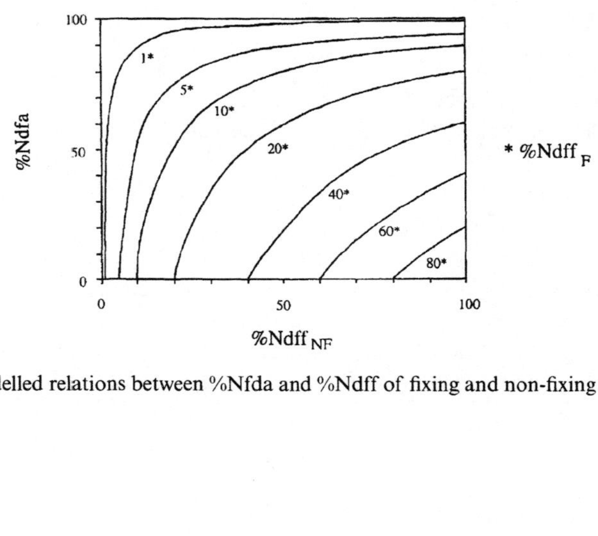

Figure 1. Modelled relations between %Nfda and %Ndff of fixing and non-fixing crops.

Table 1. Calculated values of %Ndfa from different $\%Ndff_F$ values when $\%Ndff_{NF}$ has 10% coefficient of variation

Example	$\%Ndff_F$	$\%Ndff_{NF}$	Range of %Ndfa	Differences between values
1	10	10±1	−11– 9	20
2	8	10±1	11–27	16
3	6	10±1	33–45	12
4	4	10±1	55–63	8
5	2	10±1	77–81	4

Table 2. Comparison of methods to quantify nitrogen fixation

	Conventional methods[1]	Isotope methods[2]
Simple and rapid	++	−
Useful to compare many treatments	++	−
Low cost	++	−
Non–destructive	++	−
Time-integrated estimate	+	+
%N derived from air	−	++
kg N/ha fixed	−	++

[1] Some of the conventional methods, e.g. acetylene reduction or xylem solute techniques, provide spot measurements at specific growth stages.
[2] Isotope methods provide an integrated measure of percentage and amount of N_2 fixed over the whole growing season.

To conclude, it will depend on the objectives in each case which method is the most suitable. If the objective is to screen thousands of breeding lines for nitrogen fixation one would have to use conventional methods. However, if the objective is to quantify %Ndfa or kg N/ha fixed in a few cultivars over the whole growing season, then the isotope methods would be the most suitable.

References

Danso S K A, Hardarson G, Zapata F 1993. Misconceptions and practical problems in the use of ^{15}N soil enrichment techniques for estimating N_2 fixation. Plant Soil, 152, 25–52.
Fried M, Middelboe V 1977. Measurement of amount of nitrogen fixed by legume crop. Plant Soil, 43, 713–715.
Hardarson G 1990. Use of Nuclear Techniques in Studies of Soil-Plant Relationships. IAEA, Vienna.
Hardarson G 1994. International FAO/IAEA programmes on biological nitrogen fixation. In: Graham P H, Sadowsky M J, Vance C P, eds. Symbiotic Nitrogen Fixation. Kluwer, Dordrecht, 189–202.
Hardarson G, Danso S K A 1993. Methods for measuring biological nitrogen fixation in grain legumes. Plant Soil, 152, 19–23.
Hardarson G, Zapata F, Danso S K A 1988. Dinitrogen fixation measurements in alfalfa ryegrass swards using nitrogen-15 and influence of the reference crop. Crop Sci., 28, 101–105.
McAuliffe C, Chamblee D S, Uribe-Arango H, Woodhouse Jr W W 1958. Influence of inorganic nitrogen on nitrogen fixation by legumes as revealed by ^{15}N. Agron. J., 50, 334–337.

Phillips D A, Jones M B, Foster K W 1986. Advantages of the nitrogen-15 dilution techniques for field measurements of symbiotic dinitrogen fixation in legumes. Soil Sci. Soc. Am. J., 677, 11–21.
Reichardt K, Hardarson G, Zapata F, Kirda C, Danso S K A 1987. Site variability effect on field measurement of symbiotic nitrogen fixation using the ^{15}N isotope dilution method. Soil Biol. Biochem., 19, 405–409.
Wagner G H, Zapata F 1982. Field evaluation of reference crops in study of nitrogen fixation by legumes using isotope techniques. Agron. J., 74, 607–612.

Section VIII

Actinorhizal and cyanobacterial association

37
Competitivity and persistence of *Frankia* in a managed environment

N.R. McEWAN, E.M.O GOULD, E. LUMINI, J.D. McNEILL AND
C.T. WHEELER

Abstract

Alnus rubra Bong. seedlings were inoculated in 1985 in a chemically sterilized seed bed with *Frankia* strains *HFRArI4* or *UGL013103*, with crushed nodules from the same trees that the latter strain was isolated or were left uninoculated. At transplant the following year into replicated plots established on coal-mine spoil, the inoculated plants were well nodulated and the control seedlings showed low levels of nodulation. Plants inoculated with UGL013103 showed a growth advantage over the other treatments for 2 years after out-planting but this diminished subsequently. However, after 4 years plants excavated from the uninoculated and UGL013103-inoculated plots showed similar levels of nodulation. At this time, leaf N content and nodulation were not correlated in UGL013103-inoculated plants but there was a highly significant correlation between these parameters in control plants, indicative of late nodulation of the controls on this very N-deficient site.

The movement of *Frankia* between treatments on this site facilitated investigation, 10 years after planting, of the persistence and competitivity of the two strains utilized originally to inoculate nursery seedlings. Apart from competition between these two strains, other competitors at this time were *Frankia* strains from *Alnus glutinosa*, planted outside the experimental plot 6 years after establishment, strains from the crushed nodule preparations used to inoculate alders in some plots or strains from nursery soil carried over with out-plants. The identity of *Frankia* in nodules was investigated by amplification of polymorphisms in *Frankia* DNA sequences corresponding to part of the *16S rDNA*. The technique generated a sequence-specific 'gene signature' that facilitated discrimination between two *Frankia* strains in nodules from *Alnus rubra*. Both strains were out-competed by unknown strains for occupancy of most nodule samples, but UGL013103 was the more persistent and competitive of the two strains. *Frankia*

K.A. Malik et al. (eds.), Nitrogen Fixation with Non-Legumes, 315–326.

UGL013103 was detected in 28% of nodule samples from trees inoculated originally with this strain, whereas HFRArI4 was not detected in any nodules from the specifically inoculated trees nor from other trees in the experimental plot. Strain UGL013103 was also found in 10% of samples from control plots (uninoculated prior to planting). Unidentified strains were detected in the remainder of the nodule samples.

Both strains were detected in a small number of nodules formed on plants grown in soil samples collected from alder-free areas immediately adjacent to the experimental plot. This suggests that the competitive pressures for survival of these strains, in the absence of alders as a host, may be lower than within the experimental plot.

Introduction

The ecological and economic importance of actinorhizal plants – in particular, species of the genus *Alnus* in temperate regions and of the Casuarinaceae in the tropics (Schwintzer and Tjepkema, 1990) – have encouraged the establishment of many research programmes to identify elite strains of *Frankia* and to utilize these in nursery inoculation programmes. Inoculation with selected *Frankia* strains can produce significant positive effects on plant growth in the early years following planting (Wheeler *et al.*, 1991; Prat, 1992; Lumini *et al.*, 1994). Persistence of such strains in the perennial actinorhizal nodules provides a potential for enhanced levels of N_2-fixation for several years after inoculation of seedlings with elite strains of *Frankia*. However, re-infection of new roots on the host plant, as the tree grows and older nodules decay, depends on the ability of the introduced strains to survive and to compete successfully with the indigenous population of *Frankia*. In the most effective inoculation programmes the introduced strains will survive for a long period and not be out-competed by the indigenous strains. The ubiquitous occurrence of *Frankia* in soils, even those for which there are no recent records of actinorhizal plants as components of the flora, presents particular problems for the persistence of introduced *Frankia* strains (Rodriguez Barrueco, 1968; Arveby and Huss-Danell, 1988; Paschke and Dawson, 1992). However, until recently techniques have not been available to follow the fate of strains introduced to natural or managed environments through the planting of specifically inoculated trees. Molecular techniques for DNA analysis are now being applied successfully to identify the diversity of *Frankia* strains nodulating actinorhizal plants in natural environments (Ganesh *et al.*, 1994; Sellsted, 1995; Rouvier *et al*, 1996) and in the identification of specific populations of the microsymbiont (McEwan *et al.*, 1994; Simonet *et al.*, 1994; Lumini and Bosco, 1996).

In 1986 an experimental plantation of 1-year-old, specifically inoculated *A. rubra* was established at Kyle Forest, Ayrshire, on reclaimed mine spoil that was free of *Frankia* at planting. Plots contained control plants (uninoculated in the nursery); plants inoculated with *Frankia* HFRArI4, a North American strain; with UGL013103, an elite *Frankia* strain isolated from a plantation of *A. rubra* at Lennox Forest, near Glasgow (Hooker and Wheeler, 1987; Wheeler *et al.*, 1991) and homogenized nodules also collected from *A. rubra* from Lennox Forest. Early measurements of height increment showed continued advantage of inoculation with UGL013103 for 2 years after planting (McNeill *et al.*, 1989) but this advantage gradually disappeared as the plantation aged. A possible explanation was that all trees in the plantation had become heavily nodulated due to movement of *Frankia* between plots.

Data concerning the nodulation status of specifically inoculated explants several years after establishment of this experimental plantation are presented, and form the basis for experiments in which PCR-based technology is utilized to determine the persistence and competitivity in nodulation of *Frankia* strains introduced to the site.

Materials and methods

Characteristics of Frankia

The isolation and characteristics of *Frankia* strain UGL013103 (acronym Ar125Q), the positive results from its use as inoculum for *A. rubra* in the nursery, and early data on the growth of inoculated *A. rubra* in plantations on mine-spoil reclamation areas have been described previously (Hooker and Wheeler, 1987; McNeill *et al.*, 1989; Wheeler *et al.*, 1991).

The experimental site

The plantation was located on newly reclaimed coal-mine spoil at Kyle, Ayrshire, UK. It was free of actinorhizal plants and *Alnus*-infective *Frankia* at the time of planting of specifically inoculated *Alnus rubra*, grown at the Forestry Commission nursery, Penicuik (Wheeler *et al.*, 1991). Saplings were inoculated in the nursery with one of *Frankia* strain UGL013103 (isolated from nodules from *A. rubra* from Lennox Forest, UK); *Frankia* ArI4 (DDB010210), a North American isolate kindly supplied by Dr Dwight Baker; crushed root nodules from Lennox Forest (CRN); or were not inoculated (controls). The four treatments were replicated four times in a randomized block design. Each plot consisted of 25 trees planted in rows of five trees and with 1.5 m spacing within and between rows. Each plot was separated from neighbours by a 3 m buffer which was planted with *Larix kaempferi* (Lamb.) Carr. A 3 m buffer zone was established around the block.

About 5 years after the planting of *A. rubra* in 1986, other *Alnus* species, notably *A. glutinosa*, were introduced into areas adjacent to the experimental plot. The plants were nodulated but it is probable that the strains infecting these trees are different from those introduced in the original experimental plot since they were from different nurseries.

Harvesting and analysis of plantation trees for nodulation

Four years after establishment of the plantation, five complete trees were excavated in July from each of four plots that contained control and UGL013103-inoculated trees. The leaves from the terminal 50 cm of the leader shoots from each plant were dried at 80°C to constant weight and ground for Kjeldahl analysis for total N. The roots of the trees were transported to a cold store and were then washed, the nodules removed, separated into small (>0.5 cm diameter), medium (<0.5 and >2.0 cm diameter) and large (<2.0 cm diameter categories, counted and dried at 80°C to constant weight.

Harvesting of samples for DNA analysis

Ten years after establishment of the plantation, a single nodule lobe cluster was collected from eight different trees from mid-way between trees in each of four plots of trees inoculated with UGL013103, ArI4 or crushed nodules or from uninoculated controls. DNA was isolated and analysed using the PCR-based technique below.

A soil baiting technique was used for assay of *Frankia* in soil outside the experimental plot. Using an alcohol-sterilized spade, soil was dug to a depth of 15 cm at 5 m intervals from points from 0–30 m away from the plot at the northeast and west and southeast and west corners, respectively, and transported to the laboratory in separate plastic bags.

Each soil sample was mixed with 50% perlite and placed in plastic 'Transpots' (four pots per sample). Each group of pots was isolated in a 15 cm deep bowl and surface-sterilized seeds of *A. rubra* were sown in each pot. Seedlings were thinned to three per pot after germination. Each pot received a supplement of mineral N free Crone's solution (Hooker and Wheeler, 1987). Seedlings were grown for 8 months in a heated glasshouse lit by natural daylight supplemented with 16 h light from 400 W mercury vapour lamps.

All soil-grown *A. rubra* seedlings became nodulated, but seedlings grown in perlite that were dispersed throughout the experiment remained non-nodulated.

Six nodule samples from the seedlings from each location were analysed using the PCR-based technique below.

DNA isolation

DNA was isolated from a single lobe from each nodule according to Nalin *et al.* (1995). Nodules were removed from the plant and washed in water. The periderm from single nodule lobes was removed and the lobe placed in 300 μl extraction buffer (100 mM) Tris.HCl (pH 7.0), 20 mM EDTA, 1.4 mM NaCl, 2% (w/v) hexadecyltrimethylammonium bromide (CTAB), 1% (w/v) insoluble polyvinylpyrrolidone. Peeled lobes were homogenized using a piston pestle and incubated at

65°C for 2 h. After incubation an equal volume of chloroform : isoamyl alcohol (24 : 1 v/v) was added. The two phases were mixed by inversion. The phases were separated by centrifugation in a microfuge at 13 000*g* for 20 min; 200 μl of the aqueous (upper phase) was drawn off and transferred to a fresh tube. To this an equal volume of 100 mM Tris.HCl (pH 7.0), 20 mM EDTA, 0.7% (w/v) NaCl, 10% (w/v) CTAB, which had been pre-heated to 65°C, was added. This was centrifuged further at 13 000*g* for 20 min and the aqueous phase transferred to a fresh tube. To the aqueous phase, 0.1 vol. of 3 M sodium acetate buffer (pH 5.5) and 2.5 vol. of ethanol were added. DNA was allowed to precipitate by incubating at −20°C for 45–60 min. Precipitated DNA was collected by centrifugation, DNA was dried in air and resuspended in autoclaved ultrapure water.

Amplification of DNA

DNA sequences corresponding to part of the *16S rDNA* (Normand *et al.*, 1996) were amplified by use of two primers: (i) AAGGAGGGGATCCAGCCGCA, (ii) ATGGAGAGTTTGATCATGGCTCAGGAC or ATGGAGTTTGATCAT-GGCTCAGGAC. The PCR cocktail comprised the following; 50 mM KCl, 10 mM Tris.HCl (pH 8.3), 1.5 mM $MgCl_2$, 100 μg/ml gelatin, 0.25 μM of each primer, 200 μM of each dNTP, 2.5 units of Taq polymerase per 100 μl and approximately 5 ng/ml genomic DNA. Amplification was carried out using the following cycling protocol: (95°C for 5 min, 55°C for 1 min, 72°C for 1 min) for one cycle; (95°C for 1 min, 55°C for 1 min, 72°C for 1 min) for 35 cycles and finally 72°C for 4 min.

After an initial PCR cycle for amplification of the *16S rDNA*, 10 ng of the resulting DNA was further amplified by a lower-temperature PCR method, the sequence-specific 'gene signature' (SSGS) technique, (Pena *et al.*, 1994) using only one of the second specific primers described earlier; at double its previous concentration. Other components of the PCR cocktail were as described earlier. The cycling system was: (95°C for 5 min, 32°C for 1 min, 72°C for 1 min) for one cycle; (95°C for 1 min, 32°C for 1 min, 72°C for 1 min) for 35 cycles; 72°C for 4 min. Amplified products were run on a 4% (w/v) TBE agarose gel.

Results

Nodulation and leaf nitrogen content of plantation Alnus rubra

Analysis of *Alnus rubra*, harvested from the plantation 4 years after establishment, showed no significant differences in nodulation between trees that were uninoculated in the nursery or were inoculated with UGL013103. Similarly, although apparently higher, the leaf N content of inoculated trees was not significantly different from control plants (Table 1). However, regression analysis showed that the nodule dry weight plant^{-1} was highly correlated ($r = 0.83$) with leaf N content for control plants (Figure 1a), whereas these parameters were poorly correlated ($r = 0.355$) in inoculated plants (Figure 1b).

Table 1. Nodulation and N content of *Alnus rubra* at Kyle Forest, 4 years after out-planting. Four plants were harvested for analysis from four plots per treatment. Data are means ± SE (n = 20). Nitrogen analyses are of total N in the leaves of the terminal 50 cm of the leader shoots from each plant

Treatment	Nodule dry weight (g plant-1)				No. nodules plant^{-1}			
	Small (<0.5 cm diam.)	Medium (> 0.5, < 2.0 cm)	Large (> 2.0 cm)	Total	Small (< 0.5 cm diam.)	Medium (> 0.5, < 2.0 cm)	Large (> 2.0 cm)	Leaf N (mg g^{-1} dry weight)
Uninoculated control	3.9±0.75	14.6±3.14	22.7±3.08	41.2±5.79	642±112	34.8±5.75	10.2±1.73	21.4±1.17
Inoculated UGL013103	2.1±0.43	10.2±2.27	26.0±3.44	38.3±4.42	482±95	32.0±9.27	10.6±1.26	23.7±0.92

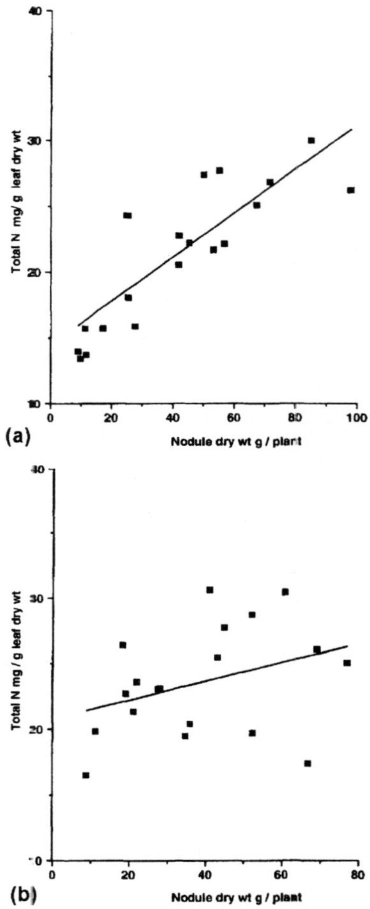

Figure 1. Relationship between nodule dry weight and the N content of the leaves of the leader shoot of UGL013103-inoculated *Alnus rubra* 4 years after planting out on reclaimed mine spoil at Kyle Forest. Five plants from each of four plots were harvested for analysis. Each data point represents the mean of duplicate N analyses of finely ground, dried leaf samples. (**a**) UGL013103-inoculated *Alnus rubra*, correlation coefficient = 0.83; (**b**) control (uninoculated) *Alnus rubra*, correlation coefficient = 0.355.

DNA analysis and the occurrence of Frankia *in the nodules of experimental plants*

The different banding patterns of Figure 2 demonstrate that the SSPP technique provides a means for distinguishing between the *Frankia* strains of interest. The technique has proved reliable when used to differentiate between HFRArI4 and UGL013103 in culture or in symbiosis with *Alnus rubra*.

The results from the application of the technique to eight nodule samples collected from trees in each of four plots of each treatment are shown in Table 2.

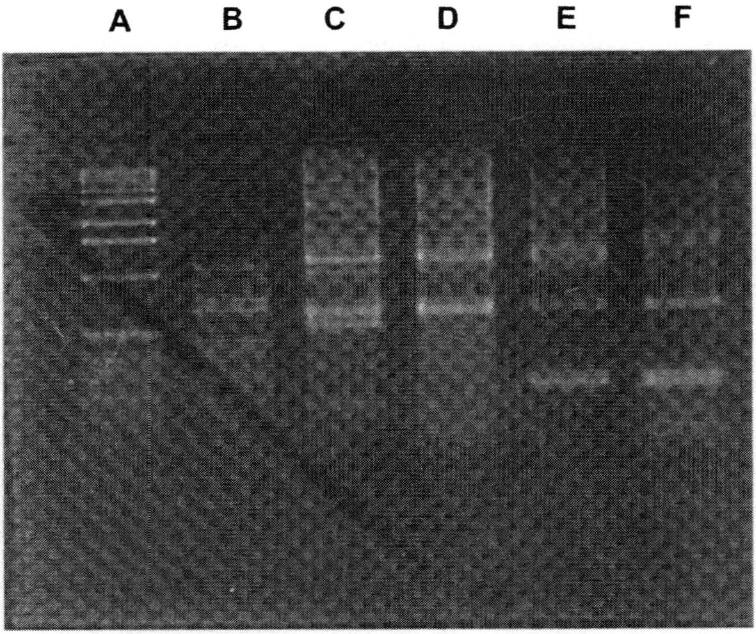

Figure 2. PCR-amplified fragments of *Frankia* DNA run on a 2% TBE agarose gel after two rounds of PCR; the first round amplifying part of the *16S rDNA* gene, the second amplication being the lower temperature amplification, using a single primer. Primer sequences were: AAGGAGGGGATC-CAGCCGCA (ii) ATGGAGAGTTTGATCATGGCTCAGGAC. **A**: BRL 1 kb size marker (517, 506, 396, 344, 298, 220, 201, 154, 134, 75 bp); **B**: unknown *Frankia* strain; **C**: symbiotic HFR ArI4; **D**: cultured HFR ArI4; **E**: symbiotic UGL 013103; **F**: cultured HFR 013103.

Patterns typical of UGL013103 were detected in 28% of the samples from trees inoculated originally with this strain. Patterns typical of UGL013103 were detected in 9% of the samples from trees inoculated originally with *Alnus rubra* crushed nodules. Patterns in nodule samples from trees inoculated originally with ArI4 and from uninoculated trees were all of unidentified strains. No nodules were harvested which showed banding patterns typical of HFRArI4, even when samples were collected from trees inoculated originally with this strain. In a more limited experiment in which four nodules from the crown of trees in plots from three treatments were examined, banding patterns typical of UGL013103 were detected in two samples from UGL013103-inoculated trees, one sample from HFRArI4-inoculated and one sample from the control trees. Again, no patterns typical of HFRArI4 were detected.

Analysis of nodules from *A. rubra* that was grown in the glasshouse in soil collected at 5 m intervals for a distance of 30 m from each of the four corners of the plot showed that 8.3% of the 168 samples gave patterns identical to UGL013103; 3% were identical to HFRArI4; and the remaining samples gave unidentified patterns. Patterns typical of UGL013103 were detected at distances

Table 2. Analysis of *Frankia* strain distribution in nodule samples from the experimental plantation of *Alnus rubra* at Kyle Forest. Strains were identified by application of the SSPP technique to samples of eight nodules from each of four replicate randomized plots established for each set of saplings. Nodules were harvested from midway between trees that were planted 1.5 m apart. Sapling sets analysed were inoculated prior to planting with either *Frankia* HFRArI4, with UGL013103, with crushed nodules (CRN) or were uninoculated (controls)

Inoculating strain	Strain identify
Control (no inoculum)	Strains unidentified in four plots
HFRArI4	Strains unidentified in four plots
CRN	Strains unidentified in three plots
	Five unidentified strains and three of UGL013103 identified in one plot
UGL013103	Strains unidentified in two plots
	Seven unidentified strains and one of UGL013103 in one plot.
	Eight of UGL013103 identified in one plot.

of up to 30 m from the plot, but HFRArI4 was detected only in samples collected at distances up to 5 m from the plot.

Discussion

Plants from control (uninoculated) chemically sterilized seed beds had no or very low levels of nodulation when planted out (Wheeler *et al.*, 1991) and infective *Frankia* was absent from the Kyle mine spoil site at planting. While a precise time framework for nodulation of the control plants cannot be given (Table 1), it is notable that inoculated plants still showed a height incremental advantage over uninoculated controls 2 years after out planting (McNeill *et al.*, 1989). Because of the absence of *Frankia* from the experimental site initially and the very low initial levels of nodulation of out-plants, it is probable that nodulation developed most extensively after year 2, as the population of *Frankia* in the soil increased. This resulted, 4 years after out-planting, in statistically similar profiles of occurrence of nodules of small, medium and large dimensions and weight in both control and UGL013103-inoculated trees (Table 1). Further support for a later, but rapid development of nodulation in the controls is suggested by the highly significant correlation between nodulation and leaf N content in these plants 4 years after out-planting, whereas these parameters were not correlated in the inoculated trees (Figures 1a,b). Positive correlations between N_2-fixation, growth and plant N content are a normal function of early growth on N-deficient media, but this relationship would be expected to break down in the field as the initial nodules age and the turnover and replacement of degenerate nodules becomes a prominent feature of the maturing root system. After 4 years the N content of leaves from leader shoots of the inoculated trees were apparently slightly (but not statistically significantly) higher than the uninoculated controls, whereas prior to out-planting inoculated seedlings of *A. rubra* had foliar N contents 2–3 times that of uninoculated control plants (Wheeler *et al.*, 1991).

The high level of nodulation of out-plants in the control plots at Kyle after 4 years (Table 1) could have arisen from infection by *Frankia* several sources, for example the very low levels of *Frankia* that were associated with controls at planting; the movement of *Frankia* within the plots, crossing the 3 m interplot barriers (this would have included a mix of strains from the crushed nodule preparations from *A. rubra* at Lennox Forest which were used to inoculate alders in some plots); or movement of *Frankia* into the plot from nodulated *Alnus glutinosa* that was planted elsewhere at Kyle during the period of the experiment. The precise source of the contamination was not determined, although the sensitive PCR-based DNA analysis technique described now renders this feasible.

The 'single specific signature' technique (Pena *et al.*, 1995), as adapted for this work offers great specificity for *Frankia* DNA and is practical for application to large numbers of samples from the field. The specificity of the primer sequences used in the initial round of PCR ensures that the only material available for amplification in the second round is from *Frankia*, while the low annealing temperature used in the second round of PCR increased the probability of detecting differences between strains.

The nodules sampled from between trees in the experimental plot were all relatively young (less than 1 cm diameter) and would have been initiated well after the trees were planted. The substrate on which the trees are growing presented particular difficulties for excavating old nodules but, as noted above, it is probable that degeneration of such nodules would have been well advanced 10 years after out-planting of *A. rubra*. Consequently, the experiments that were carried out determined the ability of *Frankia* strains released from the original nodules (i.e. those formed as a result of inoculation of nursery plants) to survive and to infect new roots in competition with other strains that were introduced into the environment. Although UGL013103 was of relatively low occurrence in the nodule samples analysed, it was detected both in nodules from plots containing trees inoculated originally with this strain and in nodules from one of the control plots. By contrast, ArI4 was not detected in any of the experimental plots, including those with specifically inoculated trees. UGL013103 was also the only strain used as inoculum to be detected in older nodules, harvested from the root crown of experimental trees (Table 2).

These data suggest that UGL013103 is a more persistent and more competitive strain than ArI4, although it is less competitive than some of the unknown strains in the environment. In this context it is of interest that ArI4 tends to be more difficult to maintain in continuous culture in the laboratory than UGL013103. However, fingerprints for both UGL013103 and ArI4 were detected in nodules initiated on seedlings grown in soil from outside the plot. Patterns typical of ArI4 were found only within 5 m of the original test plot, but UGL013103 patterns were found as much as 30 m from the test plot, which may reflect their relative ability to persist in the soil. Due to the absence of alders in the area outside the plantation from which soil samples were taken, competitive pressures for survival of ArI4 in particular may have been less than within the experimental plot. UGL013103 was detected both uphill and downhill of the experimental site, suggesting that movement in water alone is not the only means of movement of the organism. As well as the possibility of spread of spores in wind-blown dust, dispersal of *Frankia* by worms and by birds has been described (Reddell and Spain, 1991; Burleigh and Dawson, 1995).

The data show that, although UGL013103 is a highly effective strain for nodulation of, and N_2-fixation in *A. rubra* grown in controlled growth conditions or in pre-sterilized nursery beds (Hooker and Wheeler, 1987; Wheeler *et al.*, 1991) it is not strongly persistent nor competitive for infection in field conditions when a broad background of other *Alnus*-infective strains is present. As with other symbionts such as *Rhizobium*, this emphasizes the importance of developing *Frankia* inocula that contain a mix of highly effective strains, formulated to ensure that at least some will be infective, competitive and persistent under a wide range of field conditions.

Acknowledgements

This research was funded by the Natural and Environmental Research Council, UK (grant GR3/8668) and by the European Union (contract TS3/CT92/0090). Dr P. Normand, Université Claude Bernard 1, Lyon, is thanked particularly for facilitating research visits to his laboratory, for providing technical advice concerning the application of PCR techniques to *Frankia* and the sequences of the primers.

References

Arveby A S, Huss-Danell K 1988. Presence and dispersal of infective *Frankia* in peat and meadow soils in Sweden. Biol. Fert. Soils, 6, 38–44.

Burleigh S H, Dawson J O 1995. Spores of *Frankia* strain HFPCcI3 nodulate *Casuarina equisetifolia* after passage through the digestive tracts of captive parakeets (*Melopsittacus undulatus*) Can. J. Bot., 73, 1527–1530.

Ganesh G, Misra A K, Chapelon C, Normand P 1994. Morphological and molecular characterization of *Frankia* sp. isolates from nodules of *Alnus nepalensis* Don. Arch. Microbiol., 161, 152–155.

Hooker J E, Wheeler C T 1987. The effectivity of *Frankia* for nodulation and nitrogen fixation in *Alnus rubra* and *Alnus glutinosa*. Physiol. Plantarum, 70, 333–341.

Lumini E, Bosco M 1996. PCR-restriction fragment length polymorphism identification and host range of single spore isolates of the flexible *Frankia* strain UFI132715. Appl. Env. Microbiol., 62, 3026–3029.

Lumini E, Bosco M, Puppi G, Isopi R, Frattegiani M, Buresti E, Favilli F 1994. Field performance of *Alnus cordata* Loisel (Italian alder) inoculated with *Frankia* and VA-mycorrhizal strains in mine-spoil afforestation plots. Soil Biol. Biochem., 26, 659–661.

McEwan N R, Wheeler C T, Milner J J 1994. Strain discrimination of cultured and symbiotic *Frankia* by RFLP-PCR. Soil Biol. Biochem., 26, 541–545.

McNeill J D, Hollinsworth M K, Mason W L, Moffat A J, Sheppard L J, Wheeler C T 1989. Inoculation of *Alnus rubra* seedlings to improve seedling growth and forest performance. Forestry Commission Research Information Note 144.

Nalin R, Domenach A M, Normand P 1995. Molecular structure of the *Frankia* spp. *nif* intergenic spacer and design of *Frankia* genus compatible primer. Mol. Ecol., 4, 483–491

Normand P, Orso S, Cournoyer B, Jeannin P, Chapelon C, Dawson J, Evtushenko L, Misra A K 1996. Molecular phylogeny of the genus *Frankia* and related genera and emendation of the Family Frankiaceae. Int. J. Syst. Bacterial, 46, 1–9.

Paschke M W, Dawson J O 1992. *Frankia* abundance in soils beneath *Betula nigra* and other non-actinorhizal plants. Acta Oecol., 13, 407–415.

Pena S D J, Barreto G, Vago A R, Marco L D, Reinach F C, Dias-Neto E, Simpson S A J G 1994. Sequence specific 'gene signatures' can be obtained by PCR with single specific primers at low stringency. Proc. Natl. Acad. Sci. USA, 91, 1946–1949.

Prat D 1992. Effect of inoculation with *Frankia* on the growth of *Alnus* in the field. Acta Oecol., 13, 463–468.

Reddell P, Spain A V 1991. Transmission of infective *Frankia* (Actinomycetales) propagules in casts of the endogenic earthworm *Pontoscolex corethrurus* (Oligochaeta: Glossoscolecidae). Soil Biol. Biochem., 23, 775–778.

Rodriguez Barrueco C 1968. The occurrence of the root nodule endophyte of *Alnus glutinosa* and *Myrica gale* in soils. J. Gen. Microbiol., 52, 189–194.

Rouvier C, Prin Y, Reddell P, Normand P, Simonet P 1996. Genetic diversity among *Frankia* strains nodulating members of the family Casuarinaceae in Australia revealed by PCR and restriction fragment length polymorphism analysis with crushed root nodules. Appl. Env. Microbiol., 62, 979–985.

Schwintzer C R, Tjepkema J D 1990. The Biology of *Frankia* and Actinorhizal Plants. Academic Press San Diego, CA, p. 408.

Sellstedt A 1995. Specificity and effectivity in nodulation by *Frankia* on southern hemisphere actinorhiza. FEMS Microbiol. Lett., 125, 231–236.

Simonet P, Bosco M, Chapelon C, Moiroud A, Normand P 1994. Molecular characterisation of *Frankia* microsymbionts from spore+ and spore– nodules in a natural alder stand. Appl. Env. Microbiol., 60, 1335–1341.

Wheeler C T, Hollingsworth M K, Hooker J E, McNeil J D, Mason W L, Moffat A J, Sheppard L J 1991. The effect of inoculation with either cultured *Frankia* or crushed nodules on nodulation and growth of *Alnus rubra* and *Alnus glutinosa* seedlings in forest nurseries. For. Ecol. Manage., 43, 153–166.

38
Symbiotic association between *Frankia* and actinorhizal plants

NANTAKORN BOONKERD

Abstract

The actinorhizal symbioses are mutualistic relationships between the actinomycete genus *Frankia* and a number of dicotyledonous plant genera belonging to eight diverse plant families. Root nodules of actinorhizal plants induced by *Frankia* are morphologically distinct from legume nodules which are formed by rhizobia. *Frankia* is relatively conservative in its interaction with its host species, despite its broad host range. Four host-specific groups have been described; however, if one disregards cross-inoculation data from the two promiscuous host genera (*Gymnostoma* and *Myrica*), only three major groups need to be defined. There are two modes of root infections: root hair infection and intercellular penetration. *Frankia* has a mycelial growth habit consisting of two specialized structures: vesicle and sporangium. The vesicle is the multicellular structure where the enzyme nitrogenase is located. To protect this oxygen-sensitive enzyme, the vesicle is morphologically modified from vegetative cells. A multilamellar envelope surrounds it to reduce diffusion of oxygen into the structure. In the sporangium, numerous single-celled sporangiospores are borne. *Frankia* strains share sequence similarity of their nitrogenase (*nif*) genes with other nitrogen-fixing bacteria. It is also possible that they share some homology in nodulation (*nod*) genus with rhizobia. There are variations in effectiveness in N_2 fixation among *Frankia* strains within a group. Nodulated with highly effective strains of *Frankia*, the actinorhizal plants could play a major role in wood production and soil fertility improvement arising from symbiotic nitrogen fixation.

Introduction

Frankia is a genus of actinomycetes capable of forming nitrogen-fixing nodules on roots of actinorhizal plants. There are 25 genera from eight angiosperm

K.A. Malik et al. (eds.), Nitrogen Fixation with Non-Legumes, 327–331.

families identified as nitrogen-fixing actinorhizal plants (Baker and Schwintzer, 1990). Some examples of these genera are *Alnus*, *Casuarina*, *Ceanothus*, *Coriaria*, *Comptonia*, *Datisca*, *Discaria*, *Elaeagnus*, *Hippophae*, *Myrica*, *Purshia* and *Shepherdia*. Actinorhizal plants are predominantly found in temperate regions and are perenial (Baker and Selling, 1984). *Casuarina* is probably the only truly tropical actinorhizal genus.

It has been realized that some actinorhizal plants are soil-improving crops in marginal lands, poor soil conditions such as sand-dunes, mine dumps, and waste gravelly sites. Moreover, some shrub species of *Alnus*, *Ceanothus* and *Elaeagnus* have been found to be important in temperate forest management under adverse soil conditions. Many species of actinorhizal plants are woody trees having great economic impact in the wood industry fuel and wood-related products such as paper pulp.

Classification and host specificity

Unlike legume-rhizobia symbiosis, the actinorhizal association received considerable attention after the 1950s (Quispel *et al.*, 1993). Due to very slow growth (24–48 h generation time) isolation of *Frankia* has been interfered with contamination. However, at present successful isolation of *Frankia* has been made from both nodules and soil (Baker and O'Keefe, 1984). In the early period, classification of Actinomycetes into *Frankia* was based on morphology of hyphae, vesicle, spore; biochemistry; physiology and infectivity on the host plants. Since the first isolation of *Frankia* from *Comptonia* nodules in 1987 (Callaham *et al.*, 1978), numerous strains have been isolated from various plant species. Infectivity on host plants by different strains isolated from different host was studied. Baker and Schwintzer (1990) described four host-specific groups in which two promiscuous host genera, *Gymnostoma* and *Myrica* were not included: (1) strains which infect genera of the Betulaceae and Myricaceae; (2) strains which infect genera of the Casuarinaceae; (3) strains which infect genera of the Elaeagnaceae and (4) strains which cannot infect promiscuous hosts in group 3. This classification was confirmed by RFLP grouping (Nittayajarn *et al.*, 1990). At present, identification and characterization of *Frankia* by DNA probe techniques have been developed successfully by many investigators (Simonet *et al.*, 1988, 1990; Normand *et al.*, 1992; Jamann *et al.*, 1993).

Nodulation

Nodulation of actinorhizal plants can occur by either one of the two infection mechanisms, root hair infection or intercellular penetration. Root hair infection has been shown to occur in the genera *Alnus*, *Casuarina*, *Comptonia* and *Myrica* (Callaham *et al.*, 1979; Berry, 1984). Intercellular penetration has been observed only in Elaeagnaceae (Miller and Baker, 1985).

The process of root hair infection is similar to infection of legume species by rhizobia. *Frankia* colonizes the root hairs and causes their deformation in the form of branching or curling. The filaments of the bacterium are entrapped by the deformed root hairs and the cell wall of root hair is digested in an effort to enter the cell (Berry, 1984). In response to this invasion the plant cells secrete a cell wall-like matrix around the infection site to encapsulate the bacterium and maintain the integrity of the plant cell by keeping the bacterium external to the cell plasma membrane and wall. A root nodule meristem is initiated in the pericycle of the root in close proximity to the infection site (Callaham *et al.*, 1979). The invading bacterium grows through the cortical cells towards this new meristem. Upon entering the incipient root nodule the bacterium proliferates to fill the cortical cells of the nodule and then begins to differentiate vesicles, the specialized cells for nitrogen fixation. As the root nodule continues to grow and enlarge the bacterium continues to invade the new cortical cells and to maintain active nitrogen fixation. Over time the nodule meristem will divide dichotomously, resulting in a nodule with a coralloid three-dimensional structure.

The process of intracellular penetration occurs on actinorhizal plant roots which lack root hairs. *Frankia* grow on the surface of the root and, through some specific interaction, begin to digest the middle lamellar region between two adjacent epidermal cells. As the bacterium invades the intercellular spaces of the root, the plant responds by secreting a matrix to surround the bacterial filaments. The growing hyphae digest this matrix to further proliferate in the cortical intercellular region of the root. A nodule meristem develops from the pericycle and the invading bacterial hyphae grow towards this incipient nodule. The bacterium grows into the cortical region of the new nodule, remaining in the intercellular spaces until late in the development of the new nodule. Subsequently, branching hyphae invade individual cells of the nodule and become intracellular symbionts. Vesicles differentiate on the hyphae of the bacterium and nitrogen fixation commences.

Nitrogen fixation

Frankia develops within the nodule and in culture as a filamentous bacterium which differentiates into morphologically distinct forms: hyphae, swollen terminal vesicles and multicellular sporangia. The vesicles have been identified as the site for nitrogenase activity. They are similar to heterocysts of nitrogen-fixing cyanobacteria (Schultz and Benson, 1989). The formation of vesicles was repressed and acetylene reduction could not be detected in cells cultured in the presence of 2–5 mM NH_4Cl (Zhang *et al.*, 1986). This phenomenon is similar to the effect of N on nodulation and nitrogen fixation in legume-rhizobia symbiosis, *Frankia*-nodulated actinorhizal plants contribute as much fixed nitrogen to the soil as do rhizobia-nodulated legumes (Torrey, 1978).

It has been reported by many investigators (Dawson and Sun, 1981; Normand and Lalonde, 1982; Coyne, 1983; Reddell and Bowen, 1985) that *Frankia* strains varied in their effectiveness in nitrogen fixation on actinorhizal plants. Reddel and Bowen (1985) found that strains of *Frankia* highly effective on one species were frequently ineffective on the other, suggesting host-endophyte specificity. In the nodule, sporangia are frequently absent. It has been found that sporangial formation in the nodule (spore+) adversely affected the productivity of the host plants (Normand and Lalonde, 1982). VadenBosch and Torrey (1984) found that nitrogenase activity and symbiotic efficiency were higher in spore(–) than spore(+) plants.

Contribution of N fixed by *Frankia*–actinorhizal association to soil and other plants has been realized. Miller and Murray (1978) have demonstrated the beneficial effects of interplanting of 4-year-old Douglas fir with red alder. A similar result was found by growing *Ceanothus* as a companion cropping with 9–10-year old Douglas fir (Youngberg *et al.*, 1979). Gessel and Turner (1974) estimated that nodulated red alder contributed 80–200 kg N/ha per year by way of leaf fall, while non nodulated Douglas fir provided only 7–22 kg N/ha per year.

Casuarina, a tropical multipurpose tree, growing in the sand dunes and marginal land, could fix considerable amounts of N. Yoneyama et al. (1990, 1993), using the ^{15}N natural abundance method found that *Casuarina* could contribute large amounts of N through N_2 fixation, ranging from 65% to 95%. These findings were comparable to those for alder plants (Domenach *et al.*, 1989) and for *Casuarina equisetifolia* done by Sanginga *et al.* (1989), who used the total nitrogen difference method using uninoculated plants as reference. Work done by the DOA, Thailand on *C. equisetifolia* and *C. junghuhniana*, using the ^{15}N dilution method, showed that inoculated plants could fix N ranging from 47% to 94% (unpublished data).

Nitrogen fixation, the *nif* sequences, in *Frankia* share similarity with other nitrogen-fixing bacteria, especially *Anabaena*, *Azotobacter* spp. and *Klebsiella pneumoniae* (Normand *et al.*, 1988). It is also possible that they share some homology in nodulation (*nod*) genes with rhizobia (Reddy *et al.*, 1988).

References

Baker D 1987. Relationship among pure cultured strains of *Frankia* based on host specificity. Physiologia Plantarum, 70, 245–248.

Baker D, Schwintzer C R 1990. Introduction. In: Schwintzer C R, Tjepkema J D, eds. The Biology of *Frankia* and Actinorhizal Plants. Academic Press, San Diego, CA, 3–11.

Baker D, Selling E 1984. *Frankia*: new light on an actinomycete symbiont. In Ortiz-Ortiz I, Bojalil L G, Yakoleff V, eds. Biological, Biochemical and Biomedical Aspects of Actinomycetes Academic Press, New York, 563–574.

Baker D, O'Keefe D 1984. A modified sucrose fractionation procedure for the isolation of frankiae from actinorhizal root nodules and soil samples. Plant Soil, 78, 23–28.

Berry A M 1984. The actinorhizal infection process: review of recent research. In: Klusg M, Reddy C, eds. Current Perspectives in Microbial Ecology. American Society of Microbiology, Washington, 222–229.

Callaham D, Del Tredici P, Torrey J G 1978. Isolation and cultivation *in vitro* of the actinomycete causing root nodulation in *Comptonia*. Science, 199, 899–902.

Callaham D, Newcomb W, Torrey J G, Peterson R L 1979. Root hair infection in actinomycete-induced root nodule initiation in *Casuarina*, *Myrica* and *Comptonia*. Bot. Gaz., 140, 1–9.

Coyne P D 1983. Specificity between *Casuarina* species and root nodule organisms. In Midgley S J, Turnbull J W, Johnston R D, eds. *Casuarina* Ecology, Management and Utilization CSIRO, Melbourne, 205–210.

Dawson J O, Sun S H 1981. The effect of *Frankia* isolates from *Comptonia peregrina* and *Alnus crispa* on the growth of *Alnus glutionosa*, *A. cordata* and *A. incana* clones. Can J. Forest Res., 11, 758–762.

Domenach A M, Kurdal F, Bardin R 1989. Estimation of symbiotic dinitrogen fixation in alder forest by method based on natural ^{15}N abundance. Plant Soil, 118, 51–59.

Gessel S P, Turner J 1974. Litter production by red alder in Wester Washington. For. Sci., 29, 325–330.

Jamann S, Fernandez M P, Normand P 1993. Typing method for N_2-fixing bacteria based on PCR-RFLP application to the characterization of *Frankia* strains. Mol. Ecol., 2, 17–26.

Miller I M, Baker D 1985. Initiation development and structure of root nodules in *Elaeagnus angustifolia* L (Elaeagnaceae). Protoplasma, 128, 107–119.

Miller K E, Murray M D 1978. The effect of red alder on growth of Douglas-fir. In: Briggs D G, De Bell D S, Atkinson W A, eds. Utilization and Management of *Alder* USDA Forest Service. General Technical Report PHW-70, 283–306.

Nittayajarn A, Mullin B C, Baker D 1990. Screening of symbiotic frankiae for host specificity by restriction fragment length polymorphism analysis. Appl. Environ Microbiol., 56, 1172–1174.

Normand P, Lalonde M 1982. Evaluation of *Frankia* strains isolated from provenances of two *Alnus* species. Can. J. Microbiol., 28, 1133–1142.

Normand P, Simonet P, Bardin R 1988. Conservation of *nif* sequences in *Frankia*. Mol. Gen. Genet., 213, 238–246.

Normand P, Gouy M, Cournoyer B, Simonet P 1992. Nucleotide sequence of *nif* D from *Frankia alni* strain Ar13: phylogenetic inferences. Mol. Biol. Evol., 9, 495–506.

Quispel A, Rodriquez-Barrueco C, Subba Rao N S 1993. Some general considerations on symbioses of nitrogen fixing trees. In: Subba Rao N S, Rodriquez C, eds. Symbioses in Nitrogen Fixing Trees. International Science Publishers, New York, 1–32.

Reddell P, Bowen 1985. *Frankia* source affects growth, nodulation and nitrogen fixation in *Casuarina* species. New Phytol. 100.

Reddy A, Torrey J G, Hirsch A M. 1988 Isolation of *Frankia* strain HFPC cI3 nodulation genes by cross-species complementation of *Rhizobium meliloti*. In: Palacios R, Verma D P, eds. Molecular Genetics of Plant-Microbe Interactions APS Press, St Paul, MN, 213–214.

Sanginga N, Danso S K A, Bowen G D 1989. Nodulation and growth response of *Allocasuarina* and *Casuarina* species to phosphorus fertilization. Plant Soil, 118, 125–132.

Schultz N A, Benson D R 1989. Developmental potential of *Frankia* vesicles. J. of Bacteriol., 171, 6874–6877.

Simonet P, Nguyen Thi Le, DuCross E T, Bardin R 1988. Identification of *Frankia* strains by direct DNA hybridization of crushed nodules. Appl. Environ. Microbiol., 54, 2500–2503.

Simonet P, Normand P, Moiroud A, Bardin R 1990. Identification of *Frankia* strains in nodules by hybridization of polymerase chain reaction products with strain-specific oligonuclectide probes. Arch. Microbiol 153, 235–240.

Torrey J G 1978. Nitrogen fixation by actinomycete-nodulated angiosperms. Bioscience, 28, 586–591.

VandenBosch K A, Torrey J G 1984. Consequences of sporangial development for nodule function in root nodules of *Comptonia peregrina* and *Myrica gale*. Plant Physiol., 76, 556–560.

Yoneyama T, Murakami T, Boonkerd N, Wadisirisuk P, Siripin S, Kouno K 1990. Natural ^{15}N abundance in shrub and tree legumes, *Cassuina*, and non N_2 fixing plants in Thailand. Plants Soil, 128, 287–292.

Yoneyama T, Muraoka T, Murakami T, Boonkerd N 1993. Natural abundance of ^{15}N in tropical plants with emphasis on tree legumes. Plant Soil, 153, 295–304.

Youngberg C T, Wollum A G, Scott W 1979. *Ceanothus* in Douglas-fir clear cuts: Nitrogen accretion and impact on regeneration. In: Gordon J C, Wheeler C T, Perry D A, eds. Symbiotic Nitrogen Fixation in The Management of Temperate Forests. Forest Research Laboratory, Oregon State, Corrallis. Oregon, 224–233.

Zhang Z, Murry M A, Torrey J G 1986. Culture conditions influencing growth and nitrogen fixation in *Frankia* sp. HFPC c13 isolated from *Casuarina*. Plant Soil, 91, 3–15.

39
Ecophysiological studies of cyanobacteria in paddy soils of Bangladesh

M. A. HASHEM

Abstract

Ecophysiological studies of cyanobacteria isolated from three distinctly different types of paddy soils in Bangladesh were carried out. Paddy soils selected for the study were calcareous soils, saline soils and red soils. Physicochemical properties of soils had marked influences on the indigenous blue–green algal population. Wide variations in growth rate, ability to grow under heterotrophic conditions, nitrogen-fixing capacity under different conditions and tolerance to pesticides among the isolates of a genus and between different genera were observed. Heterotrophically grown cultures were more tolerant to pesticides compared to autotrophically grown cultures. Furthermore, the nitrogen fixation in heterotrophically grown cultures was 6–8 times higher than that in autotrophically grown cultures. Cyanobacteria isolated from saline soils had the capacity to reduce the electrical conductivity (salinity) measured at 4–20 ds/m. The reduction of electrical conductivity by the isolates ranged from 6% to 42.5% in liquid cultures and from 9.94% to 26.53% in soils.

Introduction

Bangladesh agriculture is predominantly rice-based, and more than 80% of the total cropped area is devoted to year-round production of rice. This crop is suitable for cultivation in soils having a wide range of characteristics, including some problem soils. Rice is largely grown under wet-land conditions, which are very conducive to the growth of cyanobacteria. In recent years Bangladesh soils have been grouped into 30 agro-ecological zones (AEZ) based on physiography, soil type, land inundation and climate. Saline soils, calcareous soils and red soils are important in Bangladesh agriculture, and management of these three soils is an important necessity from the agricultural point of view.

K.A. Malik et al. (eds.), Nitrogen Fixation with Non-Legumes, 333–344.

Cyanobacteria are being used as biofertilizers in many rice-growing countries of the world (Venkataraman, 1981). Reports are also available showing that the effect of cyanobacteria as biofertilizer is erratic and unpredictable. No reliable method is available to assure the success of inoculation with efficient cyanobacterial strains (Watanabe and Liu, 1992). Roger (1991) reported that non-indigenous strains rarely establish themselves. The growth and establishment of cyanobacteria in a given habitat are greatly influenced by physical and chemical factors (Roger and Kulasooriya, 1980). They also differ in nitrogen-fixing capacity and in their ability to thrive under different soil fertility condition (Roger and Reynaud, 1980). Hence, a study was essential to isolate cyanobacteria from different types of soils, and to assess their growth performance, nitrogen-fixation capacity and tolerance to commonly used pesticides for their subsequent use in producing biofertilizers specific to their habitats. Furthermore, we planned to study the possibility of using cyanobacteria as a biological means for the reclamation of soil salinity.

Materials and methods

Soils

Three distinctly different types of paddy soils, *viz.* saline soils, calcareous soils and red soils, located in three agro-ecological zones of Bangladesh, have been selected for the study. Soil samples were collected from the rice fields of three locations for each type of soil from 0 to 15 cm depth, and were divided into two halves – one for the physicochemical analyses of soil and the other for cyanobacterial study.

Physicochemical properties of soils

Soils were analysed for pH, organic carbon, total nitrogen, available phosphorus, sulphur, exchangeable sodium, potassium and calcium and cation exchange capacity (CEC) following standard methods (Page *et al.*, 1989; Jackson, 1962). Particle size analysis of soil was determined by the hydrometer method (Piper, 1950) and textural class was determined using Marshall's triangular coordinates of the USDA system.

Measurement of soil salinity

For determining electrical conductivity (EC) an aqueous soil extract was prepared from 1 : 5 soil to water mixtures by vacuum filtration. EC was measured by filling a EC electrode with soil extract using an 0.01 M KCl solution to calibrate the meter (Biswas and Mukherjee, 1987).

Isolation and identification of cyanobacteria

Dilution series up to 10^{-7} were made with the collected soils, inoculating 1 ml of each soil dilution into test tubes containing a modified version of the Chu 10D-N sterile growth medium (Sinclair and Whitton, 1977) and incubating at 30°C under the light of 100 W bulbs fixed 20 cm apart and 50 cm from the tubes. Unicyanobacterial cultures, obtained by gently homogenizing and plating cultures on the growth medium, were treated with cycloheximide (0.1 mg/ml) to eliminate contaminating eukaryotes, and were then freed from bacterial contamination on scored agar plates as described by Vaara *et al.*, 1979.

Growth of cultures

Cyanobacterial isolates were grown in a modified version of Chu 10D-N medium, under the light arrangement mentioned above, at a temperature of 30°C. For heterotrophic growth the autoclaved medium was supplemented with 20 mM glucose, sucrose and fructose separately, and incubated in the dark. Cultures were checked for heterotrophic growth 1 month after incubation.

Determination of the indigenous cyanobacterial population

The indigenous cyanobacterial population of soils was determined according to the method described by Prammer and Schmidt (1967).

Measurement of nitrogen fixation

For measuring nitrogen fixation the cultures were inoculated in the growth medium and incubated under the light and temperature conditions maintained during growth. Nitrogen fixation was measured according to Yoshida *et al.* (1976) after the required period of incubation.

Measurement of dry weight

The dry weight of the cyanobacterial isolates was measured by filtering the cultures and drying the samples at 85°C until a constant weight was obtained.

Results and discussions

Soil properties and indigenous cyanobacterial population

Table 1 shows the major physicochemical properties of the soils selected for the study. Cyanobacteria were recorded in all soils studied, but the population varied

Table 1. Physical and chemical properties of soils under study

Soils	pH	OM (%)	Textural class	Total N (%)	Available P (ppm)	Available S (ppm)	CEC (me/100 g)	Exchangeable K (meq/100 g)	EC (ds/m)
Saline 1	5.4	1.62	Silty clay	0.15	8.4	18.5	16.5	0.23	14.7
Saline 2	7.0	1.75	Silty clay	0.13	17.2	22.0	18.0	0.29	16.1
Saline 3	5.7	1.25	Silty clay loam	0.08	16.0	13.3	18.0	0.15	17.5
Calcareous 1	8.2	1.73	Silty loam	0.12	15.8	21.0	16.8	0.29	3.6
Calcareous 2	8.3	1.67	Silty clay loam	0.10	12.9	19.0	18.8	0.25	2.8
Calcareous 3	8.4	1.65	Clay loam	0.10	16.8	25.0	15.9	0.27	3.7
Red 1	5.2	0.95	Clay soil	0.11	6.3	11.0	14.8	0.14	2.8
Red 2	5.0	1.23	Clay soil	0.10	10.0	12.1	13.3	0.13	2.3
Red 3	5.4	1.34	Clay soil	0.04	8.4	12.5	10.3	0.12	2.9

markedly due to variations in soil properties. The cyanobacterial population in soils under study ranged from 0.4×10^4 to 22.0×10^4/g soil (Table 2). The minimum population was recorded in red soils ranging from 0.4×10^4 to 1.7×10^4/g. With respect to the number of cyanobacterial population the selected soils may be ranked as calcareous soils > saline soils > red soils.

Highly positive correlation of cyanobacterial population with the organic matter status ($r = 0.956^{***}$), soil pH ($r = 0.867^{***}$), available S ($r = 0.902^{***}$) and exchangeable K contents ($r = 0.893^{**}$) of soil was observed (Figures 1a, 1b, 1d, 1e).

A positive correlation of cyanobacterial population was also found with available P status of soil (Figure 1c). A very weak correlation was noticed between cyanobacterial population and CEC of soil (Figure 1f). Cyanobacterial population did not correlate well with total N and EC of soils. A positive correlation between soil pH and cyanobacterial population was also observed in different rice fields of Bangladesh (Biswas *et al.*, 1994) and in Japan (Roger and Kulasooriya, 1980). Ahmed *et al.* (1990) and Biswas *et al.* (1994) also reported a positive relationship between available phosphorus and cyanobacterial population in some Bangladesh soils. It appears that the growth of cyanobacteria in the rice field can be increased by supplying phosphatic fertilizer.

Cyanobacterial isolates

For each type of soil, six isolates of which three *Anabaena* and three *Nostoc* were brought under unicyanobacterial cultures for detailed study.

Heterotrophic growth

All the isolates under study were capable of growing under heterotrophic conditions (Table 3), but differences in their substrate preference were noticed. The isolates from saline soils preferred both glucose and fructose; on the other hand the isolates from red soils preferred both sucrose and fructose; the isolates from

Table 2. Indigenous cyanobacterial population in soils under study

Soils	Cyanobacterial population/g soil ($\times 10^4$)
Saline 1	4.9
Saline 2	14.0
Saline 3	2.2
Calcareous 1	11.0
Calcareous 2	22.0
Calcareous 3	14.0
Red 1	0.4
Red 2	1.1
Red 3	1.7

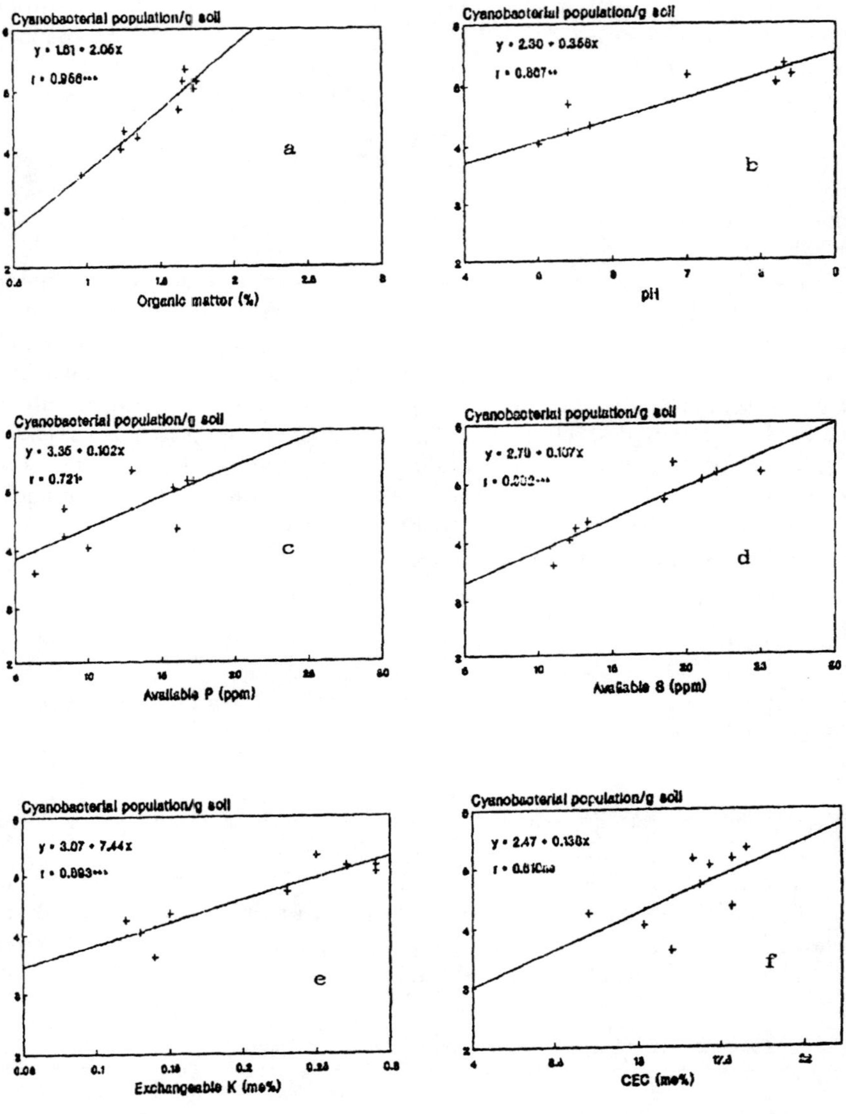

Figure 1. Relationship between indigenous cyanoacterial population (log values) and soil properties.

calcareous soils preferred only glucose. The results of this experiment also support the findings of Gallon *et al.* (1991) and Hashem *et al.* (1987), who reported that *Oscillatoria* sp. UCSB-8 can grow heterotrophically on glucose and *Oscillatoria* sp. UCSB-25 can also grow heterotrophically, but on fructose.

Table 3. Heterotrophic growth of cyanobacterial isolates on glucose, sucrose and fructose

	Substrate		
Cyanobacterial isolates	Glucose	Sucrose	Fructose
Anabaena S1	+	–	+
Nostoc S1	+	–	+
Anabaena S2	+	–	+
Nostoc S2	+	0	+
Anabaena S3	+	–	+
Nostoc S3	+	0	+
Anabaena C1	+	–	–
Nostoc C1	+	–	–
Anabaena C2	+	0	–
Nostoc C2	+	0	–
Anabaena C3	+	0	–
Nostoc C3	+	–	–
Anabaena R1	–	+	+
Nostoc R1	0	+	+
Anabaena R2	–	+	+
Nostoc R2	0	+	+
Anabaena R3	0	+	+
Nostoc R3	0	+	+

+, Positive growth; –, dead; 0, just survived.

Growth (dry weight) of cyanobacteria

Maximum growth was observed in the isolates of calcareous soils ranging from 5.5 to 7.3 mg/50 ml cultures (Table 4). The values for growth (dry weight) obtained under heterotrophic conditions were less than those obtained under autotrophic conditions (Table 4). Similar results have been reported by Stewart (1997), Carr (1973) and Rippka (1972). Nitrogen fixation was maximum in the isolates of cyanobacteria obtained from calcareous soils, followed by those from saline soils (Figures 2a, 2b, 2c). In saline soils nitrogen fixation was shown to be higher in *Anabaena* than that in *Nostoc*, but in calcareous soils nitrogen fixation in *Anabaena* and *Nostoc* was almost the same. *Anabaena* from red soils fixed nitrogen at about double the value of *Nostoc*. Nitrogen fixation by the isolates was at an increasing rate up to 50 days in all types of soils studied. The pattern of nitrogen fixation in heterotrophically grown cultures was similar to that in autotrophically grown cultures, but the rate of nitrogen fixation was about 6–8 times higher in heterotrophically grown cultures compared to that in autotrophically grown cultures (Figures 2a, 2b, 2c).

Effect of basudin on growth and nitrogen fixation by cyanobacterial isolates

An increase in growth (dry weight) was observed up to a basudin concentration of 7.5 ppm (recommended dose for field application) in all the isolates of cyanobacteria grown under autotrophic conditions (Figures 3a, 3b, 3c). The stimulatory effect of pesticides on cyanobacteria has been reported by Gangawane and

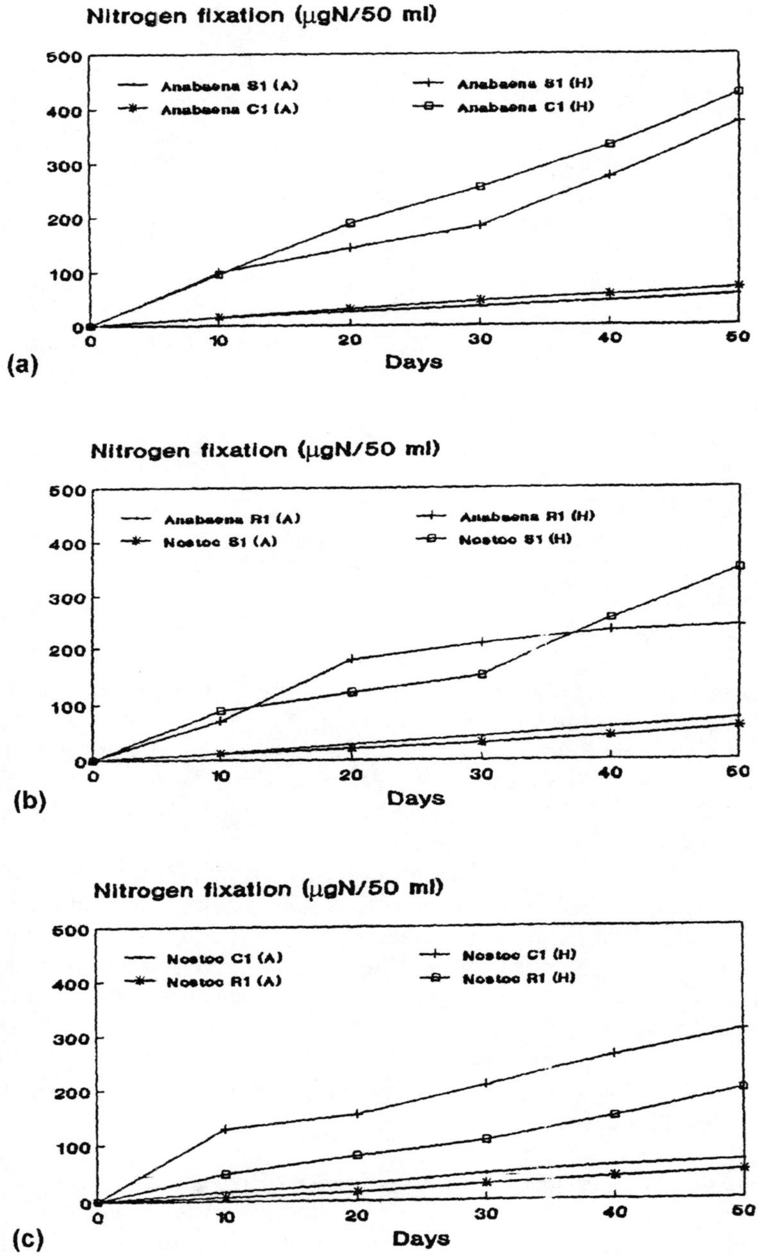

Figure 2. Nitrogen fixation by cyanobacterial isolates under autotrophic (A) and heterotrophic (H) conditions.

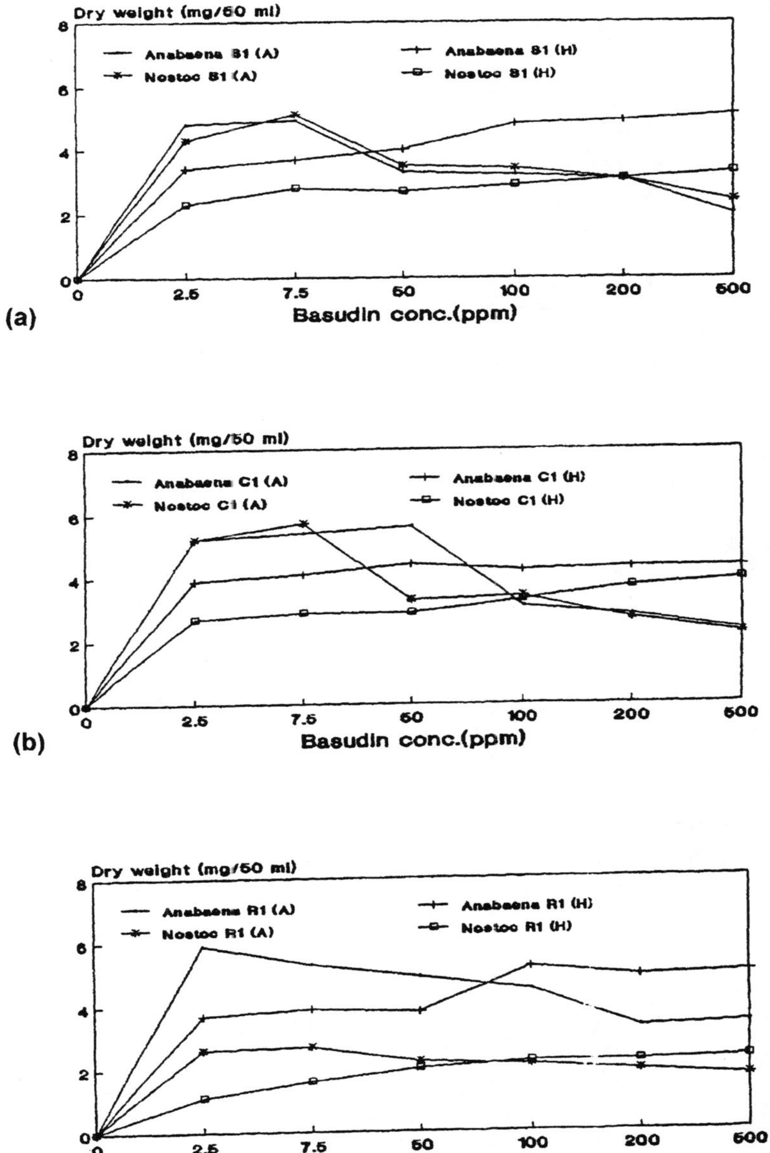

Figure 3. Effect of basudin on the growth of cyanobacterial isolates under autotrophic (A) and heterotrophic (H) conditions.

Table 4. Growth of cyanobacterial isolates under autotrophic and heterotrophic conditions after 50 days of incubation under growth conditions

	Dry weight (mg/50 ml)	
Cyanobacterial isolates	Autotrophic	Heterotrophic
Anabaena S1	4.3	2.9
Nostoc S1	3.5	1.8
Anabaena S2	4.9	3.1
Nostoc S2	3.7	2.6
Anabaena S3	5.4	3.9
Nostoc S3	2.3	1.2
Anabaena C1	6.8	4.5
Nostoc C1	5.5	2.8
Anabaena C2	7.2	4.8
Nostoc C2	6.5	3.6
Anabaena C3	6.9	3.9
Nostoc C3	7.3	4.5
Anabaena R1	5.5	3.8
Nostoc R1	2.8	1.2
Anabaena R2	6.1	3.2
Nostoc R2	2.7	1.9
Anabaena R3	3.1	1.7
Nostoc R3	2.6	1.7

Saler (1979). A decrease in growth occurred after the recommended dose. All the isolates showed a similar pattern under autotrophic conditions irrespective of soil types at different concentrations of basudin. Under heterotrophic conditions growth was slower compared to that under autotrophic conditions at lower concentrations, but an increase in growth was recorded up to the highest concentration of basudin applied (500 ppm), though at a very slow rate. All the isolates of cyanobacteria showed a similar pattern of growth under heterotrophic conditions.

Table 5. Effects of cyanobacterial isolates on changes in electrical conductivity (EC) in liquid cultures

	EC measured at				
Cyanobacterial isolates	4 ds/m	8 ds/m	12 ds/m	16 ds/m	20 ds/m
Anabaena S1	2.9	6.3	8.5	13.5	18.8
Nostoc S1	3.1	6.4	9.1	14.3	17.7
Anabaena S2	2.6	5.9	8.8	11.6	14.5
Nostoc S2	2.3	5.4	8.2	12.9	14.9
Anabaena S3	2.4	6.2	8.0	10.4	13.2
Nostoc S3	2.6	6.6	8.1	11.3	12.8

Cultures were inoculated in growth medium at electrical conductivity indicated, and incubated under growth conditions. EC was measured after 50 days of incubation.

Table 6. Effects of cyanobacterial isolates on changes in electrical conductivity (EC) in soils under laboratory conditions.

Cyanobacterial isolates	EC (ds/m): Saline 1	Percentage reduction	EC (ds/m): Saline 2	Percentage reduction	EC (ds/m): Saline 3	Percentage reduction
Anabaena S1	12.7	13.6	14.1	12.42	15.3	12.56
Nostoc S1	11.9	19.04	14.5	9.94	15.1	14.90
Anabaena S2	11.7	20.4	13.8	14.28	13.9	22.36
Nostoc S2	11.5	21.76	13.2	18.0	14.5	18.63
Anabaena S3	10.8	26.53	12.3	23.60	13.5	24.84
Nostoc S3	11.2	23.8	12.1	24.84	13.7	23.60

Cultures were inoculated in saline soils under study, kept in plastic pots and incubated under growth conditions of cyanobacteria in the laboratory. EC of soil was measured 50 days after incubation.

Effects of cyanobacteria on changes in electrical conductivity

Cyanobacterial isolates from saline soils brought down the EC of the growth medium measured at 4 to 20 ds/m (Table 5). The isolates also had the capacity to reduce the EC of the saline soils (Table 6). The inoculated cyanobacterial isolates take up sodium from the medium which is subsequently metabolized in the body cells, as a result sodium concentration decreases which might be one of the reasons for the reduction of EC by cyanobacterial isolates. Sodium and chloride uptake by cyanobacterial isolates has been reported by Gabbay-Azaria and Tel-or (1993).

References

Ahmed M, Bhuiya Z H, Mian M H, Hashem M A, Hossain M M 1990. Study on the occurrence of blue–green algae in different soils of Bangladesh under three moisture conditions. Bangladesh J. Soil Sci., 21, 80–87.

Biswas T D, Mukherjee S K 1987. Textbook of Soil Science. Tata McGraw-Hill, New Delhi, 291.

Biswas B K, Hashem M A, Islam M R, Rahman M M 1994. Influence of soil properties on the population of blue–green algae in different agro-ecological zones of Bangladesh. Bangladesh J. Crop Sci., 5, 11–18.

Carr N G 1973. Metabolic control and autotrophic physiology. In: Carr N G, Whitton B A, eds. The Biology of Blue–green Algae. Blackwell, Oxford, 39–65.

Gabbay-Azaria R, Tel-or E 1993. Mechanisms of salt tolerance in cyanobacteria In: Plant response to the environment. CRC Press, Boca Raton, FL, 000–000.

Gallon J R, Hashem M A, Chaplin A E 1991. Nitrogen fixation by *Oscillatoria* sp under autotrophic and photoheterotrophic conditions. J. Gen. Microbiol., 137, 31–39.

Gangawane L V, Saler R S 1979. Tolerance of certain fungicides by nitrogen fixing blue–green algae. Curr. Sci., 48, 306–308.

Hashem M A, Gallon J R, Chaplin A E 1987. Growth performances of two spp. of *Oscillatoria* under heterotrophic conditions. Bangladesh J. Aqua. 9, 61–67.

Jackson M L 1962. Soil Chemical Analysis. Constable, London.

Page A L, Miller R H, Keeney D R 1989. Methods of Soil Analysis, Part II, 2nd edn. American Society of Agronomics, Madison, Wisconsin, USA.

Piper C S 1950. Soil and Plant Analysis. Adelaide University Press, Adelaide, Australia.

Prammer D, Schimdt E L 1967. Algae. Experimental Soil Microbiology. Burgess, Minneapolis, Minnesota, 25–27.

Rippka R 1972. Photoheterotrophy and chemoheterotrophy among unicellular blue–green algae. Arch. Microbiol., 87, 93–98.
Roger P A 1991. Reconsidering the utilization of blue–green algae in wetland rice cultivation. In: Dutta S K and Sloger C, eds. Biological N_2 – fixation Associated with Rice Production. Oxford University Press and IBH, New Delhi, 119–141.
Roger P A, Kulasooriya S A 1980. Blue–green Algae and Rice. IRRI, Los Banos, Philippines.
Roger P A, Reynaud P A 1980. Ecology of blue–green algae in paddy fields. In: , eds. Nitrogen and Rice. IRRI, Los Banos, Philippines, 287–310.
Stewart W D P 1977. Bluegreen algae. In: Hardy R W F, Silver W S, eds. A Treatise on Dinitrogen Fixation, Section III: Biology. John Wiley, New York, 63–123.
Sinclair C, Whitton B A 1977. Influence of nitrogen source on morphology of Rivulariaceae (cyanophyta). J. Phycol., 13, 335–340.
Vaara T, Vaara M, Niemela S 1979. Two improved methods for obtaining axenic cultures of cyanobacteria. Appl. Env. Microbiol., 38, 1011–1014.
Venkataraman G S 1981. Blue–green algae for rice production. FAO Soils Bulletin. No. 46, Rome.
Watanabe I, and Liu C C. Improving nitrogen fixing systems and integrating them into sustainable rice farming. Plant Soil, 141, 57–67.
Yoshida S, Forno D A, Cook J H, Gomez K A 1976. Laboratory Manual for Physiological Studies of Rice. IRRI, Los Banos, Philippines.

40
Cyanobacterial root nodules in *Tribulus terrestris* L. (Zygophyllaceae)

A. MAHMOOD AND M. ATHAR

Abstract

This chapter is a histological study of the nodular swellings found on the roots of *Tribulus terrestris* L., a xerophytic non-leguminous plant. Anatomically the nodular swellings found on this species show similarities to the tissue pattern of non-leguminous nodules. The cytology of the proliferated tissue is described. Two types of endophytes were observed in the cortical cells of the nodule. Cells of the outer cortex were bigger, showing large intercellular spaces in which clumps of rhizobia (*Bradyrhizobium*) could be seen. The invasion pathway of bacteria appears to be by 'crack entry', and their cell-to-cell movement was both intercellular and intracellular. Cells of the inner cortex were smaller and were heavily colonized by a Cyanobacterium. This has been described as a new taxon and is designated as *Newmania karachiensis* Mahmood and Athar gen. nov et sp. nov. The occurrence of a blue–green alga as endophyte in the root nodules of an angiosperm is being reported for the first time.

Introduction

The general assumption that *Rhizobium*-induced nodules are restricted to legumes only was maintained for many years, until Trinick (1973) discovered that nodules in certain Ulmaceae are induced by a *Rhizobium* species. Three species of *Parasponia* have been found to bear nodules (Gresshoff *et al.*, 1984). Root nodules have also been reported in certain Zygophyllaceae (Medan and Tortosa, 1983; Gresshoff *et al.*, 1984; Athar and Mahmood, 1985). Nodular growths on the roots of *Tribulus* spp. contained a fungus (Issatschenko, 1913) while Sabet (1946), Mostafa and Mahmoud (1951) and Athar and Mahmood (1972) described *Rhizobium* in the nodules of *Tribulus alatus* and *Tribulus terrestris*, respectively. On

K.A. Malik et al. (eds.), Nitrogen Fixation with Non-Legumes, 345–350.

the other hand some workers (Allen and Allen, 1950; Sahibzada and Khan, 1973; Gresshoff *et al.*, 1984) failed to find any bacterial occupant inside the nodular growths on the roots of *Tribulus* spp. On the basis of these conflicting reports the nature of the endophyte in the nodules of *T. terrestris* was reinvestigated.

Materials and methods

The nodules of *Tribulus terrestris* used in the present study were collected from plants growing at Karachi University campus area. Nodules were washed thoroughly with running tap water and fixed in formalin–acetic acid–alcohol in the ratio of 5 : 5 : 90, and embedded in paraffin wax (m.p. 56°C). Serial longitudinal sections 8–10 μm thick were cut with a rotary microtome. Paraffin was removed with xylol (Jensen, 1962) and the sections were stained with fuchsin methyl green following Lawrence *et al.* (1967). The photographs were taken under ordinary light and phase-contrast illumination using a Nikon Biophot microscope.

Results and discussion

Nodules were found both on main as well as lateral roots of *T. terrestris*. The nodules occurred singly as well in branched forms and averaged 1.5–2.0 mm in diameter. Nodules were loosely attached to the root system and could be detached easily. They were white to yellowish in colour, similar to the colour of the root. Field observations demonstrated that maximum nodulation occurred in sandy soils and sandy gravels.

Anatomy of nodules

The nodule in a radial longitudinal section could be differentiated into an epidermis which was replaced by a periderm in older nodules. The nodule cortex formed the bulk of the nodular tissue. It could be differentiated into outer and inner cortex. The cells of the outer cortex were larger than the cells of the inner cortes. These cells contained large intercellular spaces in which bacteria could be seen. The cells of the inner cortex were smaller, more or less rectangular in shape, arranged in radial files, did not show intercellular spaces and always took deeper stain than the cells of the outer cortex. The cells of the inner cortex were heavily colonized by a Cyanobacterium. The cortical cells situated at the base of the nodule contained starch. The nodules were similar in morphology to those of root nodules on non-legumes, i.e. they had a central vascular system and the infected tissue occupied the cortical cells on both sides of the stele. Internal infection threads such as described in nodules of *Leucaena leucocephala* (Chen and Rolfe, 1988; Iqbal and Mahmood, 1992) could be seen.

The invasion path way of the bacteria appeared to be by 'crack entry' (Chandler *et al.*, 1982), and their cell-to-cell movement was both intercellular and intracellular. Bacterial infection through root hairs reported by Sabet (1946) in *Zygophyllum coccineum* L. was not observed. From the comparison of nodular growths on roots of *T. cistoides* (Allen and Allen, 1950) with those on *T. terrestris*, it can be said that the latter showed close similarities with a non-leguminous root nodule.

It is worth mentioning that a new system of angiosperm phyllogenies, based on DNA sequence comparisons, revealed a markedly different relationship among nodulated plants and indicated a more coherent group than has previously been thought. Zygophyllaceae fits in this scheme, and finds a place along with other non-leguminous nodule-bearing families of angiosperms (Swensen and Mullin, 1996).

Nodule endophyte

Two types of endophytes were observed:

1. *Bacteria isolated from the nodules of T. terrestris.* Bacteria occupied the cells of the outer cortex both intercellularly and intracellularly. The cells of the outer cortex were large with big intercellular spaces in which clumps of rhizobia could be seen. The bacterial isolates have been characterized as rhizobia (*Bradyrhizobium*) on the basis of their colonial morphology, staining reactions (Athar and Mahmood, 1972) and on the basis of cross-infection of legume genera, viz. *Vigna, Arachis* and *Sesbania*, with bacterial isolates from *T. terrestris* nodules (Athar, 1980). Cross-infection of leguminous plants such as *Trifolium* and *Arachis* with bacterial isolates from *Tribulus alatus* nodules has been reported by Mostafa and Mahmoud (1951).
2. *Cyanobacteria isolated from the nodules of T. terrestris.* The nodular tissue of the outer as well as the inner cortex were occupied by cyanobacterial cells, but the cells of the inner cortex were heavily colonized. Each host cell contained one algal cell. The organism was unicellular, globose (3-) 6.5–13.0 μm in diameter, cell wall thick, bilayered, central pseudovacuole. Nannocytes (plannocytes or baeocytes) were present which were highly refractile under the microscope (Figure 1). They were rod-shaped as well as circular in outline. This is a new taxon and is designated as *Newmania* gen. nov. et N. *karachiensis* sp. nov. after the late Dr I. V. Newman, Department of Botany, University of Sydney, Australia for his outstanding research contributions in the field of apical biology.

Description

Newmania karachiensis Mahmood and Athar gen. nov. et sp. nov resembles the genera *Cynechocystis* and *Aphanocapsa* in being spherical and cells single (-pairs) (Desikacharry, 1959) but differs in a distinct sheath, unequal cell divisions once

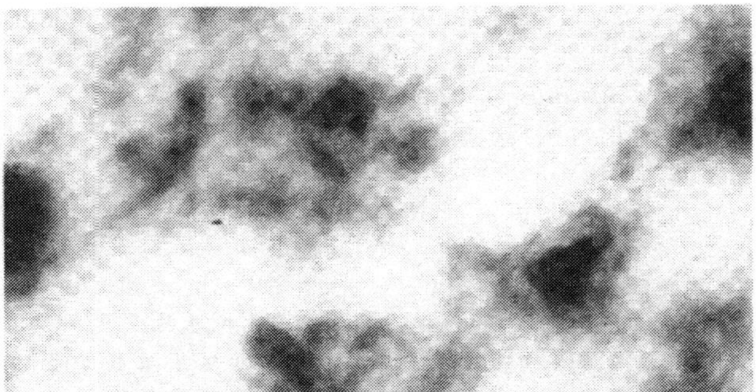

Figure 1. A single cell of *Newmania karachiensis* showing a bilayered cell wall (c) and nannocytes (n) in a ring around the central pseudovacuole (v) (phase contrast × 4000).

or twice forming two or three cells (Figure 2), a central pseudovacuole, presence of nannocytes and in endophytic habit. *N. karachiensis* resembles *A. zanardinii* (Hauck.) Hansgrig in cell dimensions (10–16) μm) but differs in thickness of sheath (Tilden, 1910).

Description: *Newmania karachiensis* Mahmood and Athar gen. nov et sp. nov. unicellularis, globosis (3-) 6.5–13 μm diam., crassis paries, bistralis, cum vacuole centralis; nannocytes/plannocytes/baeocytes refractis, bacilliformis vel ambitus circularis. Holotype: Karachi University, Botany Department. Leg M. Athar, 4-9-1984, KUH. 17.

The entry of the cyanobacterium inside the nodule could occur at the nannocyte stage. We have observed nannocytes in the xylem vessels of *T. terrestris* nodules, which in the presence of bacteria could enter the cortical cells of the nodule. It has been reported that bacteria degrade the cell walls and facilitate the entry of blue–greens into the host tissue (Towata, 1985a).

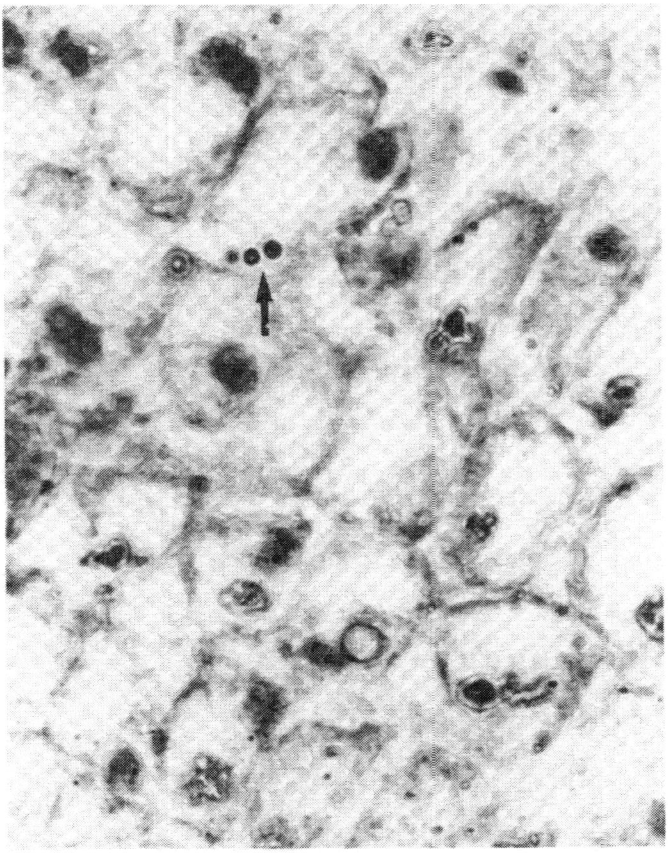

Figure 2. An enlarged view of a portion of the inner cortex of a *Tribulus terrestris* nodule showing cell configurations of *Newmania karachiensis* and the pattern of cell division (arrow) (× 4000).

The occurrence of a unicellular cyanobacterium colonizing the nodule tissues of an angiosperm is being reported for the first time. Known description of a cyanobacterium living as endophyte in angiosperms is *Gunnera–Nostoc* symbiosis where *Nostoc* lives intracellularly in the cortical tissue of the host plant (Silvester, 1976) and *Trifolium–Nostoc* association where *Nostoc* was found occasionally in the root nodules of the host plant (Subba Rao, 1977).

Acknowledgements

The authors express their sincere gratitude to Prof. Dr M. Nizamuddin for identification of the cyanobacterium, and helpful suggestions throughout the research. The authors thankfully acknowledge Dr J. Pirzada, Institute of Marine Biology, University of Karachi, for useful discussions.

References

Allen E K, Allen O N 1950. The anatomy of nodular growths on roots of *Tribulus cistoides* L. Proc. Soil Soc. Am., 14, 179–183.

Athar M 1980. Investigations on non-leguminous nodule bearing families of angiosperms from Karachi area with particular reference to Zygophyllaceae. MPhil thesis. University of Karachi.

Athar M, Mahmood A 1972. Root nodules in some members of Zygophyllaceae growing at Karachi University Campus. Pak. J. Bot., 4, 209–210.

Athar M, Mahmood A 1981. Extension of *Rhizobium* host range to Zygophyllaceae. In: Gibson A H, Newton D E, eds. Current Perspectives in Nitrogen Fixation. Australian Academy of Sciences, Canberra, 481.

Athar M, Mahmood A 1985. Nodulation and nitrogen fixation by *Tribulus terrestris* under natural conditions. Pak. J. Agric. Res., 6, 101–106.

Chandler M R, Date R A, Roughley R J 1982. Infection and root-nodule development in *Stylosanthes* species by *Rhizobium*. J. Exp.t. Bot., 33, 47–57.

Chen H, Rolfe B G 1988. *Rhizobium* infection of *Leucaena leucocephala* via the formation of infection threads in curled root hairs. J. Plant Physiol, 332, 379–382.

Desikacharry T V 1959. Cyanophyta. Indian Council of Agricultural Research, New Delhi.

Gresshoff P M, Newton S, Mohapatra S S, Scott K P, Mowitt S, Price, G D, Blender G L, Shine J, Rolfe G B 1984. Symbiotic nitrogen fixation involving *Rhizobium* and non-legume *Parasponia*. In: Veegar C, Newton W, eds. Advances in Nitrogen Research. Martinus Niijhoff, the Hague, 483–489.

Iqbal R, Mahmood A 1992. Structural studies on root nodules of *Leucaena leucocephala* with particular reference to the infection process. Pak. J. Bot., 24, 142–152.

Issatschenko B L 1913. Uber die Wurzelknollchen bei *Tribulus terrestris* L. Bull. Jord. Bot. St. Peterson, 13, 23–31 (in Russian, with German summary).

Jensen W A 1962. Botanical Histochemistry, Principles and Practice. W H Freeman, San Francisco.

Lawrence D B, Schoenike R E, Quispel A, Bond G 1967. The role of *Dryas drummondii* in vegetation development following ice recession at Glacier Bay, Alaska, with special reference to its nitrogen fixation by root nodules. J. Ecol., 55, 793–813.

Medan P D, Tortosa R D 1983. Nodulos radicales en *Larrea* (Zygophyllaceae). Bol. Soc. Argentina Bot., 22, 221–236.

Mostafa M A, Mahmoud M A 1951. Bacterial isolates from root nodules of Zygophyllaceae. Nature, Lond., 167, 446.

Sabet Y S 1946. Bacterial root nodules in the Zygophyllaceae. Nature, Lond., 157, 656–657.

Sahibzada M A, Khan A G 1973. Starch-filled nodular hypertrophies in *Tribulus terrestris*. Pak. J. Bot., 5, 87–88.

Silvester W B 1976. Endophyte adaptation in *Gunnera-Nostoc* symbiosis. In: Nutman P S, ed. Symbiotic Nitrogen Fixation in Plants, IBP-7. Cambridge University Press, Cambridge, 521–528.

Subba Rao N S 1977. Soil Microorganisms and Plants Growth. Oxford University Press and IHB Publishing Co., New Delhi.

Swensen S M, Mullin B C 1996. The impact of molecular systematics on hypothesis for the evolution of root nodule symbiosis. Seventh International Symposium on BNF with Non-legumes. 16–21 October, Faisalabad, Pakistan (abstract).

Tilden J 1910. Minnesota Algae, vol. I. Minnesota University Press, Minneapolis, Minnesota.

Trinick M J 1973. Symbiosis between *Rhizobium* and the non-legume, *Trema aspera* (*Parasponia cannabina*). Nature. 244, 459–460.

Towata E M 1985a. Mucilage glands and cytochemical ultrastructural analyses of the *Gunnera Kealensis/Nostoc* symbiosis. Bot. Gaz., 146, 56–62.

Posters

41 Poster

Use of *Azolla* as biofertilizer to enhance crop yield in a rice–wheat cropping system under mild climate

SIKANDER ALI, NAIMA HAMID, DILROSH KHAN AND KAUSER A. MALIK

Abstract

A field study was conducted to observe the effect of *Azolla* on rice yield, N uptake and its residual effect on subsequent wheat crop. *Azolla* developed into a thick mat due to the mild temperature prevailing during the rice season. *Azolla* was grown as intercrop with rice in field conditions and was used as cover, and also as incorporated treatment alone and in combination with a low dose of urea. As compared to unfertilized control, increase in rice grain yield was 28% for *Azolla* cover, 53% for *Azolla* incorporated and 71% for *Azolla* incorporated +30 kg N/ha treatment. The grain yield of wheat, grown after rice harvest and without any fertilizer application, was significantly higher for *Azolla* than control and chemical-N treatments. The results of this field study indicated that *Azolla* can successfully be used for increasing crop yield in a rice–wheat cropping system.

Introduction

Rice and wheat are the most important staple food crops in the world. The yields of rice and wheat in South Asia have been steadily declining; thus serious efforts are needed to increase the yield of these crops (IRRI, 1992). Because of the high prices of chemical fertilizers, the farmers of developing countries can barely afford to use expensive chemical fertilizers, and thus cannot achieve optimum yield. *Azolla* has been used for centuries as a biofertilizer in rice (Lumpkin and Plucknett, 1982), and is still considered a potential nitrogen biofertilizer in a rice-based cropping system (Peoples *et al.*, 1995). The usefulness of *Azolla* for increasing rice yield has been reported in various countries (Watanabe, 1987; Kumarasinghe and Eskew, 1993). However, little information is available on its use in a rice–wheat cropping system; therefore the present study was carried out

K.A. Malik et al. (eds.), Nitrogen Fixation with Non-Legumes, 353–357.

to investigate the usefulness of *Azolla* for enhancing the grain yield of these important crops.

Materials and methods

The field experiment was carried out at Swat, located in the rice-growing northern valley of Pakistan, having a maximum temperature of 35°C during the rice season (Choudhry, 1978). The study was conducted in 3 m × 5 m plots, arranged in a randomized complete block design. The soil was flooded for a few days before rice transplanting, and rice (var. JP-5) was transplanted at three seedlings/hill with 20 × 20 cm hill spacing. A mixed culture of six strains of *Azolla, viz. A. pinnata* (local), *A. filiculoides, A. caroliniana, A. microphylla, A. pinnata var. pinnata and Rong Ping* (hybrid) was inoculated in *Azolla* nursery, and farmyard manure was applied at 0.5 kg/m^2 to encourage its multiplication. Twenty-five days after rice transplanting (DAT) a mixed inoculum of above-described six *Azolla* strains was inoculated at 200 g (f.w)/m^2, into *Azolla* experimental plots (Table 1), and 5 g P_2O_5/m^2 (superphosphate) was also applied to all plots to supply phosphorus to rice, and enhance *Azolla* growth. For prevention of insect attack to *Azolla* and rice, Furadan granules (at 60 mg carbofuran/m^2) were broadcast 25 DAT to all plots. Urea at 30 kg N/ha was applied 38 DAT to nitrogen alone and N+*Azolla* treatment plots.

After *Azolla* inoculation it was allowed to grow without mixing into soil for *Azolla* cover treatments; while for *Azolla*-incorporated treatments nearly half of the *Azolla* mat was incorporated manually into the soil and the remainder was left to grow again until the next incorporation. Two such incorporations, i.e. 60 and 73 DAT, were made during the period of rice growth. Rice plants were harvested at maturity and then dried at 70°C for dry weight and total N analysis.

Table 1. Effect of *Azolla* and urea on dry-matter yield of rice, cv. JP-5

Treatment	Straw (kg/ha)	Grain (kg/ha)	Straw+grain (kg/ha)
T1. Control	5542^{c}	3917^{d}	9458^{c}
	0	*0*	*0*
T2. *Azolla* cover	6000^{bc}	5000^{c}	11000^{bc}
	8	*28*	*16*
T3. *Azolla* incorporated	7188^{abc}	6000^{ab}	13188^{ab}
	30	*53*	*39*
T4. 30 kg N/ha	7375^{ab}	6125^{ab}	13500^{ab}
	33	*56*	*43*
T5. *Azolla* cover+30 kg N/ha	6250^{bc}	5625^{bc}	11583^{bc}
	13	*44*	*22*
T6. *Azolla* incorporated +30 kg N/ha	8108^{a}	6683^{a}	14792^{a}
	46	*71*	*56*

Means followed by the same letter are not statistically different at 5% *p*. Figures in italics show percentage increase over control.

After harvesting rice, the soil was ploughed to mix the left-over *Azolla* and rice stubble into the soil. To investigate the residual effect of *Azolla*, wheat (var. Khyber 87) was sown without applying any fertilizer.

Results and discussion

Azolla developed into a thick mat in the nursery and experimental plots, which may be due to the mild temperature (20–35°C) prevailing during the rice growth period; as for most of the *Azolla* species, a temperature around 25°C is considered to be optimum for its growth and nitrogen fixation (Lumpkin and Plucknett, 1982).

As compared to unfertilized control the increase in rice grain yield was 28% for *Azolla* cover, 53% for *Azolla* incorporated and 71% for *Azolla* incorporated +30 kg N/ha (Table 1). Similarly an increase in rice yield due to *Azolla* cover and/or incorporation was also reported by Watanabe (1987) and Kulasooriya *et al.* (1994), while the use of *Azolla* along with chemical-N fertilizer was found to be better than chemical fertilizer for rice yield by Yanni *et al.* (1994). As compared to control, the total nitrogen accumulated in rice plants (straw+grain) was 71% while in rice grain it was 82% higher, for *Azolla* incorporated +30 kg N/ha (Table 2). An improvement in N uptake and fertilizer-N use efficiency in the presence of *Azolla* was also reported from different countries (Kumarasinghe and Eskew, 1993).

The application of *Azolla* not only increased rice yield, but its residual effect on wheat straw and grain yield was significantly higher as compared to control for *Azolla* treatments (Table 3). As compared to unfertilized control the increase in wheat straw yield was 106% for *Azolla* cover, 108% for *Azolla* cover+30 kg N/ha and 143% for *Azolla* incorporated +30 kg N/ha (Table 3). Similarly, the wheat grain yield was 70% and 93% higher for *Azolla* incorporated (T3) and

Table 2. Effect of *Azolla* and urea on nitrogen uptake in rice, cv. JP-5

Treatment	Straw (kg/ha)	Grain (kg/ha)	Straw+grain (kg/ha)
T1. Control	26.44^{c}	35.61^{c}	62.05^{d}
	0	*0*	*0*
T2. *Azolla* cover	30.80^{bc}	46.65^{b}	77.46^{cd}
	16	*31*	*25*
T3. *Azolla* incorporated	37.01^{ab}	55.49^{ab}	92.50^{abc}
	40	*56*	*49*
T4. 30 kg N/ha	42.14^{a}	55.89^{ab}	98.04^{ab}
	59	*57*	*58*
T5. *Azolla* cover+30 kg N/ha	33.23^{abc}	54.37^{b}	87.60^{bc}
	26	*53*	*41*
T6. *Azolla* incorporated +30 kg N/ha	41.39^{a}	64.88^{a}	106.27^{a}
	57	*82*	*71*

Means followed by the same letter are not statistically different at 5% *p*. Figures in italics show percentage increase over control.

Table 3. Residual effect of *Azolla* and urea on dry-matter yield of wheat, cv. Khyber-87

Treatment (given to rice)	Straw (kg/ha)	Grain (kg/ha)	Straw+grain (kg/ha)
T1. Control	2273[d]	1740[c]	4013[d]
	0	*0*	*0*
T2. *Azolla* cover	4690[b]	2640[b]	7330[b]
	106	*52*	*83*
T3. *Azolla* incorporated	2817[c]	2960[b]	5777[c]
	24	*70*	*44*
T4. 30 kg N/ha	2933[c]	2750[b]	5683[c]
	29	*58*	*42*
T5. *Azolla* cover+30 kg N/ha	4723[b]	2990[b]	7713[b]
	108	*72*	*92*
T6. *Azolla* incorporated +30 kg N/ha	5517[a]	3363[a]	8880[a]
	143	*93*	*121*

Means followed by the same letter are not statistically different at 5% *p*. Figures in italics and smaller font show percentage increase over control.

Azolla incorporated + N treatments (T6) respectively (Table 3). The wheat biomass (straw+grain) was maximum for *Azolla* incorporated + 30 kg N/ha (T6) followed by *Azolla* cover treatments, i.e. T2 and T5 (Table 3). Kolhe and Mitra (1987) have also reported a better residual effect of *Azolla* on the subsequent wheat crop, while a higher residual effect of *Azolla* and *Sesbania* than urea was observed on the following rice crop by Watanabe and Ventura (1992). The combined effect of conserved soil N during the rice season due to nitrogen fixation by *Azolla* and mineralization of unused *Azolla* during wheat growth, might have led to a higher residual effect for *Azolla* than control or chemical-N application, as a similar residual effect is expected for a leguminous crop on the subsequent crop (Peoples *et al.*, 1995).

The field study indicated that *Azolla* alone and/or in combination with a low dose of chemical-N can be used quite successfully to increase grain yield in a rice-wheat cropping system under mild climate conditions.

Acknowledgement

We are grateful to the Pakistan Science Foundation for funding the research work.

References

Choudhry M B 1978. Rice soils of Pakistan. In: IRRI, eds. Soil and Rice. International Rice Research Institute (IRRI), PO Box 933, Manila, Philippines, 147–161.

IRRI 1992. IRRI Reporter 4/92. International Rice Research Institute (IRRI), PO Box 933, Manila, Philippines, (pamphlets).

Kolhe S S, Mitra B N 1987. Effect of *Azolla* as an organic source of nitrogen in rice-wheat cropping system. J Agro. Crop Sci., 159, 212–215.

Kulasooriya S A, Seneviratne G, Wijesundera C 1994. An evaluation of some factors that limit the widespread use of *Azolla* in rice cultivation. In: Hegazi N A, Fayez M, Monib M, eds. Nitrogen Fixation with Non-Legumes. American University in Cairo Press, Cairo, 469–474.

Kumarasinghe K S, Eskew D L (eds) 1993. Isotopic Studies of *Azolla* and Nitrogen Fertilization of Rice. Kluwer, Dordrecht, 145 p.

Lumpkin T A, Plucknett D L 1982. *Azolla* As a Green Manure: Use and Management in Crop Production. Westview Press, Boulder, Co., 230 p.

Peoples M B, Herridge D F, Ladha J K 1995. Biological nitrogen fixation: an efficient source of nitrogen for sustainable agricultural production? Plant Soil, 174, 3–28.

Watanabe I 1987. Summary report of the *Azolla* program of the International Network on Soil Fertility and Fertilizer Evaluation for Rice. In: IRRI eds. *Azolla* Utilization. International Rice Research Institute (IRRI), PO Box 933 Manila, Philippines, 197–205.

Watanabe I, Ventura W 1992. Long-term effects of *Azolla* and *Sesbania* on rice yield and fertility of tropical wetland rice soils. In: Proceedings, International Symposium on Paddy Soils. Chinese Academy of Science, Nanjing, China, 331–342.

Yanni Y G, Shalaan S N, Haddad M E 1994. Potential role of *Azolla* as green manure for rice in Nile Delta under different levels of inorganic fertilization. In: Hegazi N A, Fayez M, Monib M, eds. Nitrogen Fixation with Non-legumes. American University in Cairo Press, Cairo, 127–132.

PLANT AND SOIL

Contents Volume 194, Nos. 1–2 July 1997

OPPORTUNITIES FOR BIOLOGICAL NITROGEN FIXATION IN RICE AND OTHER NON-LEGUMES

Papers presented at the Second Working Group Meeting of the Frontier Project on Nitrogen Fixation in Rice held at the National Institute for Biotechnology and Genetic Engineering (NIBGE) Faisalabad, Pakistan, 13–15 October 1996

Guest Editors: J.K. Ladha, F.J. de Bruijn and K.A. Malik

Preface vii

Introduction: Assessing opportunities for nitrogen fixation in rice – a frontier project
J.K. Ladha, F.J. de Bruijn and K.A. Malik 1–10

Fertilizers and biological nitrogen fixation as sources of plant nutrients: Perspectives for future agriculture
O.C. Bøckman 11–14

Isolation of endophytic diazotrophic bacteria from wetland rice
W.L. Barraquio, L. Revilla and J.K. Ladha 15–24

Isolation of endophytic bacteria from rice and assessment of their potential for supplying rice with biologically fixed nitrogen
J.R. Stoltzfus, R. So, P.P. Malarvithi, J.K. Ladha and F.J. de Bruijn 25–36

Association of nitrogen-fixing, plant-growth-promoting rhizobacteria (PGPR) with kallar grass and rice
K.A. Malik, R. Bilal, S. Mehnaz, G. Rasul, M.S. Mirza and S. Ali 37–44

Occurrence, physiological and molecular analysis of endophytic diazotrophic bacteria in gramineous energy plants
G. Kirchhof, V.M. Reis, J.I. Baldani, B. Eckert, J. Döbereiner and A. Hartmann 45–55

***Azoarcus* spp. and their interactions with grass roots**
B. Reinhold-Hurek and T. Hurek 57–64

Biological nitrogen fixation in non-leguminous field crops: Facilitating the evolution of an effective association between *Azospirillum* and wheat
I.R. Kennedy, L.L. Pereg-Gerk, C. Wood, R. Deaker, K. Gilchrist and S. Katupitiya 65–79

Rhizobial communication with rice roots: Induction of phenotypic changes, mode of invasion and extent of colonization
P.M. Reddy, J.K. Ladha, R.B. So, R.J. Hernandez, M.C. Ramos, O.R. Angeles, F.B. Dazzo and F.J. de Bruijn 81–98

Natural endophytic association between *Rhizobium leguminosarum* bv. *trifolii* and rice roots and assessment of its potential to promote rice growth
Y.G. Yanni, R.Y. Rizk, V. Corich, A. Squartini, K. Ninke, S. Philip-Hollingsworth, G. Orgambide, F. de Bruijn, J. Stoltzfus, D. Buckley, T.M. Schmidt, P.F. Mateos, J.K. Ladha and F.B. Dazzo 99–114

Interactions of rhizobia with rice and wheat
G. Webster, C. Gough, J. Vasse, C.A. Batchelor, K.J. O'Callaghan, S.L. Kothari, M.R. Davey, J. Dénarié and E.C. Cocking **115–122**

Interactions between bacterial diazotrophs and non-legume dicots: *Arabidopsis thaliana* as a model plant
C. Gough, J. Vasse, C. Galera, G. Webster, E. Cocking and J. Dénarié **123–130**

Root morphogenesis in legumes and cereals and the effect of bacterial inoculation on root development
B.G. Rolfe, M.A. Djordjevic, J.J. Weinman, U. Mathesius, C. Pittock, E. Gärtner, K.M. Ride, Z. Dong, M. McCully and J. McIver **131–144**

Strategies for increased ammonium production in free-living or plant associated nitrogen fixing bacteria
R. Colnaghi, A. Green, L. He, P. Rudnick and C. Kennedy **145–154**

Genetics of *Azospirillum brasilense* with respect to ammonium transport, sugar uptake, and chemotaxis
A. Van Dommelen, E. Van Bastelaere, V. Keijers and J. Vanderleyden **155–160**

Chitin recognition in rice and legumes
G. Stacey and N. Shibuya **161–169**

The role of phytohormones in plant-microbe symbioses
A.M. Hirsch, Y. Fang, S. Asad and Y. Kapulnik **171–184**

The impact of molecular systematics on hypotheses for the evolution of root nodule symbioses and implications for expanding symbioses to new host plant genera
S.M. Swensen and B.C. Mullin **185–192**

***Nif* gene transfer and expression in chloroplasts: Prospects and problems**
R. Dixon, Q. Cheng, G.-F. Shen, A. Day and M. Dowson-Day **193–203**

Enhancing biological nitrogen fixation: An appraisal of current and alternative technologies for N input into plants
S. Shantharam and A.K. Mattoo **205–216**

Author index **217**

Instructions to authors **219–221**

Developments in Plant and Soil Sciences

50. P.J. Randall, E. Delhaze, R.A. Richards and R. Munns (eds.): *Genetic Aspects of Plant Mineral Nutrition*. 1993 ISBN 0-7923-2118-9
51. K.S. Kumarasinghe and D.L. Eskew (eds.): *Isotopic Studies of Azolla and Nitrogen Fertilization of Rice*. 1993 ISBN 0-7923-2274-6
52. F.A. Bliss and G. Hardarson (eds.): *Enhancement of Biological Nitrogen Fixation of Common Baen in Latin America*. 1993 ISBN 0-7923-2451-X
53. M.A.C. Fragoso and M.L. van Beusichem (eds.): *Optimization of Plant Nutrition*. 1993 ISBN 0-7923-2519-2
54. N.J. Barrow (ed.) *Plant Nutrition – From Genetic Engineering to Field Practice*. 1993 ISBN 0-7923-2540-0
55. A.D. Robson (ed.): *Zinc in Soils and Plants*. 1993 ISBN 0-7923-2631-8
56. A.D. Robson, L.K. Abbott and N. Malajczuk (eds.): *Management of Mycorrhizas in Agriculture, Horticulture and Forestry*. 1994 ISBN 0-7923-2700-4
57. P.H. Graham, M.J. Sadowsky and C.P. Vance (eds.): *Symbiotic Nitrogen Fixation*. 1994 ISBN 0-7923-2781-0
58. F. Baluška, M. Čiamporová, O. Gašparíkova and P.W. Barlow (eds.): *Structure and Function of Roots*. 1995 ISBN 0-7923-2832-9
59. J. Abadía (ed.): *Iron Nutrition in Soils and Plants*. 1995 ISBN 0-7923-2900-7
60. P.S. Curtis, E.G. O'Neill, J.A. Teeri, D.R. Zak and K.S. Pregitzer (eds.): *Belowground Responses to Rising Atmospheric CO_2*. Implications for Plants, Soil Biota, and Ecosystem Processes. 1994 ISBN 0-7923-2901-5
61. P.C. Struik, W.J. Vredenberg, J.A. Renkema and J.E. Parlevliet (eds.): *Plant Production on the Threshold of a New Century*. 1994 ISBN 0-7923-2903-1
62. L.O. Nilsson, R.F. Hüttl and U.T. Johansson (eds.): *Nutrient Uptake and Cycling in Forest Ecosystems* (CEC/IUFRO Symposium, Halmstad, Sweden, 1993). 1995 ISBN 0-7923-3030-7
63. H.P. Collins, G.P. Robertson and M.J. Klug (eds.): *The Significance and Regulation of Soil Biodiversity* (International Symposium, East Lansing, 1993). 1995 ISBN 0-7923-3138-9
64. R.A. Date, N.J. Grundon, G.E. Rayment and M.E. Probert (eds.): *Plant-Soil Interactions at Low pH: Principles and Management* (3rd International Symposium, Brisbane, 1993). 1995 ISBN 0-7923-3198-2
65. J.K. Ladha and M.B. Peoples (eds.): *Management of Biological Nitrogen Fixation for the Development of More Productive and Sustainable Agricultural Systems*. (15th Congress of Soil Science, Acapulco, Mexico, 1994) 1995 ISBN 0-7923-3413-2; Pb 0-7923-3414-0
66. C. Rodriguez-Barrueco (ed.): *Fertilizers and Environment*. 1996 ISBN 0-7923-3729-8
67. B. Buerkert, B.E. Allison and M. von Oppen (eds.): *Wind Erosion in Niger*. Implications and Control Measures in a Millet-based Farming System. 1996 ISBN 0-7923-3885-5
68. O. van Cleemput, G. Hofman and A. Vermoesen (eds.): *Progress in Nitrogen Cycling*. 1996 ISBN 0-7923-3962-2
69. N. Ahmad (ed.): *Nitrogen Economy in Tropical Soils*. 1996 ISBN 0-7923-4094-9
70. M. Rahman, A. Kumar Podder, C. van Hove, Z.N. Tahmida Begum, T. Heulin and A. Hartmann (eds.): *Biological Nitrogen Fixation Associated with Rice Production*. 1996 ISBN 0-7923-4197-X

Developments in Plant and Soil Sciences

71. D. Rosen, E. Tel-Or, Y. Hadar and Y. Chen (eds.): *Modern Agriculture and the Environment*. 1997 ISBN 0-7923-4295-X
72. G.H. Elkan and R.G. Upchurch (eds.): *Current Issues in Symbiotic Nitrogen Fixation*. 1997 ISBN 0-7923-4367-0
73. H.M. Anderson, P.W. Barlow, D.T. Clarkson, M.B. Jackson and P.R. Shewry (eds.): *Plant Roots – From Cells to Systems*. 1997 ISBN 0-7923-4369-7
74. Zhu Zhao-liang, Wen Qi-xiao and J.R. Freney (eds.): *Nitrogen in Soils of China*. 1997 ISBN 0-7923-4372-7
75. J.K. Ladha, F.J. de Bruijn and K.A. Malik (eds.): *Opportunities for Biological Nitrogen Fixation in Rice and Other Non-Legumes* (2nd Working Group Meeting, Faisalabad, Pakistan, 1996). 1997 ISBN 0-7923-4514-2
76. R.W. Bell and B. Rerkasem (eds.): *Boron in Soils and Plants* (International Symposium, Chiang Mai, Thailand, 1997). 1997 ISBN 0-7923-4705-6
77. B. Dell, P.H. Brown and R. Bell (eds.): *Boron in Soils and Plants: Reviews*. 1997 ISBN 0-7923-4729-3
78. T. Ando, K. Fujita, T. Mae, H. Matsumoto, S. Mori and J. Sekiya (eds.): *Plant Nutrition for Sustainable Food Production and Environment* ISBN 0-7923-4796-X
79. K.A. Malik, M.S. Mirza and J.K. Ladha (eds.): *Nitrogen Fixation with Non-Legumes*. 1998 ISBN 0-7923-4873-7

Kluwer Academic Publishers – Dordrecht / Boston / London